WITHDRAWN

PROGRESS IN

Nucleic Acid Research and Molecular Biology

Volume 33

PROGRESS IN
Nucleic Acid Research and Molecular Biology

edited by

WALDO E. COHN
Biology Division
Oak Ridge National Laboratory
Oak Ridge, Tennessee

KIVIE MOLDAVE
University of California
Santa Cruz, California

Volume 33

1986

ACADEMIC PRESS, INC.
Harcourt Brace Jovanovich, Publishers

Orlando San Diego New York Austin
Boston London Sydney Tokyo Toronto

COPYRIGHT © 1986 BY ACADEMIC PRESS, INC.
ALL RIGHTS RESERVED.
NO PART OF THIS PUBLICATION MAY BE REPRODUCED OR
TRANSMITTED IN ANY FORM OR BY ANY MEANS, ELECTRONIC
OR MECHANICAL, INCLUDING PHOTOCOPY, RECORDING, OR
ANY INFORMATION STORAGE AND RETRIEVAL SYSTEM, WITHOUT
PERMISSION IN WRITING FROM THE PUBLISHER.

ACADEMIC PRESS, INC.
Orlando, Florida 32887

United Kingdom Edition published by
ACADEMIC PRESS INC. (LONDON) LTD.
24–28 Oval Road, London NW1 7DX

LIBRARY OF CONGRESS CATALOG CARD NUMBER: 63-15847

ISBN 0–12–540033–0

PRINTED IN THE UNITED STATES OF AMERICA

86 87 88 89   9 8 7 6 5 4 3 2 1

Contents

ABBREVIATIONS AND SYMBOLS ix

SOME ARTICLES PLANNED FOR FUTURE VOLUMES xiii

Expression of Plasmid-Coded Mutant Ribosomal RNA in E. coli: Choice of Plasmid Vectors and Gene Expression Systems

Rolf Steen, David K. Jemiolo, Richard H. Skinner, John J. Dunn, and Albert E. Dahlberg

I.	Expression of Mutant Ribosomal RNA from Wild-Type Promoters, P1 and P2	3
II.	Expression of Mutant Ribosomal RNA from Plasmids with Inducible Promoters	8
III.	Specific Labeling of Cloned rDNA Genes	13
IV.	Conclusion	17
	References	18

The Ubiquitin Pathway for the Degradation of Intracellular Proteins

Avram Hershko and Aaron Ciechanover

I.	Structure of Ubiquitin and of Its Conjugate with Histone	19
II.	Structure and Organization of Ubiquitin Genes	21
III.	Discovery of the Role of Ubiquitin in Protein Breakdown	23
IV.	Enzymatic Reactions in the Formation of Ubiquitin–Protein Conjugates	25
V.	Breakdown of Proteins Conjugated with Ubiquitin	30
VI.	Ubiquitin–Protein Lyases	33
VII.	Recognition of Protein Structure by the Ubiquitin System: Role of the α-Amino Group	35
VIII.	Proposed Sequence of Events in the Ubiquitin Proteolytic Pathway	40
IX.	Involvement of tRNA in Ubiquitin-Mediated Protein Breakdown	41
X.	Evidence for Ubiquitin-Dependent Proteolysis in Various Cells	43
XI.	Possible Roles of Ubiquitin in Histone Modification	48
XII.	Concluding Remarks	51
	References	53
	Addendum	301

DNA Polymerase-α: Enzymology, Function, Fidelity, and Mutagenesis

Lawrence A. Loeb, Philip K. Liu, and Michael Fry

I.	Identification of DNA Polymerase-α	58
II.	Biochemical Characteristics of DNA Polymerase-α	60
III.	Auxiliary Activities Associated with DNA Polymerase-α	72
IV.	Roles of DNA Polymerase-α in Replication and Repair	78
V.	Chromosomal Localization of the Gene for DNA Polymerase-α	83
VI.	Role of DNA Polymerase-α in the Fidelity of DNA Synthesis	84
VII.	Mutants in DNA Polymerase-α	95
VIII.	Prospects for Cloning the Gene for DNA Polymerase-α	99
	References	101

Replication of Superhelical DNAs *in Vitro*

Kenneth J. Marians, Jonathan S. Minden, and Camilo Parada

I.	Initiation	116
II.	Elongation	129
III.	Termination and Segregation of Daughter Molecules	131
IV.	Conclusions	137
	References	138

Aspects of the Growth and Regulation of the Filamentous Phages

Wilder Fulford, Marjorie Russel, and Peter Model

I.	The Elements of the f1 Self-Regulatory Circuit	143
II.	The Role of Thioredoxin in Phage Assembly	156
III.	Concluding Remarks	163
	References	164

Roles of Double-Strand Breaks in Generalized Genetic Recombination

Franklin W. Stahl

I.	The Role of Phage λ in Recombination Studies	169
II.	The Lytic Cycle of Phage λ	172
III.	Recombination of Nonreplicated λ Chromosomes	173

IV.	Recombination by λ's Red System	174
V.	E. coli's RecBC Pathway	186
VI.	Red and RecBC as Models for Meiotic Recombination	191
	References	192

Regulation of Protein Synthesis by Phosphorylation of Ribosomal Protein S6 and Aminoacyl-tRNA Synthetases

J. A. Traugh and A. M. Pendergast

I.	Phosphorylation of Ribosomal Protein S6	196
II.	Phosphorylation of Aminoacyl-tRNA Synthetases	210
III.	Coordinate Regulation of Protein Synthesis	223
	References	225

The Primary DNA Sequence Determines *in Vitro* Methylation by Mammalian DNA Methyltransferases

Arthur H. Bolden, Cheryl A. Ward, Carlo M. Nalin, and Arthur Weissbach

I.	Characterization of DNA Methyltransferases	232
II.	Methylation of Oligodeoxynucleotides	236
III.	*De Novo* and Maintenance Methylation Sites	239
IV.	Inhibitors of Methyltransferases	244
V.	Summary	249
	References	249

The Interferon Genes

Charles Weissmann and Hans Weber

I.	Types, Effects, and Properties of Interferons	251
II.	Analysis of the Interferon System by Recombinant DNA Technology	253
III.	The IFN-α Genes	255
IV.	The IFN-β Genes	276
V.	The IFN-γ Genes	280
VI.	The Evolution of the IFN Gene Family	283
VII.	Conclusions	291
	References	293

INDEX 303

Abbreviations and Symbols

All contributors to this Series are asked to use the terminology (abbreviations and symbols) recommended by the IUPAC-IUB Commission on Biochemical Nomenclature (CBN) and approved by IUPAC and IUB, and the Editor endeavors to assure conformity. These Recommendations have been published in many journals (1, 2) and compendia (3) in four languages and are available in reprint form from the Office of Biochemical Nomenclature (OBN), as stated in each publication, and are therefore considered to be generally known. Those used in nucleic acid work, originally set out in section 5 of the first Recommendations (1) and subsequently revised and expanded (2, 3), are given in condensed form (I–V) below for the convenience of the reader. Authors may use them without definition, when necessary.

I. Bases, Nucleosides, Mononucleotides

1. *Bases* (in tables, figures, equations, or chromatograms) are symbolized by Ade, Gua, Hyp, Xan, Cyt, Thy, Oro, Ura; Pur = any purine, Pyr = any pyrimidine, Base = any base. The prefixes S–, H_2, F–, Br, Me, etc., may be used for modifications of these.

2. *Ribonucleosides* (in tables, figures, equations, or chromatograms) are symbolized, in the same order, by Ado, Guo, Ino, Xao, Cyd, Thd, Ord, Urd (Ψrd), Puo, Pyd, Nuc. Modifications may be expressed as indicated in (1) above. Sugar residues may be specified by the prefixes r (optional), d (=deoxyribo), a, x, l, etc., to these, or by two three-letter symbols, as in Ara-Cyt (for aCyd) or dRib-Ade (for dAdo).

3. *Mono-, di-, and triphosphates of nucleosides* (5′) are designated by NMP, NDP, NTP. The N (for "nucleoside") may be replaced by any one of the nucleoside symbols given in II-1 below. 2′-, 3′-, and 5′- are used as prefixes when necessary. The prefix d signifies "deoxy." [Alternatively, nucleotides may be expressed by attaching P to the symbols in (2) above. Thus: P-Ado = AMP; Ado-P = 3′-AMP] cNMP = cyclic 3′:5′-NMP; Bt_2cAMP = dibutyryl cAMP, etc.

II. Oligonucleotides and Polynucleotides

1. Ribonucleoside Residues

(a) Common: A, G, I, X, C, T, O, U, Ψ, R, Y, N (in the order of I-2 above).

(b) Base-modified: sI or M for thioinosine = 6-mercaptopurine ribonucleoside; sU or S for thiouridine; brU or B for 5-bromouridine; hU or D for 5,6-dihydrouridine; i for isopentenyl; f for formyl. Other modifications are similarly indicated by appropriate *lower-case* prefixes (in contrast to I-1 above) (2, 3).

(c) Sugar-modified: prefixes are d, a, x, or l as in I-2 above; alternatively, by *italics* or **boldface** type (with definition) unless the entire chain is specified by an appropriate prefix. The 2′-O-methyl group is indicated by *suffix* m (e.g., -Am- for 2′-O-methyladenosine, but -mA- for 6-methyladenosine).

(d) Locants and multipliers, when necessary, are indicated by superscripts and subscripts, respectively, e.g., -m_2^6A- = 6-dimethyladenosine; -s^4U- or -^{4}S- = 4-thiouridine; -ac^4Cm- = 2′-O-methyl-4-acetylcytidine.

(e) When space is limited, as in two-dimensional arrays or in aligning homologous sequences, the prefixes may be placed *over the capital letter*, the suffixes *over the phosphodiester symbol*.

2. Phosphoric Residues [left side = 5′, right side = 3′ (or 2′)]

(a) Terminal: p; e.g., pppN... is a polynucleotide with a 5′-triphosphate at one end; Ap is adenosine 3′-phosphate; C > p is cytidine 2′:3′-cyclic phosphate (1, 2, 3); p < A is adenosine 3′:5′-cyclic phosphate.

(b) Internal: hyphen (for known sequence), comma (for unknown sequence); unknown sequences are enclosed in parentheses. E.g., pA-G-A-C(C_2,A,U)A-U-G-C > p is a sequence with a (5') phosphate at one end, a 2':3'-cyclic phosphate at the other, and a tetranucleotide of unknown sequence in the middle. (**Only codon triplets should be written without some punctuation separating the residues.**)

3. Polarity, or Direction of Chain

The symbol for the phosphodiester group (whether hyphen or comma or parentheses, as in 2b) represents a 3'-5' link (i.e., a 5'... 3' chain) unless otherwise indicated by appropriate numbers. "Reverse polarity" (a chain proceeding from a 3' terminus at left to a 5' terminus at right) may be shown by numerals or by right-to-left arrows. Polarity in any direction, as in a two-dimensional array, may be shown by appropriate rotation of the (capital) letters so that 5' is at left, 3' at right when the letter is viewed right-side-up.

4. Synthetic Polymers

The complete name or the appropriate group of symbols (see II-1 above) of the repeating unit, **enclosed in parentheses if complex or a symbol,** is either (a) preceded by "poly," or (b) followed by a subscript "n" or appropriate number. **No space follows "poly"** (2, 5).

The conventions of II-2b are used to specify known or unknown (random) sequence, e.g., polyadenylate = poly(A) or A_n, a simple homopolymer;

poly(3 adenylate, 2 cytidylate) = poly(A_3C_2) or $(A_3,C_2)_n$, an *irregular* copolymer of A and C in 3:2 proportions;

poly(deoxyadenylate-deoxythymidylate) = poly[d(A-T)] or poly(dA-dT) or $(dA-dT)_n$ or $d(A-T)_n$, an *alternating* copolymer of dA and dT;

poly(adenylate,guanylate,cytidylate,uridylate) = poly(A,G,C,U) or $(A,G,C,U)_n$, a random assortment of **A, G, C,** and **U** residues, proportions unspecified.

The prefix copoly or oligo may replace poly, if desired. The subscript "n" may be replaced by numerals indicating actual size, e.g., $A_n \cdot dT_{12-18}$.

III. Association of Polynucleotide Chains

1. *Associated* (e.g., H-bonded) chains, or bases within chains, are indicated by a *center dot* (not a hyphen or a plus sign) separating the *complete* names or symbols, e.g.:

$$\text{poly(A)} \cdot \text{poly(U)} \quad \text{or} \quad A_n \cdot U_m$$
$$\text{poly(A)} \cdot 2\,\text{poly(U)} \quad \text{or} \quad A_n \cdot 2U_m$$
$$\text{poly(dA-dC)} \cdot \text{poly(dG-dT)} \quad \text{or} \quad (dA-dC)_n \cdot (dG-dT)_m.$$

2. *Nonassociated* chains are separated by the plus sign, e.g.:

$$2[\text{poly(A)} \cdot \text{poly(U)}] \rightarrow \text{poly(A)} \cdot 2\,\text{poly(U)} + \text{poly(A)}$$
$$\text{or} \quad 2[A_n \cdot U_m] \rightarrow A_n \cdot 2U_m + A_n.$$

3. Unspecified or unknown association is expressed by a comma (again meaning "unknown") between the completely specified chains.

Note: In all cases, each chain is completely specified in one or the other of the two systems described in II-4 above.

IV. Natural Nucleic Acids

RNA	ribonucleic acid or ribonucleate
DNA	deoxyribonucleic acid or deoxyribonucleate
mRNA; rRNA; nRNA	messenger RNA; ribosomal RNA; nuclear RNA
hnRNA	heterogeneous nuclear RNA
D-RNA; cRNA	"DNA-like" RNA; complementary RNA

ABBREVIATIONS AND SYMBOLS

mtDNA	mitochondrial DNA
tRNA	transfer (or acceptor or amino-acid-accepting) RNA; replaces sRNA, which is not to be used for any purpose
aminoacyl-tRNA	"charged" tRNA (i.e., tRNA's carrying aminoacyl residues); may be abbreviated to AA-tRNA
alanine tRNA or tRNA^{Ala}, etc.	tRNA normally capable of accepting alanine, to form alanyl-tRNA, etc.
alanyl-tRNA or alanyl-tRNA^{Ala}	The same, with alanyl residue covalently attached. [*Note:* fMet = formylmethionyl; hence tRNA^{fMet}, identical with tRNA$_f^{Met}$]

Isoacceptors are indicated by appropriate subscripts, i.e., tRNA$_1^{Ala}$, tRNA$_2^{Ala}$, etc.

V. Miscellaneous Abbreviations

P_i, PP_i	inorganic orthophosphate, pyrophosphate
RNase, DNase	ribonuclease, deoxyribonuclease
t_m (not T_m)	melting temperature (°C)

Others listed in Table II of Reference 1 may also be used without definition. No others, with or without definition, are used unless, in the opinion of the editor, they increase the ease of reading.

Enzymes

In naming enzymes, the 1984 recommendations of the IUB Commission on Biochemical Nomenclature (4) are followed as far as possible. At first mention, each enzyme is described *either* by its systematic name *or* by the equation for the reaction catalyzed *or* by the recommended trivial name, followed by its EC number in parentheses. Thereafter, a trivial name may be used. Enzyme names are not to be abbreviated except when the substrate has an approved abbreviation (e.g., ATPase, but not LDH, is acceptable).

REFERENCES

1. *JBC* **241**, 527 (1966); *Bchem* **5**, 1445 (1966); *BJ* **101**, 1 (1966); *ABB* **115**, 1 (1966), **129**, 1 (1969); and elsewhere.†
2. *EJB* **15**, 203 (1970); *JBC* **245**, 5171 (1970); *JMB* **55**, 299 (1971); and elsewhere.†
3. "Handbook of Biochemistry" (G. Fasman, ed.), 3rd ed. Chemical Rubber Co., Cleveland, Ohio, 1970, 1975, Nucleic Acids, Vols. I and II, pp. 3–59.
4. "Enzyme Nomenclature" [Recommendations (1984) of the Nomenclature Committee of the IUB]. Academic Press, New York, 1984.
5. "Nomenclature of Synthetic Polypeptides," *JBC* **247**, 323 (1972); *Biopolymers* **11**, 321 (1972); and elsewhere.†

Abbreviations of Journal Titles

Journals	Abbreviations used
Annu. Rev. Biochem.	ARB
Annu. Rev. Genet.	ARGen
Arch. Biochem. Biophys.	ABB
Biochem. Biophys. Res. Commun.	BBRC
Biochemistry	Bchem
Biochem. J.	BJ
Biochim. Biophys. Acta	BBA
Cold Spring Harbor	CSH

Cold Spring Harbor Lab.	CSHLab
Cold Spring Harbor Symp. Quant. Biol.	CSHSQB
Eur. J. Biochem.	EJB
Fed. Proc.	FP
Hoppe-Seyler's Z. physiol. Chem.	ZpChem
J. Amer. Chem. Soc.	JACS
J. Bacteriol.	J. Bact.
J. Biol. Chem.	JBC
J. Chem. Soc.	JCS
J. Mol. Biol.	JMB
J. Nat. Cancer Inst.	JNCI
Mol. Cell. Biol.	MCBiol
Mol. Cell. Biochem.	MCBchem
Mol. Gen. Genet.	MGG
Nature, New Biology	Nature NB
Nucleic Acid Research	NARes
Proc. Nat. Acad. Sci. U.S.	PNAS
Proc. Soc. Exp. Biol. Med.	PSEBM
Progr. Nucl. Acid. Res. Mol. Biol.	This Series

Some Articles Planned for Future Volumes

DNA Polymerase
  F. J. BOLLUM

The Structural and Functional Basis of Collagen Gene Diversity
  P. BORNSTEIN

UV-Induced Crosslinks in Nucleoprotein Structure Investigations
  E. I. BUDOWSKY

Hormonally Regulated Eukaryotic Genes
  R. W. HANSON

Translational Control in Eukaryotic Protein Synthesis
  J. W. B. HERSHEY

Messenger RNA Capping Enzymes from Eukaryotic Cells
  Y. KAZIRO

Foreign Gene Expression in Plant Cells
  P. F. LURQUIN

Translocation of mRNA
  W. E. G. MÜLLER

Structure and Organization of the Genome of *Mycoplasma capricolum*
  S. OSAWA

Intermediates in Homologous Recombination Promoted by *recA* Protein
  CHARLES RADDING

Oligonucleotide-Directed Site-Specific Mutagenesis
  U. L. RAJBHANDARY

Early Signals and Molecular Steps in the Mitogenic Response
  E. ROZENGURT

Chloroplast DNA Genes: Structure and Transcription
  K. K. TEWARI

Damage to Mammalian DNA by Ionizing Radiation
  J. F. WARD

Expression of Plasmid-Coded Mutant Ribosomal RNA in *E. coli*: Choice of Plasmid Vectors and Gene Expression Systems

ROLF STEEN,*
DAVID K. JEMIOLO,*
RICHARD H. SKINNER,*
JOHN J. DUNN,† AND
ALBERT E. DAHLBERG*

* Section of Biochemistry
Division of Biology and Medicine
Brown University
Providence, Rhode Island 02912
† Biology Department
Brookhaven National Laboratory
Upton, New York 11973

Rapid advances in the field of molecular genetics have now made it possible to construct mutations of almost any type in a cloned gene. Recently these powerful methods have begun to be applied to cloned *E. coli* ribosomal DNA (rDNA) to explore the structure and function of ribosomal RNAs (rRNA). The *rrn*B operon of *E. coli* was initially cloned into the multicopy plasmid pBR322 by Noller *et al.* (1). The first mutants using this plasmid, pKK3535, were deletions in 16-S and 23-S rRNA (2–4). The plasmid was linearized by digestion with a restriction enzyme, followed by limited treatment with the exonuclease *Bal* 31. Subsequently, point mutations have been produced by bisulfite (5, 6) and ethyl methanesulfonate mutagenesis (7, 8) and by synthetic oligonucleotide-directed mutagenesis (9).

From this early work, it was apparent that many mutations caused drastic reductions in cell growth rates (2). In some cases, mutations in certain regions of the operon could not be cloned, presumably because the gene product was lethal to the cells. To solve the problem of studying lethal mutations, we developed or adopted a number of additional vector systems. The description of these vectors comprises the first half of this chapter and includes a plasmid with low-copy-number and plasmids in which the wild-type promoters for the $rrnB^1$ operon, P1 and P2, are replaced by inducible promoters, either the lambda promoter, P_L, or a T7 late-promoter. These systems have allowed us to clone otherwise lethal mutations and to study their expression.

A second major problem encountered in the study of ribosomal RNA mutants has to do with analysis of the mutant rRNA set in the background of host-coded rRNA. The expression of the rRNA operon is an involved and complex process. The operon is transcribed as a long primary transcript containing 16-S, 23-S, and 5-S rRNA. This primary transcript must be cleaved, methylated, and bound to 52 different ribosomal proteins for gene expression to be complete. Because processing involves these many complex steps, which have not been reproduced *in vitro,* it is necessary to express the cloned genes *in vivo.* This then raises the problem of distinguishing the cloned gene transcript from wild-type rRNA transcribed from the seven rRNA operons on the host chromosome. We address this problem in the second half of this chapter as we describe two procedures that accomplish the specific labeling of cloned rRNA genes.

Different plasmid vectors and expression systems must be used to answer different questions about rRNA. In what follows, we describe the advantages and disadvantages of each system as it applies to the

---

[1] Glossary:

rrnB: A transcriptional unit (operon) coding for 16-S, 23-S, and 5-S ribosomal RNA (in that order). Transcription of the *rrnB* operon is initiated at two promoters (P1 and P2) located in front of the gene for 16-S and terminated at two terminators (T1 and T2) located after the gene for 5-S rRNA.

NR1: A 90 kb plasmid described by Taylor and Cohen (*13*). There are two copies of the plasmid per chromosome in *E. coli.* The plasmid confers resistance to chloramphenicol and streptomycin.

P1, P2, P_L: P1 and P2 are the two natural promoters initiating transcription of the *rrn*B operon. P_L is a repressible promoter from bacteriophage lambda that replaces P1 and P2 in front of the *rrn*B operon in plasmid pNO2680 (*16*).

LacUV5: The lactose operon in *E. coli* coding for β-galactosidase, permease, and acetylase. This operon is transcribed from an inducible promoter that can be activated by lactose or iPrSGal (IPTG).

BL21/DE3: A lambda lysogen of *E. coli* ($r_B^- m_B^- rif^S$) in which the prophage carries a copy of the gene for bacteriophage T7 RNA polymerase under the control of the *Lac*UV5 promoter.

pEMBL9$^+$: A 4 kb high-copy number plasmid carrying the origin of replication both for plasmid replication and for bacteriophage f1. The plasmid confers resistance to ampicillin.

***Col*E1:** A group of plasmids all derived from the *E. coli* plasmid *Col*E1.

CSR603: An *E. coli* strain (*rec, uvr, phr*) unable to repair UV-damaged DNA.

HB101: An *E. coli* strain (*rec, str, pro*).

p23S: The precursor of 23-S rRNA.

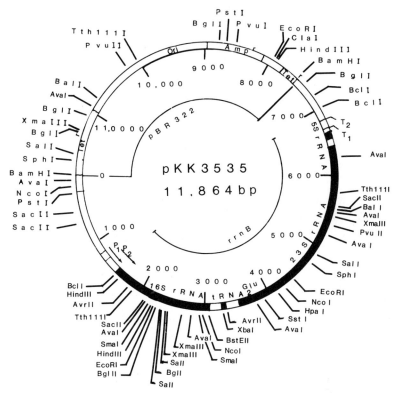

Fig. 1. Plasmid pKK3535 (1).

analysis of the structure and function of these large and complex macromolecules, the rRNAs.

I. Expression of Mutant Ribosomal RNA from Wild-Type Promoters, P1 and P2

A. Mutant rRNA Expressed from a High-Copy-Number Plasmid, pKK3535

The initial phase of our work with ribosomal RNA mutagenesis employed the plasmid pKK3535 (Fig. 1). This plasmid was constructed by Noller et al. (1) as a derivative of the high-copy-number plasmid pBR322, with the rrnB operon of E. coli inserted at a unique BamHI restriction site (see Fig. 1). The rrnB operon contains two tandem promoters, P1 and P2. The plasmid-borne operons are under

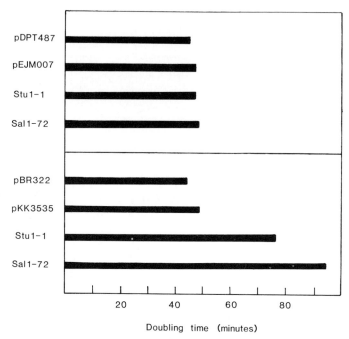

FIG. 2. Doubling times of *E. coli* HB101 strains containing different plasmids. Cell growth rates were measured as described in 2.

stringent control (*10*). Thus the plasmid-borne genes are probably regulated by the same control mechanisms as the host rDNA operons. However, because of the high-copy-number, products from the plasmid-borne genes account for a large fraction of the total cellular pool of rRNA. Plasmid-coded wild-type transcripts may represent close to 50% of the total rRNA, although the level of plasmid-coded mutant transcript may be much reduced in some cases due to instability of the products of the mutant genes (*11*).

We have constructed a number of deletion mutations in the plasmid-borne *rrn*B operon that greatly affect growth rate (*2*). We have also identified deletions that have no effect on cell growth (*5*). Examples of the former are deletion mutations within 16-S rRNA of 371 bases (*Stu*I-1, between two *Stu*I sites) and of 53 bases (*Sal*I-72, between two *Sal*I sites). Each of these causes a significant increase in cell doubling time (see Fig. 2). While we are unsure of the exact reason(s) for this effect, several possible mechanisms can be considered. The most interesting possibility involves rRNAs with small deletions, such as the 53-base deletion in *Sal*I-72 that may result in an rRNA product that is impaired in function. It may be assembled into a ribosomal subunit defective in one or more of the processes involved in translation: initiation, elongation, or termination. Large deletion mutant transcripts may affect cell growth by depleting the cell's resources to produce totally nonfunctional ribosomes. Additionally, they may lack binding sites for ribosomal proteins that regulate their own expression (autogenous regulation, see *12*) thus altering the regulation of expression of ribosomal proteins for host-coded rRNA as well. Translation factors and enzymes involved in rRNA processing (methylases and RNases) may also be tied up with the nonfunctional ribosomal particles. The intracellular turmoil produced by these mutants must be considerable!

In addition to deletion mutants, we have constructed point mutations in plasmid pKK3535 by several techniques, including bisulfite-induced (*5, 6*) and oligonucleotide-directed mutagenesis (*9*). In some cases the mutants had little or no effect on cell growth, while in other cases we failed to recover mutant plasmids, presumably because the products were lethal. For example, in a study of bisulfite-induced mutations near the 3' end of 16-S rRNA, mutations were commonly found in variable regions of the rRNA sequence (e.g., the stem structure 1409–1491, see Fig. 3), but were rarely isolated in highly conserved regions around 1400 and 1500 (*6*) We were also unsuccessful in producing an oligonucleotide-directed mutation in the mRNA-binding (Shine–Dalgarno) region of 16-S rRNA in plasmid pKK3535. A transition involving C to U at position C1538, within the mRNA binding region, was produced in a cloned fragment in phage M13, but could not be cloned back into pKK3535. Thus plasmid pKK3535 has certain limits to its usefulness in the study of rRNA mutants. While the inability to isolate mutants in certain regions of the rRNA is informative, it provides only indirect evidence of the functional importance of these regions.

In some, if not all cases, the lethal or slow-growth phenotype of the rRNA mutation might be dependent on gene dosage. To determine this, we have cloned mutants into a low-copy-number plasmid that contains the *E. coli* promoters, P1 and P2, and is thus controlled by the cell, much like pKK3535. Using both this system and pKK3535, we can test for the dependence of slow growth (or no growth) phenotype on gene dosage.

B. Mutant rRNA Expressed from a Low-Copy-Number Plasmid, pEJM007

The plasmid pDPT487 is a derivative of the low-copy-number plasmid NR1 (*13*). It carries two genes, chloramphenicol acetyltransferase (EC 2.3.1.28) and streptomycin 3″-adenylyltransferase

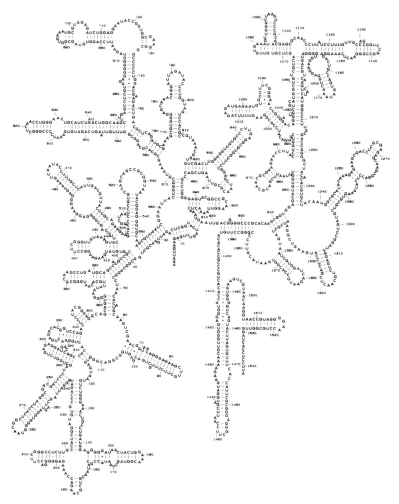

Fig. 3. Secondary structure of 16-S rRNA (26).

(EC 2.7.7.47), which confer antibiotic resistance to cells harboring the plasmid, and it is maintained at two copies per cell. Although much of our work with this plasmid is quite recent, it has already been very useful in demonstrating the effect of gene dosage of mutant rRNA on phenotype. Plasmid pDPT487 has a unique *Bam*HI restriction site that enabled us to clone the entire *rrn*B operon, either with or without a mutation, directly from pKK3535 in a single step producing plasmid pEJM007 (Fig. 4, Jemiolo unpublished data). The plasmid-coded

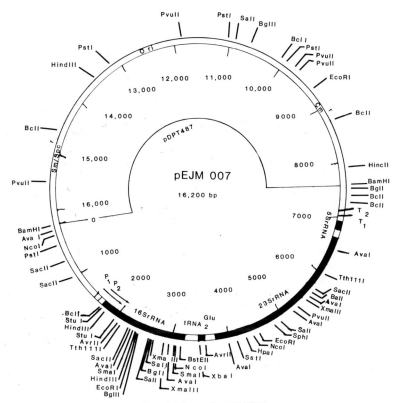

FIG. 4. Plasmid pEJM007.

rRNA remains, therefore, under stringent control because promoters P1 and P2 are present. Examples of mutants cloned into this low-copy-number plasmid include the two deletion mutations in 16-S rRNA (*Stu*I-1, and *Sal*I-72 with deletions of 371 and 53 bases, respectively) described above and already characterized in pKK3535. Each mutant greatly reduces the cell growth rate when present in high-copy-number (Fig. 2). However, when these mutants are cloned into the low-copy-number plasmid and cells are transformed with the mutant plasmids, the growth rates are the same as cells with a wild-type plasmid (see Fig. 2). Thus the effect of these mutants on cell growth is very dependent on gene dose.

It had been shown that the mutant rRNA produced from the *Stu*I-1 plasmid is not processed and assembled into subunits found in 70-S ribosomes (*11*). (The method for determining this is described in Section III of this chapter.) This information, together with the strong relationship between slow-growth phenotype and high-plasmid copy-number, is consistent with the hypothesis that the StuI-1 deletion effects cell growth by disrupting the levels of components involved in protein synthesis (factors, ribosomal proteins, etc.) rather than by interfering directly with the translation process per se. In contrast, mutant 16-S rRNA produced from the SalI-72 plasmid is not very stable; nevertheless, 30 to 40% of what is labeled in maxicells is processed and is found in 70-S ribosomes (14).

A strong dependence on gene dosage was also noted with the C-to-U mutation at position CI538 in 16-S rRNA (see Fig. 3), at the mRNA binding (Shine–Dalgarno) site described above. Whereas this mutant is lethal in the multicopy plasmid pKK3535, it has little effect on cell growth-rate when cloned into pEJM007. This strong dependence on gene dosage might be explained by the fact that this mutation gives rise to functioning ribosomes that are altered in their recognition of Shine–Dalgarno sequences in mRNA during initiation. The mRNAs are translated with greater or lesser frequency depending on their complementarity to the mutant Shine–Dalgarno sequence in the 16-S rRNA (Jacob, Santer, and Dahlberg, unpublished data).

We are now searching for rDNA mutations that do display a slow growth phenotype in plasmid pEJM007 (weakly dependent or independent of gene dosage). These mutations could be particularly interesting as they might involve vital steps in the process of translation. For example, a mutant ribosome affecting elongation might be relatively insensitive to copy number if it can retard the rate of normal ribosomes behind it as it moves slowly along the mRNA. A similar hypothesis has been proposed to explain the dominance of the streptomycin-sensitive phenotype in hybrid cells containing both streptomycin-sensitive and streptomycin-resistant ribosomes (15). Selection of gene-dosage-independent rRNA mutants will represent an important application of this low-copy-number plasmid in future studies.

II. Expression of Mutant Ribosomal RNA from Plasmids with Inducible Promoters

A. A Conditional rRNA Gene Expression System Utilizing the Lambda-P$_L$ Promoter in pNO2680

In order to clone mutations giving rise to gene products that are lethal to the cell, we have employed the conditional rRNA expression system developed by R. Gourse in Nomura's laboratory (16). In this

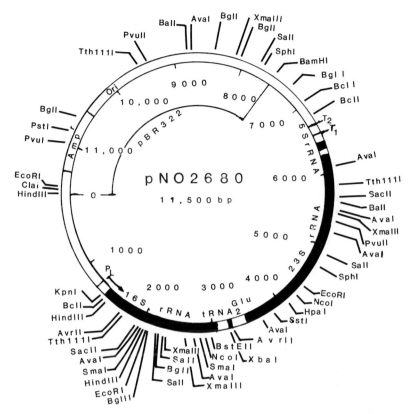

FIG. 5. Plasmid pNO2680 (16).

system, the wild-type promoters, P1 and P2, are replaced by the lambda promoter P_L in the plasmid pNO2680 (Fig. 5). This plasmid can be introduced into cells harboring a gene for the P_L repressor protein. Under these conditions, the P_L promoter is repressed and the *rrn*B operon is not expressed. Conditional expression from P_L can be achieved by utilizing a temperature-sensitive variant of the repressor protein, CI857. This protein functions normally as a repressor at low temperatures (e.g., 30°C) but undergoes thermal denaturation at elevated temperatures (e.g., 42°C). At 42°C, the protein no longer binds to P_L and expression from this promoter can occur.

Using the P_L promoter in place of the wild-type promoters offers several advantages. First, as described above, expression from the promoter is conditional. This allows one to clone in the absence of expression at permissive temperatures (permissive for repressor function) and then "turn on" expression at the nonpermissive temperature (42°C). In addition a graded response of expression can be achieved by varying the temperature between these two extremes. Second, the gene for the repressor protein can be employed in several ways. For example, in the *E. coli* strains K5637, M5219, and N4830, the gene for CI857 is integrated into the host chromosome and these cells produce repressor protein at the level of a single-gene dose. Alternatively, the repressor gene may be introduced into cells on the multicopy plasmid pCI857 (17). This plasmid carries the gene for the repressor protein CI857 and it belongs to a different incompatibility group than plasmid pNO2680. Therefore, the two plasmids can be maintained stably together. The plasmid-coded repressor protein is produced in larger quantities and repression is very efficient. The two-plasmid system also allows one to test for lethal mutations by simply cloning pNO2680 into the same strain either with or without pCI857 at 30°C (permissive temperature for repression). Lethal mutations result in low transformation frequencies in the absence of repressor.

We have used plasmid pNO2680 to clone successfully and to express the Shine–Dalgarno mutation (C to U at position 1538) in 16-S rRNA (Jacob, Santer, and Dahlberg, unpublished data). Since the P_L promoter is independent of the mechanism that modulates host rRNA expression, the cloned rRNA can be overproduced. It has been estimated that up to 80% of *de novo* rRNA synthesis is from plasmid-coded genes (16). The temperature shift (heat shock) does not appear to impede production of plasmid-coded ribosomes, but the cell growth-rate does level off after several hours at 42°C.

B. A Conditional rRNA-Gene-Expression System Utilizing a T7 Late-Promoter in pAR3056

Although the P_L promoter allows one to clone otherwise detrimental mutations, the use of P_L with the conditional repressor does present some experimental difficulties. It has been our experience that under certain conditions repression is not complete. The P_L promoter is a substrate for *E. coli* RNA polymerase, and some expression does occur (at 30°C) when the repressor gene is in a low-copy-number plasmid. Although this problem can be circumvented by placing the gene in a high-copy-number plasmid, one runs into the difficulty of inefficient derepression at elevated temperatures. In addition, precise temperature control is difficult to achieve, and the temperature extremes place certain constraints on cell growth (slow growth at 30°C and erratic growth at 42°C).

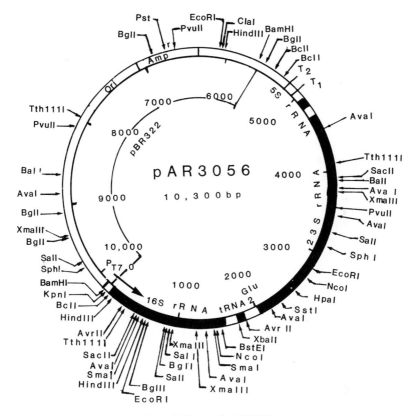

FIG. 6. Plasmid pAR3056.

These problems are avoided when the P_L promoter is replaced by a T7 late-promoter. In the plasmid pAR3056 (Fig. 6), the structural gene of the *rrn*B operon is fused to a T7 late-promoter. This promoter is not recognized by the *E. coli* RNA-polymerase (EC 2.7.7.6) but rather is transcribed specifically by the T7 RNA-polymerase (EC 2.7.7.6). (During T7 infections, one of the early genes, transcribed by the host RNA-polymerase, is the gene for the T7 RNA-polymerase. This enzyme is then used to transcribe late genes.) Thus, in the absence of T7 RNA-polymerase, expression is completely absent.

To express genes from a T7 promoter, the T7 RNA-polymerase must be present. In the *E. coli* strain BL21/DE3, the structural gene for T7 RNA-polymerase has been integrated into the chromosome under control of the *lac*UV5 promoter. The *lac*UV5 promoter is normally repressed but can be derepressed by the addition of isopropyl thiogalactoside (iPrSGal or IPTG). Thus expression of T7 promoter regulated genes in this strain is dependent on production of T7 RNA-polymerase which, in turn, is dependent on chemical induction by iPrSGal.

Mutations in the rRNA structural gene in pAR3056 can be propagated by cloning the plasmid in cells lacking T7 RNA-polymerase. Under these conditions the cloned gene is completely silent. Alternatively, mutations in pAR3056 can be cloned in *E. coli* BL21/DE3 (r_B^-, m_B^-, rif^S) in the absence of iPrSGal. It is known (18) that the *lac*UV5 promoter is weakly expressed in the absence of iPrSGal (0.5% of full induction). This low level of expression probably results in the production of innocuous levels of mutant rRNA. For example, when the 53-nucleotide deletion in the central region of 16-S rRNA (*Sal*I-72) was constructed in pAR3056, no effect on growth could be detected under repressed conditions (absence of inducer). However, when cells were grown in its presence (0.5 mM), the doubling time increased from 35 to 120 minutes for the mutant compared to 70 minutes for BL21/DE3 harboring pAR3056. (For a comparison with the doubling time of *Sal*I-72 in pKK3535; see Fig. 2). A mutation (A to U) at position 814 in 16-S rRNA, lethal in pKK3535, has also been cloned and expressed successfully in this plasmid (Santer and Dahlberg, unpublished data). The expression of this mutant also causes a dramatic increase in doubling time.

We have constructed another plasmid, pGQ15 (Fig. 7), derived from pEMBL9$^+$ (19), that also contains the structural gene for *rrn*B regulated by the T7 promoter. Plasmid pEMBL9$^+$ has the origin of replication of the single-stranded DNA-producing-phage f1 inserted into a *Col*E1-derived plasmid carrying ampicillin resistance. Under normal conditions, the plasmid replicates from the *Col*E1 origin of replication, but upon infection with phage f1, a *trans*-acting phage-coded replication function initiates production of single-stranded DNA from the f1 origin of replication on pGQ15. Single-stranded DNA is packaged and exported into the medium, from which it can easily be isolated and used as a template for sequencing reactions and mutagenesis protocols. Because the complete *rrn*B structural gene is present and regulated by the T7 promoter, subcloning of mutations made on cloned fragments can be circumvented. Site-specific mutations can be constructed directly in the gene in the absence of gene expression. We are currently using this plasmid to make a number of mutations at the site of cleavage of 23-S rRNA by the cytotoxic nucleolytic α-sarcin (20). The integrity of this region of rRNA seems to be of critical importance for both the structure and the function of the large ribosomal subunit (21).

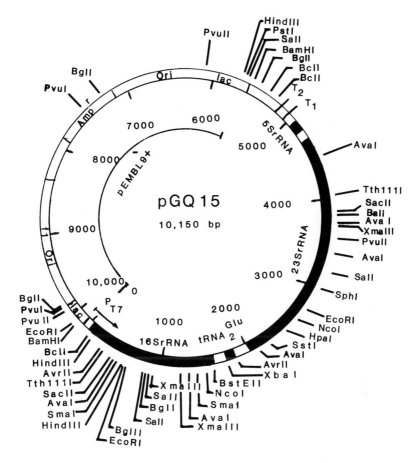

Fig. 7. Plasmid pGQ15.

III. Specific Labeling of Cloned rDNA Genes

Transcription is only the first step in the expression of plasmid-coded rRNA. The precursor rRNA transcript must undergo cleavage and methylation as well as binding to ribosomal proteins to be assembled into a functioning ribosomal subunit. While rRNA transcripts with large deletions can be separated from host rRNA by gel electrophoresis or other physical methods, smaller deletions and point mutations require other methods for analysis. We have developed two general methods for studying mutations, both of which specifically label plasmid-coded rRNA independent of host rRNA.

A. A Maxicell System Induced by UV

Bacteria carrying plasmids with the rRNA operon express both the plasmid-coded and host-coded rRNA genes. In order to study the expression of plasmid-coded rRNA in the absence of host-coded rRNA synthesis, we modified (*11*) the maxicell system developed by Sancar *et al.* (*22*). The principle behind this system is to exploit the difference in target size between the plasmid and host chromosome for damage by UV. In the *E. coli* strain CSR603 (*rec*A1, *uvr*A6, *phr*-1), DNA damaged by UV irradiation cannot be repaired, and most of the chromosomal DNA becomes degraded over a period of several hours. The plasmids contained in such strains largely escape UV-induced damage because of their small target size in comparison with the host chromosome. They continue to replicate and to be transcribed within the "maxicells," which permits specific expression of plasmid-coded rRNA genes. [Note: While the original maxicell used a strain deficient in UV-damage repair pathways, cells wild-type for these functions (e.g., *E. coli* HB101, *rec, str, pro*) can also be used. Because the repair mechanisms require light to function, these cells can be grown in the dark or in opaque flasks and fail to repair UV-damaged DNA.]

The development of maxicell expression of plasmid-coded rRNA in the absence of host-coded rRNA synthesis made it possible to study the effects of mutations on processing and assembly into ribosomal subunits. Certain modifications of the original procedure (*22*) were necessary to achieve optimum phosphate incorporation into plasmid-coded rRNA. It was particularly important that the UV fluence be reduced 10-fold. Under these conditions, synthesis of rRNA was entirely dependent on the presence of plasmid pKK3535. In addition, the maxicells continued to synthesize proteins coded for by the chromosome, including ribosomal proteins and rRNA processing enzymes. These proteins, in turn, permit the plasmid-coded 16-S and 23-S rRNAs to be processed and assembled into mature ribosomal subunits that combine to form 70-S ribosomes.

Using a mutation in pEJM007, we can illustrate some of these points. When cells are transformed with the plasmid containing the 371-nucleotide deletion mutation in 16-S rRNA (*Stu*I-1), and then pulse-labeled with [^{32}P]phosphate, both host and plasmid-coded rRNAs incorporate ^{32}P. Figure 8 shows both host-coded 17-S and 16-S and the slightly faster migrating mutant 16-S rRNA (Δ371), as well as 23-S rRNA transcribed from both the host and plasmid operons (lane 1). Since the RNA was pulse-labeled, both the precursor (17-S) and the mature (16-S) rRNAs are present. When maxicells are prepared,

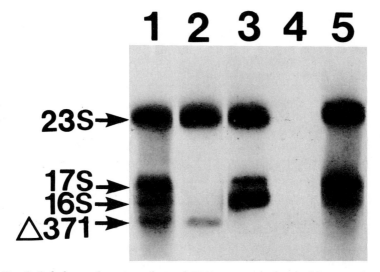

FIG. 8. Gel electrophoretic analysis of rRNA transcripts from wild-type and mutant plasmids labeled for 30 minutes in growing cells (lane 1) and 10 hours in UV-induced maxicells (lane 2–5) (2, 11). The plasmids from which rRNA samples were transcribed are as follows: lanes 1 and 2, pEMJ007 (with StuI-1 mutation); lane 3, pEMJ007 (with wild-type rrnB); lane 4, pDPT487 (parent of pEMJ007, no rrnB operon); lane 5, pEMJ007 (with SalI-72).

the RNA pattern is quite different; in the region of 16-S rRNA, only the plasmid-coded Δ371 transcript is labeled (lane 2). Thus it can be seen that the UV irradiation has completely abolished the synthesis of rRNA from the host chromosome without affecting the production of plasmid-coded rRNA. This is also demonstrated in lane 4 by the absence of labeling in maxicells containing a control plasmid, pDPT487, which lacks the rrnB operon, compared with those containing pEJM007 (lane 3). In addition plasmid-coded rRNAs in maxicells are normally processed to mature 16-S and 23-S rRNAs (lane 3), but one case in which processing is slowed and 17-S-like rRNA accumulates is the 53-nucleotide deletion mutation in 16-S rRNA, SalI-72, shown in lane 5.

An explanation for the greater sensitivity to UV of host-coded rDNA compared to other genes (e.g., ribosomal-protein genes) may be that low UV fluences result in "nicking" and "relaxation" rather than degradation of the host chromosome. It has been shown that transcription of rDNA is particularly sensitive to inhibition of DNA gyrase [DNA topoisomerase (ATP-hydrolyzing), EC 5.99.1.3], indicating that superhelicity of rDNA is required for it to be expressed efficiently (23,

24). The proposed relaxation of host-coded rDNA could account for the plasmid-specific transcription of rRNA in maxicells while the cells maintain the ability to synthesize host-coded proteins.

Although the maxicell technique was first modified for use with the multicopy-plasmid pKK3535, it has been adapted without modification to the low-copy-number plasmid system. In addition, it can be used to label transcripts from the P_L promoter in plasmid pNO2680. Cells containing this plasmid, with the P_L promoter controlled by the temperature-sensitive repressor-protein, can be propagated at 30°C and then irradiated with UV. After a suitable recovery period (up to 5 hours) the temperature is shifted to 42°C, to initiate transcription from the P_L promoter, at which time [^{32}P]phosphate is added to label specifically the plasmid-coded rRNA. These rRNA transcripts are processed and correctly assembled into subunits at 42°C.

B. A Maxicell-like System Induced Chemically

The cloning system based on a T7 promoter can also be used to label specifically plasmid-coded transcripts. This technique is not dependent on UV irradiation to stop host transcription. Rather it takes advantage of the fact that T7 RNA-polymerase, unlike *E. coli* RNA-polymerase, is resistant to rifampicin. After BL21/DE3 cells are treated with iPrSGal for 45 minutes, sufficient T7 RNA polymerase is produced that further transcription of this gene can be terminated by the addition of rifampicin. Within 5 minutes all transcription of host genes by *E. coli* RNA polymerase ceases and specific labeling of plasmid-coded transcripts can be achieved by addition of [^{32}P]phosphate.

The rRNA transcribed from the T7-regulated operon is processed and assembled into ribosomal subunits for at least 35 minutes after addition of rifampicin. This indicates that all of the necessary processing enzymes and ribosomal proteins are available during this time. As there are no large pools of ribosomal proteins in *E. coli*, they must be synthesized *de novo* from stable mRNAs. Because ribosomal proteins control their own synthesis by autogenous regulation (*12*), we assume that the high rate of rRNA synthesis from the plasmid accounts for the continued translation and stability of the ribosomal protein mRNAs. Preliminary data show that there is greater incorporation of [^{35}S]methionine into iPrSGal-induced cells than into noninduced cells. Finally, if the level of available ribosomal protein mRNA varies with growth rate, this may account for the observation that the plasmid-coded rRNA is processed when the inducer is added to cells in the late, but not in the early or mid-log phase of growth.

A mutation (A to U) at position 814 in 16-S rRNA (see Fig. 3), which is lethal in pKK3535, has been cloned into pAR3056, as mentioned above. Maxicell analysis of this mutant rRNA shows (Santer, Dahlberg, and Steen, unpublished) that it is processed and assembled into subunits that associate with 50-S subunits to form 70-S ribosomes. Preliminary data indicate that the lethal nature of the mutant subunit may arise from its involvement in some phase of the process of translation. This mutant is the first of many mutants that we are now cloning into pAR3056, as the T7-maxicell system has several advantages over the UV-maxicell system. The T7 system is also much simpler, requiring only the addition of chemicals rather than UV irradiation. It is much faster: 1.5 rather than 15 hours. It gives a greater yield of plasmid-coded rRNA; liters of T7 maxicells may be used in contrast to irradiation of 10 ml. Equally significant is the high level of transcription by T7 RNA-polymerase. The only negative aspect of the T7 system is the reliance on stable mRNA to provide ribosomal proteins for processing and assembly of the rRNA.

The T7 plasmid pAR3056 also offers the potential for *in vitro* transcription of the *rrn*B operon to produce large amounts of mutant rRNA. Recently we have demonstrated this by transcribing the *Bam*HI–*Pvu*I DNA fragment from plasmid pAR3056 (see Fig. 6). This DNA template, which covers the entire *rrn*B operon, is very efficiently transcribed *in vitro* by T7 RNA-polymerase to produce full length transcripts. (Two transcripts are actually produced, corresponding to termination at T1 and the "run-off" transcript.) Although the 5' end of the primary transcript from the T7 promoter differs from that of the wild-type transcript (because transcription with T7 RNA-polymerase starts within the promoter, subsequent processing *in vitro* with RNase III (EC 3.1.26.3) produces identical p23-S and 17-S precursor rRNAs (Dunn, unpublished data).

IV. Conclusion

We have limited the coverage of material in this chapter primarily to a description of work from our laboratories. We have attempted to provide an update of current methods, including both plasmid vectors and gene expression systems, now available for studying the expression of plasmid-coded mutant rRNA in *E. coli*. This field is still at a very early stage of development, but information about the structural and functional roles of rRNA is accumulating rapidly. Just a few years ago we did not know if *E. coli* could survive if transformed with plasmids containing rRNA mutants. Fortunately for us (and for them), they can survive. By applying the many mutagenic techniques now available to the plasmids and gene expression systems described in this chapter, we are beginning to gain an understanding about one of the most complex and yet most primitive macromolecules in living systems (25).

Acknowledgments

We are grateful to W. Jacob and M. Santer who shared their unpublished data with us. We also want to thank G. Q. Pennable and E. J. Montaro for numerous discussions and help during preparation of this chapter, R. Gourse and D. Taylor for sending us their plasmids pNO2680 and pDPT487, and P. Makosky for technical assistance. Part of this work was supported by an American Cancer Society Grant IN-45Y to Brown University (D.K.J.) and a National Institutes of Health Grant GM19756 (A.E.D.). Research carried out at Brookhaven National Laboratory was under the auspices of the United States Department of Energy.

References

1. J. Brosius, T. J. Dull, D. D. Sleeter and H. F. Noller, *JMB* **148**, 107 (1981).
2. R. L. Gourse, M. J. Stark and A. E. Dahlberg, *JMB* **159**, 397 (1982).
3. C. Zweib and A. E. Dahlberg, *NARes* **12**, 7135 (1984).
4. R. H. Skinner, M. J. Stark and A. E. Dahlberg, *EMBO J.* **4**, 1605 (1985).
5. C. Zweib and A. E. Dahlberg, *NARes* **12**, 4361 (1984).
6. D. K. Jemiolo, C. Zwieb and A. E. Dahlberg, *NARes* **13**, 8631 (1985).
7. L. Mark, C. Sigmund and E. Morgan, *J. Bact.* **155**, 989 (1983).
8. C. Sigmund and E. Morgan, *PNAS* **79**, 5602 (1983).
9. H. U. Goringer, R. Wagner. W. F. Jacob, A. E. Dahlberg and C. Zwieb, *NARes* **12**, 6935 (1984).
10. R. L. Gourse, M. J. Stark and A. E. Dahlberg, *Cell* **32**, 1347 (1983).
11. M. J. Stark, R. L. Gourse and A. E. Dahlberg, *JMB* **159**, 416 (1982).
12. M. Nomura, R. L. Gourse and G. Baughman, *ARB* **53**, 75 (1984).
13. D. P. Taylor and S. N. Cohen, *J. Bact.* **137**, 92 (1979).
14. M. J. Stark, R. J. Gregory, R. L. Gourse, D. L. Thurlow, C. Zweib, R. A. Zimmermann and A. E. Dahlberg, *JMB* **178**, 303 (1984).
15. A. E. Dahlberg, E. Lund and N. O. Kjeldgaard, *JMB* **78**, 627 (1973).
16. R. L. Gourse, Y. Takebe, R. Sharrock and M. Nomura, *PNAS* **82**, 1069 (1985).
17. E. Remaut, H. Tassao and W. M. Fiers, *Gene* **22**, 103 (1983).
18. W. Gilbert and B. Muller-Hill, *PNAS* **56**, 1891 (1962).
19. L. Dente, G. Cesareni and R. Cortese, *NARes* **11**, 1645 (1983).
20. Y. Endo and I. G. Wool, *JBC* **257**, 9056 (1982).
21. I. G. Wool, *TIBS* **9**, 14 (1984).
22. A. Sancar, N. D. Clark, J. Griswold, W. J. Kennedy and W. D. Rupp, *JMB* **148**, 63 (1981).
23. H. L. Yang, K. Heller, M. Gellert and G. Zubay, *PNAS* **76**, 3304 (1979).
24. B. A. Oostra, A. J. van Vliet, G. Ab and M. Gruber, *J. Bact.* **148**, 782 (1981).
25. H. F. Noller and C. R. Woese, *Science* **212**, 403 (1981).
26. R. R. Gutell, B. Weiser, C. R. Woese and H. F. Noller, *This Series* **32**, 155 (1985).

The Ubiquitin Pathway for the Degradation of Intracellular Proteins

> Avram Hershko and
> Aaron Ciechanover
>
> Unit of Biochemistry
> Faculty of Medicine
> Technion-Israel Institute of
>   Technology
> Haifa, 31096, Israel

Though the dynamic turnover of cellular proteins was discovered almost half a century ago (1), the mechanisms of intracellular protein breakdown have begun to be elucidated only in the past few years. It is now clear that this breakdown is carried out by several separate systems that may operate under different physiological conditions. When cells are deprived of nutrients, certain hormones, or serum growth factors, lysosomal autophagy is greatly accelerated. Lysosomal proteolysis may be of massive proportions under deprivation conditions, but it appears to be essentially nonselective, i.e., it degrades mainly "resident," long-lived cellular proteins (2). Other cellular proteins, such as abnormal proteins, short-lived normal proteins, regulatory enzymes, and long-lived proteins in normally growing cells, are degraded by nonlysosomal, energy-dependent proteolytic systems. Some of the latter types of selective protein breakdown are carried out by the ubiquitin proteolytic pathway.

Since our last extensive review on this topic (2), sufficient progress has been achieved to warrant a new summary. The present review is intended to discuss recent developments in our understanding of the mode of action and biological functions of the ubiquitin system. Some earlier results, described in previous reviews (2–4), are mentioned only briefly to provide background information. General aspects of intracellular protein breakdown are described in several earlier reviews (5–7) and in a recent symposium (8).

I. Structure of Ubiquitin and of Its Conjugate with Histone

Ubiquitin is a small polypeptide ($M_r = 8500$) present in apparently all eukaryotic cells. It was originally isolated (9) in the characteriza-

```
                          5                           10
    Met - Gln - Ile - Phe - Val - Lys - Thr - Leu - Thr - Gly -
                          15                          20
    Lys - Thr - Ile - Thr - Leu - Glu - Val - Glu - Pro - Ser -
                          25                          30
    Asp - Thr - Ile - Glu - Asn - Val - Lys - Ala - Lys - Ile -
                          35                          40
    Gln - Asp - Lys - Glu - Gly - Ile - Pro - Pro - Asp - Gln -
                          45                          50
    Gln - Arg - Leu - Ile - Phe - Ala - Gly - Lys - Gln - Leu -
                          55                          60
    Glu - Asp - Gly - Arg - Thr - Leu - Ser - Asp - Tyr - Asn -
                          65                          70
    Ile - Gln - Lys - Glu - Ser - Thr - Leu - His - Leu - Val -
                          75
    Leu - Arg - Leu - Arg - Gly - Gly
```

FIG. 1. Amino-acid sequence of ubiquitin from higher eukaryotes. In the original sequence determinations of ubiquitin (13, 14), the carboxyl-terminal Gly-Gly residues were missed, presumably due to a proteolytic artifact (57).

tion of polypeptide hormones of the calf thymus. Initial claims that ubiquitin induces lymphocyte differentiation and stimulates adenylate cyclase (9, 10) have not been confirmed (11, 12). The term "ubiquitin" was coined following the observation that the polypeptide is found (as determined by radioimmunoassay) in all living cells examined (9).

The primary structure of ubiquitin is shown in Fig. 1. It is apparently the most highly conserved of known proteins. The sequence of ubiquitin is identical in organisms as diverse as cattle (13), man (14), trout (15), toad (16), and insects (17). In yeast ubiquitin, the amino-acid sequence of which was deduced from the nucleotide sequence of the cloned gene (18), only three residues (out of 76) differ from the sequence of human ubiquitin.

The crystal structure of ubiquitin has recently been elucidated at 2.8 Å resolution (19). It is a compact, globular protein with a hydrophobic core and considerable secondary structure. These structural features account for the previously observed high resistance of ubiquitin to heat denaturation (20), or to treatments with acid, alkali, or denaturants (21). The amino-terminal residue is buried, but the carboxyl-terminal Gly-Gly protrudes and has considerable freedom of motion. This feature is presumably required for the linkage of the carboxyl-terminal glycine to the amino groups of proteins (see below).

The extraordinary conservation of ubiquitin in evolution indicates some basically important cellular functions. There are two processes in which ubiquitin is known to be involved, histone modification and intracellular protein breakdown. In both cases, ubiquitin is linked to protein amino groups. The structure of the conjugate of ubiquitin with histone 2A (formerly called protein A24 and now termed uH2A) has been elucidated by Busch, Goldknopf, and co-workers and was summarized in previous reviews by these authors (22–24). In this conjugate, a single molecule of ubiquitin is linked through its carboxy-terminal glycine residue to the ε-amino group of lysine-119 of histone 2A by an isopeptide linkage (23).

In cultured animal cells, about 10% of histone 2A (25) and 1–1.5% of histone 2B (26) are in the form of their conjugates with ubiquitin. All variants of H2A and H2B are modified to similar extents (25, 26), indicating no selectivity of ubiquitin attachment with regard to these histone variants. Histone uH2A is an integral component of nucleosomes, as shown by its presence in purified core nucleosome particles (27, 28) and by its ability to become assembled with DNA in a manner similar to that of core nucleosome histones (28). Either one or both H2A molecules in the same nucleosome can be replaced by uH2A (29, 30). Cross-linkage experiments show that the interaction of H2A with H2B in solution (28) or the proximity of H2A to H1 in chromatin (31) is not affected by the attachment of ubiquitin to H2A. When purified uH2A is reconstituted into core particles in place of H2A, no changes are observed in the rate and pattern of DNase I digestion, the size of DNA produced by micrococcal nuclease action, or in the binding of HMG (high mobility group) proteins 14 and 17 to the core particles (30). It has been concluded that substitution of uH2A in place of H2A has little effect at the level of the mononucleosome. However, it is possible that the attachment of ubiquitin of histones affects interactions between nucleosomes, or between the nucleosome and some other chromatin constituents.

Possible functions of ubiquitin attachment to histones are discussed in Section XI.

II. Structure and Organization of Ubiquitin Genes

The gene of ubiquitin, recently elucidated in various organisms by several groups of investigators, has a unique organization. In yeast there are six adjacent copies of ubiquitin genes linked "head-to-tail," without intervening termination signals (18). The last ubiquitin-coding repeat is followed by an extra asparagine residue, before the translation "stop" codon. These findings indicate that ubiquitin is synthesized as a polyprotein in which the carboxyl-terminus of one ubiquitin residue is linked to the amino-terminus of the next. The precursor is then presumably cleaved at the peptide linkages between ubiquitin residues by specific enzyme(s). A similar organization of multiple and contiguous ubiquitin genes occurs in *Xenopus laevis* (16), chicken (32), and human (33) cells. In addition, it has been pointed out (32) that a repetitive 228-nucleotide sequence found in an ecdysone puff region of *Drosophila melanogaster* (34) is a tandem array of ubiquitin genes. No spacer regions between ubiquitin repeats nor intervening sequences within coding regions have been found in any of the reported cases. Since this unique gene structure is conserved from yeast to man, it apparently confers some as yet unknown functional advantage.

Though the general structure of polyubiquitin genes has been conserved in evolution, there are species differences in the number of ubiquitin repeats, number of genomic loci, and nature of the amino-acid residue(s) following the carboxyl-terminus of the last ubiquitin repeat. The number of ubiquitin-coding repeats in a single locus has been estimated as 5–6, 6, 9, >12, and >15 in chicken, yeast, man, *Xenopus*, and *Drosophila*, respectively (32, 18, 33, 16, 34). While in yeast all ubiquitin-coding repeats are confined to a single genomic locus, multiple ubiquitin mRNAs of different sizes have been observed in *Xenopus* (16), *Drosophila* (34), and human (33, 35) cells. That these may be transcribed from different genomic loci, containing different numbers of ubiquitin copies, is indicated by the finding that at least two types of ubiquitin loci exist in the human genome, one coding for a single ubiquitin and another for nine ubiquitin repeats (33). In addition, considerable polymorphism has been reported in sizes of ubiquitin mRNAs in different stocks of *Drosophila* (34) or at different stages of development of *Xenopus laevis* (16).

Species differences also exist with regard to the extra amino acid at the carboxyl-terminus of the final ubiquitin unit, which is asparagine in yeast (18), tyrosine in chicken (32), and valine in the human sequence (33). The function of this extra residue is not known, though it has been suggested that it may serve to prevent the conjugation of the unprocessed polyubiquitin precursor (18). However, in the clone isolated from *Xenopus laevis*, there is no extra amino acid, and the termination codon follows immediately that of Gly^{76} of the final ubiquitin coding unit (*16*). A still different case was reported for a human liver cDNA clone, in which a single ubiquitin coding unit is joined to an 80-amino-acid, carboxyl-terminal extension (*35*). The carboxyl-terminal polypeptide has a high proportion of basic amino acids, and it has been proposed that its function may be to transport the ubiquitin precursor to the nucleus (*35*).

It is noteworthy that the structure of the avian ubiquitin gene (*32*) was elucidated in the course of a study on genes of "heat-shock" proteins. The heat-shock response is the induction of a specific set of proteins when cells are exposed to elevated temperature or to a variety of agents such as ethanol, heavy metals, oxidants or amino-acid analogues (see ref. *36–38* for reviews). It is generally assumed that the heat-shock proteins protect the cell from these damaging conditions, but the specific function of most heat-shock proteins is not known. To study the heat-shock genes of avian cells, Bond and Schlesinger (*32*) selected clones containing heat-inducible mRNA sequences from a cDNA library; one of these clones contained the ubiquitin genes. The level of ubiquitin mRNA was about 5-fold higher in heat-shocked cells than in nonshocked cells, and a marked increase in the rate of ubiquitin synthesis was observed in heat-shocked cells (*32*). These findings indicate that ubiquitin is a heat-shock protein. The authors suggested that an increased activity of the ubiquitin proteolytic pathway may be required to remove abnormal or misfolded proteins that may accumulate at high temperature. A relationship between the heat-shock response and protein breakdown had been suggested by the observation that heat-shock proteins are induced by amino-acid analogues (*39*), which produce rapidly degradable abnormal proteins. It is notable that the formation of the ATP-dependent *lon* protease in *E. coli* (see Section V) is under the control of the heat-shock regulatory gene *htp*R (*40–42*). Formation of abnormal proteins by a variety of methods induces the synthesis of heat-shock proteins in *E. coli* (*43*), as well as in eukaryotic cells. It might be interesting to examine whether some of the heat-shock proteins in eukaryotic cells are enzymes of the ubiquitin pathway.

III. Discovery of the Role of Ubiquitin in Protein Breakdown

The role of ubiquitin in protein breakdown was uncovered in our studies, in part in collaboration with Irwin Rose, on the mode of action of an ATP-dependent proteolytic system from reticulocytes. The energy dependence of intracellular protein breakdown has long been known (*44, 45*), but the underlying mechanisms remained obscure.

An ATP-dependent proteolytic system from reticulocyte lysates was first described by Etlinger and Goldberg (46). We tried to fractionate this system, with the aim of isolating the component(s) responsible for its ATP requirement. Initially, reticulocyte lysates were fractionated on DEAE-cellulose into two crude fractions, fraction I, which is not adsorbed, and fraction II, which contains all proteins adsorbed and eluted with high salt. The aim of this fractionation had been to remove hemoglobin (present in fraction I), but we found that most ATP-stimulated proteolytic activity was lost in fraction II, as compared to crude lysates. ATP-dependent proteolytic activity could be recovered, however, by the addition of fraction I to fraction II (20). The active component in fraction I is a small, heat-stable polypeptide (20). It was first termed APF-1 (ATP-dependent proteolysis Factor 1). Following its purification (47), it was found to be similar to ubiquitin (48).

The covalent conjugation of ubiquitin to proteins was also discovered by accident. We first thought that the polypeptide might be an activator or a regulatory subunit of some enzyme of the proteolytic system present in fraction II. To look for such a possible association, purified ubiquitin was radioiodinated and incubated with crude fraction II from reticulocytes in the presence or absence of ATP. A dramatic ATP-dependent binding of ^{125}I-ubiquitin to reticulocyte proteins was observed upon gel-filtration chromatography (49). We were very astonished, however, to find that a covalent amide linkage was formed, as shown by the stability of the "complex" to acid, alkali, hydroxylamine, and boiling with dodecyl sulfate and mercaptoethanol (49). Only then did we consider the possibility that ubiquitin may be linked to protein substrates (rather than to an enzyme of the system), thus providing a marking mechanism for protein breakdown. Indeed, we found that ubiquitin is conjugated to exogenous proteins, which are good substrates for the ATP-dependent proteolytic system (50). Upon the conjugation of ubiquitin with exogenous protein substrates, multiple bands were observed by SDS-polyacrylamide-gel electrophoresis, which apparently consisted of several molecules of ubiquitin linked to a single molecule of the protein (50). Based upon these findings, it was proposed that the conjugation of ubiquitin with proteins is an obligatory intermediary event in protein breakdown, and that an enzyme system exists that specifically degrades proteins conjugated to ubiquitin (50). Direct evidence for such a conjugate-degrading enzyme system was found several years later (ref. 51 and see Section V).

IV. Enzymatic Reactions in the Formation of Ubiquitin–Protein Conjugates

A. Activation of Ubiquitin

Our next task was to delineate the enzymatic reactions in the formation of ubiquitin–protein conjugates. We first concentrated on the activation reaction, which must precede amide bond formation. A specific ubiquitin-activating enzyme was detected in reticulocyte extracts by a ubiquitin-dependent $^{32}PP_i$–ATP exchange reaction (52). The enzyme also catalyzes ubiquitin-dependent AMP–ATP exchange reaction, and binding of ubiquitin to the enzyme by a thioester linkage was observed (52). A two-step reaction sequence was proposed to account for these observations, in which first a ubiquitin-adenylate is formed (with the displacement of PP_i from ATP) and then the adenylate is transferred to a thiol site of the enzyme (with the release of AMP). This model was subsequently confirmed by direct methods (53–55, and see below). The activated amino-acid residue of ubiquitin is its carboxyl-terminal glycine, indicated by its specific labeling upon reductive cleavage of the intermediate with [^{3}H]borohydride (56). This finding also cleared up some previous confusion about the carboxyl-terminus, since Arg^{74} was reported as the carboxyl-terminal residue of ubiquitin (13, 14). Subsequently, it was shown (57) that active ubiquitin has the carboxyl-terminal sequence Arg^{74}-Gly^{75}-Gly^{76} and that the loss of the carboxyl-terminal Gly^{75}-Gly^{76} was a proteolytic artifact.

The ubiquitin-activating enzyme was purified to near homogeneity by "covalent-affinity" chromatography (58). In this technique, a crude extract is applied to a column of ubiquitin-Sepharose in the presence of Mg-ATP. Matrix-bound ubiquitin is activated by the activating enzyme, and a covalent thioester linkage is formed between the enzyme and ubiquitin-Sepharose. Nonspecifically bound proteins are removed with a high-salt wash, and the enzyme is specifically eluted with AMP and PP_i, which reverse the activation reaction. By this procedure, a nearly homogeneous preparation of ubiquitin-activating enzyme is obtained in a single step. The enzyme is a dimer of two identical subunits of 105,000 (58).

Using the affinity-purified preparation, the mechanisms of the ubiquitin-activating enzyme were studied in detail by Rose and co-workers. Direct evidence for the formation of the ubiquitin-adenylate was provided by the finding that, following incubation with [2,8-^{3}H]ATP, tritium radioactivity became linked to ubiquitin, as shown by its precipitation with trichloroacetic acid and by the size of the product on gel-filtration chromatography (53). Without acid precipitation, ubiquitin-AMP was tightly bound to the activating enzyme at a 1 : 1 molar ratio to enzyme subunit. At the same time, two equivalents of ubiquitin were bound to each enzyme subunit. These findings were explained by a reaction sequence in which, following the transfer of activated ubiquitin from the adenylate site to the thiol site, a second ubiquitin undergoes reaction at the adenylate site, while the first ubiquitin is still bound to the thiol site:

$$\text{ATP} + \text{Ub} + \text{E}_{\text{SH}} \rightleftharpoons \text{E}_{\text{SH}}^{\cdot \text{AMP-Ub}} + \text{PP}_i \quad (1)$$

$$\text{E}_{\text{SH}}^{\cdot \text{AMP-Ub}} \rightleftharpoons \text{E}_{\text{S-Ub}} + \text{AMP} \quad (2)$$

$$\text{E}_{\text{S-Ub}} + \text{ATP} + \text{Ub} \rightleftharpoons \text{E}_{\text{S-Ub}}^{\cdot \text{AMP-Ub}} + \text{PP}_i \quad (3)$$

The molar ratio ubiquitin : AMP : enzyme = 2 : 1 : 1 indicates that the product of reaction (3) is the main intermediate, which accumulates under the conditions employed. This interpretation was supported by the finding that two equivalents of $^{32}\text{PP}_i$ were released from $[\gamma\text{-}^{32}\text{P}]\text{ATP}$ per equivalent of enzyme subunit in the initial "burst" due to the formation of enzyme-bound intermediates. That the second ubiquitin is bound as a thioester was supported by the observation that, following treatment of the enzyme with iodoacetamide (which inactivates the thiol site without affecting the adenylate site), the molar ratio of ubiquitin : AMP : enzyme was 1 : 1 : 1 (53). Further experiments with isolated ubiquitin adenylate (prepared by acid precipitation of enzyme-bound intermediate) were in complete accord with its expected role in the above reaction mechanism (55). Thus, the addition of fresh enzyme to isolated ubiquitin-AMP caused the "burst" release of free AMP in an amount equal to fresh activating enzyme added [reaction (2)]; following the initial "burst" release, the enzyme-catalyzed slower release of AMP in the presence of DTT required a functional thiol site, as shown by its prevention by iodoacetamide treatment of enzyme [reaction (2)]; and isolated ubiquitin-AMP was converted to ATP and ubiquitin in the presence of PP_i and Mg^{2+} [reversal of reaction (1)] (55). Kinetic examination of the details of the partial reactions indicated that reaction (1) is an ordered process in which the binding of ATP precedes that of ubiquitin, as suggested by the inhibition of PP_i–ATP exchange by excess ubiquitin (54). Similarly, AMP–ATP exchange was inhibited by excess PP_i, which was taken as evidence that PP_i is released before AMP. Equilibrium and kinetic constants of the various steps were determined by the isotope exchange technique (54).

It was initially suggested that the energy-rich thioester bond between ubiquitin and the thiol site of the enzyme preserves the activated state of the acyl group of ubiquitin and serves as a donor for the formation of the final amide linkage (52). That this is indeed the case was supported by the following experiment (53). The intermediate was formed with a limiting amount of ubiquitin relative to the activating enzyme. Under these conditions, most of ubiquitin was converted to thioester, and only a small fraction remained as adenylate. ATP was removed with hexokinase and deoxyglucose; upon addition of fraction II, the formation of the ubiquitin–protein conjugates was observed in the absence of ATP. A similar experiment with iodoacetamide-treated activating enzyme (in which the formation of thioester, but not of adenylate, is blocked) did not result in conjugate formation. Both experiments suggest that the thioester of ubiquitin is the donor for conjugate formation (53).

B. Transfer of Activated Ubiquitin for Conjugation

Following isolation of the ubiquitin-activating enzyme, we found that the purified activating enzyme does not carry out ubiquitin–protein conjugation by itself (58). It could be assumed, therefore, that activated ubiquitin is linked to amino groups of protein by the action of another enzyme, which would be the ubiquitin–protein ligase. However, we found that at least two other enzymes participate in this process. These were termed E_2 and E_3 (E_1 is the ubiquitin-activating enzyme). E_2 and E_3 were partially purified from fraction II by affinity chromatography on ubiquitin-Sepharose (59). E_3 is noncovalently bound to the affinity column and does not require the presence of ATP for binding. On the other hand, the binding of E_2 to ubiquitin-Sepharose requires ATP and E_1. These enzymes are also eluted from the affinity column under different conditions: E_3 can be eluted with high salt or by raising the pH to 9.0, while E_2 is not displaced with the high-salt wash, but can be eluted with increased concentrations of a thiol compound. These findings suggest that E_2 is bound to the ubiquitin column by a covalent thioester linkage (see also below). E_2 and E_3 were further purified by gel-filtration chromatography of the affinity-column eluates. Two forms of E_2 were observed, with apparent molecular sizes of 35,000 and 250,000, while E_3 eluted as a single peak with an apparent molecular weight of around 300,000 (59).

We found that all three enzymes (E_1, E_2, and E_3) are absolutely required for the conjugation of ubiquitin to substrates of the proteolytic system, such as lysozyme or oxidized ribonuclease. All three enzymes are also required for the reconstitution of the protein breakdown system in the presence of the unadsorbed fraction of the affinity column, ubiquitin, and ATP. That indeed the same enzymes are required for ubiquitin conjugation and protein breakdown was indicated by the coincidence of these two activities across peak fractions of gel filtration columns (59).

The available evidence indicates that E_2 has a ubiquitin-carrier function: it accepts activated ubiquitin from E_1 and then transfers it for amide bond formation with proteins. The following sequence of reactions was proposed (59):

$$E_1\text{-SH} + ATP + Ub \longrightarrow E_1\text{-S-Ub} + AMP + PP_i \qquad (4)$$

$$E_1\text{-S-Ub} + E_2\text{-SH} \longrightarrow E_2\text{-S-Ub} + E_1\text{-SH} \qquad (5)$$

$$E_2\text{-S-Ub} + \text{Protein} \xrightarrow{E_3} E_2\text{-SH} + \text{Protein-Ub} \qquad (6)$$

The first indication for such a role of E_2 was the observation (cited above) that the binding of E_2 to ubiquitin-Sepharose requires E_1 and ATP. One possible explanation was that E_2 replaces E_1 (covalently bound to the ubiquitin column) by a thioester transacylation process (step 5). This interpretation was supported by the finding that E_2 is rapidly inactivated by iodoacetamide, but can be protected against iodoacetamide inactivation by prior incubation with E_1, ATP, and ubiquitin (59). This suggests that an E_1-mediated binding of ubiquitin to an essential thiol site of E_2 protects this site from the sulfhydryl-blocking agent. Finally, transfer of activated ubiquitin from E_1 to E_2 could be directly observed in the following experiment (59). ^{125}I-Ubiquitin was incubated with E_1 and ATP to form the ^{125}I-ubiquitin-E_1 thioester. Following the addition of a large excess of unlabeled ubiquitin (which effectively lowered the specific radioactivity of residual-free ^{125}I-ubiquitin), E_2 was added, and the reaction products were separated by SDS-polyacrylamide gel electrophoresis under conditions that prevent the breakdown of thioesters (at 4°C and in the absence of mercaptoethanol). There was a pronounced loss of E_1-bound ^{125}I-ubiquitin, with a corresponding transfer of radioactivity to four smaller E_2-ubiquitin thioesters. In experiments of similar design, the further transfer of activated ubiquitin from E_2-ubiquitin thioesters to amide linkages with proteins could be observed, a reaction that required the presence of E_3 (59).

At present, we do not know much about the mode of action of E_3. It might be the ligase which catalyzes amide-bond formation between ubiquitin and amino groups of protein substrates. However, E_3-independent transfer of ubiquitin from E_2 to certain proteins that are not substrates for proteolysis has recently been observed (60, and see below). E_3 may have a structural role, allowing interaction between E_2 and some specific proteins, or it may have some other, as yet unknown function.

An important related question is which component of the ubiquitin–protein ligase system (E_2 or E_3) contains recognition site(s) for protein substrates. It would be surprising (though not impossible) for one enzyme to recognize a wide variety of different proteins. E_3 has not yet been extensively purified, but there is no indication at present for more than one species of E_3 in reticulocyte extracts. On the other hand, we observed multiple species of E_2 and suggested that the different subspecies of E_2 may represent a family of enzymes of related function, but of different specificities (59).

The functional heterogeneity of some E_2 species has recently been investigated (60). Polyacrylamide gel electrophoresis in dodecyl sulfate of the E_2-containing thiol eluate of the ubiquitin-Sepharose affinity column showed five proteins, with subunit molecular weights of 32,000, 24,000, 20,000, 17,000, and 14,000. These proteins were termed bands 1 to 5, in order of decreasing molecular size. The proteins were partially resolved by gel-filtration chromatography. E_2 activity to stimulate protein breakdown (in the presence of E_1, E_3, and the unadsorbed fraction of the affinity column) coincided with band 5. The same peak also contained E_2 activity required for the E_3-dependent conjugation of ubiquitin to substrates of the proteolytic system, such as reduced and carboxymethylated serum albumin, creatine phosphokinase, or oxidized ribonuclease. On the other hand, E_2 activities that transfer ubiquitin to primary amines or to basic proteins (such as histones or cytochrome c) eluted in two peaks: one coinciding with band 5 and another with bands 1 to 3. These latter two transfer activities did not require E_3. It is noteworthy that the E_3-independent conjugation of ubiquitin to histones and cytochrome c produces predominantly the monoubiquitin derivative, in contrast to the multiubiquitin conjugates of protease substrates produced by the E_3-dependent reaction (60). These results indicate that at least some E_2 species are specific for certain types of proteins, and it appears reasonable to assume that some of the E_3-independent conjugation reactions are involved in functions of ubiquitin in protein modification, rather than in protein breakdown. The problem of whether only one type of

E_2 species is involved in protein breakdown requires further resolution studies.

V. Breakdown of Proteins Conjugated with Ubiquitin

The experiments described above established that ubiquitin conjugation is required for protein breakdown, but did not necessarily indicate that conjugation with the protein *substrate* is an obligatory event. It was possible, for example, that a protease is activated by ubiquitin conjugation. Another possibility was suggested by Speiser and Etlinger (61), who proposed that ATP and ubiquitin act by repressing an inhibitor of an ATP-independent protease. However, we found that the crude inhibitor preparation used by these authors contained a positively required factor of the ubiquitin system that could be separated from two types of protease inhibitors by gel-filtration chromatography (62). The "inhibitors" are endogenous protein substrates which compete with the labeled exogenous substrate, as shown by kinetic competition and by the finding that inhibition can be abolished by incubation of protease with inhibitor. It was concluded that the "inhibitor" cannot have a regulatory role in the ubiquitin proteolytic system (62). That the protease and its inhibitor are not related to the ubiquitin pathway was also concluded by Katznelson and Kulka (63), who found that proteins with blocked amino groups are not degraded by the ATP-ubiquitin proteolytic system, but are efficiently broken down by the "inhibitor-free" protease from reticulocyte extracts.

For direct examination of the notion that ubiquitin–protein conjugates are intermediates in protein breakdown, it was necessary to demonstrate the existence of an enzyme system that preferentially degrades proteins conjugated to ubiquitin, but not unconjugated proteins. To look for such a system, we isolated ubiquitin–protein conjugates (labeled in their protein moiety) to serve as substrates for the partial system (51). As the enzyme source, we used the part of reticulocyte fraction II which is not adsorbed to the ubiquitin affinity column ("affinity-unadsorbed" fraction) and which contains all the enzymes of the system except for the three enzymes involved in the formation of ubiquitin–protein conjugates. In another of the many surprises which we had in the course of this work, we found that Mg-ATP is absolutely required for degradation of the protein moiety of lysozyme–ubiquitin conjugates to acid-soluble products (51). Control experiments indicated that the effect of ATP on conjugate breakdown is different from its action on the degradation unconjugated proteins (where it is required for ubiquitin conjugation as well). Thus, free lysozyme was not degraded by the affinity-unadsorbed fraction unless E_1, E_2, E_3, and ubiquitin were added, whereas the degradation of lysozyme–ubiquitin conjugates required the supplementation of ATP alone. The addition of a specific antibody directed against ubiquitin (which reacts with free ubiquitin, but not with protein-conjugated ubiquitin) inhibited the breakdown of lysozyme in the complete system, but had no influence on the degradation of lysozyme–ubiquitin conjugates. These results indicate that the ATP-dependent degradation of ubiquitin conjugates is independent of ubiquitin conjugation, whereas the breakdown of unconjugated proteins does require ubiquitin conjugation. This conclusion was supported by the finding that the addition of a large excess of unlabeled lysozyme decreased the release of acid-soluble radioactivity from free lysozyme (by its isotopic dilution), but had no effect on the breakdown of ubiquitin-conjugated lysozyme. This shows that the free protein is not a direct substrate for the ATP-dependent system that degrades ubiquitin-conjugated proteins.

Examination of the nucleotide requirement of the conjugate-degrading system showed that ATP is most effective and that Mg^{2+} is absolutely required for ATP action. Nonhydrolyzable analogs of ATP, substituted at either α–β or β–γ positions by methylene or imido groups, are ineffective. This suggests (but does not prove) that hydrolysis of ATP may be required for breakdown of the conjugates. Of various nucleotides tested, only CTP could partially replace ATP. It is interesting to note that CTP cannot replace ATP in the ubiquitin activation reaction, and it also cannot replace ATP in the complete proteolytic system (degradation of unconjugated proteins in the presence of ubiquitin). Therefore, CTP specifically replaces ATP only at the level of conjugate breakdown (51).

All the above experiments indicate that there is a second site at which ATP is involved in the ubiquitin proteolytic pathway. In addition to its action in the formation of ubiquitin–protein conjugates, ATP is involved in the proteolytic breakdown of the conjugates.

Results from other laboratories are consistent with such a second role of ATP in the ubiquitin pathway. Hough and Rechsteiner (64) reported that the degradation of lysozyme–ubiquitin conjugates in reticulocyte lysates is stimulated by ATP. Rapoport and co-workers (65) have estimated ATP hydrolysis accompanying the ubiquitin-dependent degradation of mitochondrial proteins in reticulocyte extracts, and found that more than one ATP is consumed per peptide bond cleaved. Since this value is much higher than that expected for ubiquitin–protein conjugation, it was suggested that most ATP consumption may be required for the breakdown of ubiquitin conjugates (65). The amount of ATP consumed in a futile cycle consisting of ubiquitin conjugation and isopeptidase action (see Section VI) cannot be estimated by this method.

The above findings provided strong evidence that conjugation of ubiquitin to the proteolytic substrate is an intermediary process in protein breakdown. They also raised a new enigma concerning the role of ATP in conjugate breakdown. ATP-dependent proteases in *Escherichia coli* (66, 67) and mammalian mitochondria have been described (68, 69). The protease from *E. coli* is the product of the *lon* gene (70, 71); purified to apparent homogeneity, it has an ATPase activity (72). Hydrolysis of ATP to ADP + P_i is stimulated by (but not absolutely dependent upon) the addition of protein substrates (72). Though the mechanism of this enzyme has not yet been elucidated, a likely possibility is that peptide bond cleavage is coupled to ATP hydrolysis, as is the case with the ATP-requiring 5-oxoprolinase (EC 3.5.2.9) (73). The ATP-dependent conjugate-degrading system might be analogous to the *E. coli* enzyme, but it is definitely not similar, since it appears to be specific for ubiquitin-conjugated proteins, whereas the *lon* protease acts on free proteins. Furthermore, while the *lon* protease degrades proteins to small peptides by itself, the ATP-dependent conjugate-degrading system from reticulocytes has several components. Our fractionation studies indicates that the affinity-unadsorbed fraction contains at least three separable components, all of which are absolutely required for the breakdown of lysozyme–ubiquitin conjugates in the presence of ATP (unpublished results). The function of these factors is not known at present.

Another question raised by these studies was the structural nature of the conjugates that are substrates for the degrading system. We noted that high-molecular-weight lysozyme–ubiquitin conjugates are better substrates than low-molecular-weight conjugates. Furthermore, we (51) and Rechsteiner and co-workers (74) observed that the apparent molecular weight of lysozyme–ubiquitin conjugates is higher than expected from the number of amino groups in the protein. Thus, lysozyme has seven amino groups (six lysines and the α-amino group), and the calculated M_r of ubiquitin$_7$–lysozyme is 74,000, yet more than 12 bands of ubiquitin–lysozyme conjugates are formed, the higher of which have apparent molecular masses above 100 kDa. An explanation for the structure of the "super-high"-molecular-weight conjugates was provided by our recent observation that their formation requires free amino groups of ubiquitin (75). When the amino groups of ubiquitin are blocked by reductive methylation, it may be efficiently conjugated to lysozyme, but the higher-molecular-weight conjugates are not formed. The simplest explanation is that the high-molecular-weight conjugates of native ubiquitin contain polyubiquitin structures, in which one molecule of ubiquitin is linked to an amino group of another molecule of ubiquitin (75). Cyanogen bromide cleavage experiments suggested that linkages are mainly to ε-amino groups of ubiquitin (75), and thus the posttranslationally formed polyubiquitin chains differ from the biosynthetic "head-to-tail" polyubiquitin precursor (18).

We next asked whether the formation of polyubiquitin chains in ubiquitin–protein conjugates is obligatory for protein breakdown to occur. This does not seem so, since reductively methylated ubiquitin stimulates protein breakdown with fraction II at about half the rate obtained with native ubiquitin. Furthermore, isolated conjugates of lysozyme with methylated ubiquitin are efficiently degraded by the affinity-unadsorbed fraction in the presence of Mg-ATP. It was concluded that the formation of polyubiquitin chains is not obligatory for protein breakdown, though it may accelerate the rate of this process (75).

VI. Ubiquitin–Protein Lyases

In the course of the study on the ATP-dependent breakdown of ubiquitin-conjugated proteins (51), it became apparent that high-molecular-weight ubiquitin–protein conjugates are subject to degradation in reticulocyte extracts by two alternative pathways. In the presence of ATP, the protein moiety of the conjugates is degraded to acid-soluble products, as described above. On the other hand, in the absence of ATP the same, large conjugates are also rapidly degraded (51). This ATP-independent process is accompanied by the accumulation of smaller ubiquitin–protein conjugates and of free and undegraded protein (51). In earlier experiments we found that, upon incubation of ubiquitin–protein conjugates with reticulocyte extracts in the absence of ATP, free and reusable ubiquitin is released (50). It is evident, therefore, that, in the absence of ATP, the linkage between ubiquitin and protein in the conjugates is cleaved by ubiquitin–protein lyases present in reticulocyte extracts. At least part of these may be isopeptidases (which cleave the glycyl-ε-NH_2-lysine isopeptide linkage), but since other linkages are possible (see Section VII), the more general term of "ubiquitin–protein lyases" is preferred. That the ATP-independent ubiquitin–protein lyases act on the same high-molecular-weight conjugates that are substrates for the ATP-dependent degradation system is indicated by an experiment in which ^{125}I-lysozyme–ubiquitin conjugates were first incubated with reticulocyte extract in the absence of ATP and then ATP was supplemented. It was found that the first incubation in the absence of ATP abolished nearly completely the subsequent ATP-dependent release of acid-soluble radioactivity, indicating that the two systems compete for common ubiquitin–protein conjugate substrates (*51*).

Ubiquitin–protein lyases that act on multiubiquitinated proteins have not yet been sufficiently characterized. Preliminary experiments indicate that reticulocyte extracts contain at least three lyases that act on high-molecular-weight conjugates; their apparent molecular masses are 450,000, 300,000, and 100,000 Da (E. Leshinsky and A. Hershko, unpublished results). The cellular role of ubiquitin–protein lyases is unknown. It has been suggested that some lyases may have a corrective function, releasing incorrectly ubiquitinated proteins not destined for degradation (*50*). For this model, it will have to be further assumed that incorrectly ubiquitinated proteins become exposed to the action of lyases following their rejection from the ubiquitin conjugation machinery, since lyases do not appear to have high specificity for certain types of ubiquitin–protein conjugates. Such a compartmentalization between the two competing systems would also prevent the uncontrolled action of a futile cycle of ubiquitin conjugation and release. All these ideas remain to be examined.

Another role of ubiquitin–protein lyases might be to release ubiquitin linked to small peptides or to the ε-amino group of lysine. Such a process has been suggested to occur at the final stages of the ubiquitin proteolytic pathway (*50*). It is not known whether lyases that cleave high-molecular-weight ubiquitin–protein conjugates can also act on ubiquitin linked to small residues. On the other hand, a specific hydrolase that acts only on linkages between ubiquitin and a small molecule was recently described. This enzyme was originally described by Rose and Warms (*76*) as a thioesterase cleaving the linkage between the carboxyl-terminus of ubiquitin and simple thiols, such as dithiothreitol or glutathione. Such compounds are formed by the attack of thiols on ubiquitin thioesters of E_1 and E_2 (*76*). Subsequently, it was found that the same enzyme can cleave amide linkages between ubiquitin and small amines, such as lysine (ε-NH_2), glycine methyl ester, spermidine, and hydroxylamine (*77*). These derivatives can be formed by the E_3-independent transfer of ubiquitin from E_2 to amines (*60*). On the other hand, this hydrolase has no significant action on conjugates of ubiquitin with proteins such as lysozyme or cytochrome c. It was suggested that the hydrolase may function at the final stages of the proteolytic pathway. Alternatively, or in addition, it may act to "rescue" ubiquitin from its possible linkage with thiols or amines present in cells at high concentrations, such as glutathione or polyamines. The hydrolase from reticulocytes or erythrocytes can easily be purified, due to its high affinity for ubiquitin-Sepharose and its stability to trichloroacetic acid. It is a monomer of about 30,000 Da (77).

An additional possible role of at least some ubiquitin–protein lyases is to reverse protein modification due to ubiquitin attachment. As described below (Section XI), the reversible conjugation of ubiquitin to histones seems to be involved in histone modification, rather than breakdown. Andersen et al. (78) first described an enzyme activity in rat liver nucleoli that releases ubiquitin and histone from uH2A. The site of the cleavage was identified as the isopeptide linkage by the analysis of the carboxyl-terminus of released ubiquitin (79). Isopeptidases that cleave uH2A are present in the cytosol of all eukaryotic cells examined, but not in E. coli (80). It is not clear whether the cytosolic enzyme is similar to that observed in nuclei. In metaphase cells, there is some increase in nuclear isopeptidase activity, but more than 98% of total activity is still found in the cytosol (80). The enzyme from calf thymus has been partially purified; it binds to affinity columns containing histones, but not to ubiquitin-Sepharose (81). It is not known whether this isopeptidase is specific for ubiquitinated histones, or whether it can also act on other ubiquitin–protein conjugates. It is different from the reticulocyte enzymes that release ubiquitin from the high-molecular-weight conjugates, in that the latter enzymes do bind to the ubiquitin affinity column (E. Leshinsky and A. Hershko, unpublished results).

VII. Recognition of Protein Structure by the Ubiquitin System: Role of the α-Amino Group

A problem of central importance is what determines the specificity of the ubiquitin conjugation system, so that certain proteins are conjugated and are committed for degradation, while other proteins are not. Intracellular protein breakdown is a highly specific and selective process. Since proteolysis per se is an exergonic reaction, one can see the "justification" for the great expenditure of energy in the ubiquitin proteolytic pathway only if energy is used to obtain specificity. We are therefore intrigued by the question as to which specific features of protein structure are recognized by the ubiquitin conjugation system. Clearly, it cannot be the availability of an ε-amino group of any lysine residue, since most lysines are exposed on the surface of most native proteins. Our recent study, done in collaboration with Irwin Rose, indicates that the availability of a free α-amino group on the amino terminal residue of a protein is an important feature that determines its degradation by the ubiquitin system (82).

We began to realize the special role of the α-amino group when we examined the effects of the selective modification of amino groups of proteins on their degradation by the ubiquitin system. Other investigators had previously used complete blocking of the amino groups of proteins to distinguish between ubiquitin-dependent and ubiquitin-independent proteolytic pathways (83, 84). We compared the effects of increasing degrees of modification of amino groups of lysozyme by reductive methylation and carbamoylation at pH 6 and found that, in the latter treatment, the breakdown of lysozyme was inhibited at much lower extents of amino-group modification than in the case of reductive methylation (82). Though different effects of different protein modifications are subject to different interpretations, one possible interpretation is that under the conditions employed for carbamoylation, a specific amino group of the protein that is more important for ubiquitin conjugation than the other amino groups is blocked. That this may be the α-amino group was suggested by the information that, when carbamoylation with cyanate is carried out at a slightly acidic pH, the α-amino group reacts about 100-fold faster than the ε-amino group, due to the lower pK_a of the former. In the case of hemoglobin, even more selective methods have been worked out to carbamoylate the amino-terminal α-amino groups (with negligible modification of ε-amino groups), because of its potential use in the treatment of sickle-cell anemia. We therefore repeated the experiment, using selectively N^α-carbamoylated globin chains, and found that in this case, too, degradation by the ubiquitin system was prevented, even though practically all ε-amino groups remained free (82).

We next examined the influence of selective blocking of α-amino groups of proteins on their conjugation with ubiquitin. We found that N^α-carbamoylation of globin or lysozyme greatly decreased their conjugation with ubiquitin. The formation of high-molecular-weight conjugates containing several molecules of ubiquitin was most drastically inhibited. This is compatible with the notion that conjugation of ubiquitin to the amino-terminus may precede ε-amino conjugation in the formation of high-molecular-weight ubiquitin–protein conjugates. It seems, however, that the requirement for a free α-amino group is specific to a pathway leading to the formation of high-molecular-weight ubiquitin–protein conjugates committed for degradation. Thus, the formation of the monoubiquitin derivative of histone 2A does not require a free α-amino group, since this histone has a blocked amino-terminus (23).

Our next question was whether proteins with blocked amino-termini can be made into substrates by creating new α-amino groups. This could be done by the addition of poly(amino-acid) side-chains with N-carboxyamino acid anhydride. By this process, the ε-amino group of the lysine residue to which the poly(amino-acid) is attached is replaced by the α-amino group of the side-chain. We indeed found that polyalanylation of lysozyme derivatives that had previously been carbamoylated at their α-amino groups restored to a large extent their susceptibility to degradation by the ubiquitin proteolytic system (82).

Further experiments indicated that a free α-amino group in the absence of ε-amino groups is sufficient for degradation by the ubiquitin system. This could be examined for the case of guanidinated proteins, since quanidination with O-methylisourea blocks ε-amino groups, but not the α-amino group. Rechsteiner and co-workers first observed that guanidinated proteins are degraded in reticulocyte lysates by an ATP-dependent process (74), and we found that the degradation of guanidinated proteins requires ubiquitin and the three ubiquitin-conjugating enzymes (82). In fraction II from reticulocytes, guanidinated lysozyme is degraded at about 25% of the rate of unmodified lysozyme. It might be, therefore, that conjugation of ubiquitin to ε-amino groups serves to accelerate the rate or to increase the affinity of the substrate to some component of the proteolytic system.

In all the above experiments, we used chemical modification techniques to explore the role of the α-amino group. To examine whether naturally occurring modification of amino-termini of proteins has a similar effect, we determined the degradation of some N^α-acetylated proteins by the ubiquitin proteolytic system from reticulocytes. Many cellular proteins have acetylated amino-termini, but the function of N^α-acetylation is not known (85). We found that N^α-acetylated cytochrome c and enolase from mammalian tissues are not degraded by the ubiquitin system, while their nonacetylated counterparts from yeast are good substrates (82). None of the naturally occurring N^α-acetylated proteins tested were degraded by the ubiquitin system. On the other hand, not all proteins with free amino-termini are good substrates. It might be that in the latter cases, the unblocked α-amino group is buried and is thus not available to the action of the ubiquitin system.

The cumulative evidence from the above experiments led to the suggestion (82) that the exposure of an amino-terminal α-amino group may be a requisite signal for protein degradation by the ubiquitin system. As opposed to the ε-amino groups of lysine residues, which are mostly exposed, amino-termini in many native proteins are buried within the protein structure. The exposure of a buried amino-terminus may be the consequence of several types of alterations in protein structure, such as denaturation or subunit dissociation. It may even be imagined that specific modulators or metabolites may control the rate of degradation of particular proteins by inducing conformational changes leading to the exposure of their amino-termini. This hypothesis remains to be examined by direct methods.

An interesting exception to the above generalizations has recently been reported by Gregori and colleagues (86). These investigators found that calmodulin from bovine brain is not degraded by the ATP, ubiquitin-dependent proteolytic system from reticulocytes, while calmodulin from *Dictyostelium discoideum* is degraded at a fairly rapid rate. Mammalian calmodulins are N^α-acetylated, but *Dictyostelium* calmodulin also has a blocked amino-terminus (87); the nature of the blocking group has not been identified. Examination of the formation of conjugates of ubiquitin with the various calmodulins (in the presence of fraction II from reticulocytes) showed that mammalian calmodulin is not conjugated, while *Dictyostelium* calmodulin forms a prominent monoubiquitin derivative. No higher molecular weight conjugates of *Dictyostelium* calmodulin were observed even in the presence of hemin, which inhibits conjugate-degrading enzymes. The site of the conjugation of ubiquitin to *Dictyostelium* calmodulin was tentatively identified as lysine-115, by analysis of cyanogen bromide fragments of the labeled conjugate. The authors noted that mammalian calmodulin contains a trimethyllysine residue at the same position, and proposed that methylation of this residue may protect mammalian calmodulin against degradation by the ubiquitin system (86). It might be very interesting to find out what is unique about the structural environment of lysine-115, as opposed to those of the seven other lysine residues in *Dictyostelium* calmodulin. Another point of interest is that two general "rules," which we observed for a variety of protein substrates (i.e., requirement for a free α-amino group and for the formation of high-molecular-weight conjugates) do not apply in the case of *Dictyostelium* calmodulin, suggesting a possible connection between the two requirements. It might be that several ubiquitin ligation systems exist for various types of proteins, one that requires a free α-amino group and the formation of high-molecular-weight conjugates, and another that does not.

The above observations may pertain to the question as to how some N^α-acetylated proteins are broken down within cells. The idea that N^α-acetylation may protein proteins against degradation was first suggested by Jörnvall (88). On the other hand, N^α-acetylated and non-acetylated proteins of cultured cells turn over at similar rates (89). Though there are some questions about the specificity of the methods used in this study, other reports also indicate that some proteins blocked in α-amino groups or in all amino groups, can be degraded in intact cells. Katznelson and Kulka (83) reported that proteins completely blocked in their amino groups are not degraded in reticulocyte extracts, but are in hepatoma cells following microinjection. These authors concluded that microinjected proteins are degraded by an energy-dependent nonlysosomal system that differs from the ubiquitin pathway (83). Other investigators also observed fairly rapid degradation of N^α-blocked proteins following microinjection, such as N^α-carbamoylated hemoglobin (90) or N^α-acetylated cytochrome c (63). Tanaka et al. (84) observed that though the modification of all amino groups of proteins reduces their rate of degradation in reticulocyte lysates, the residual degradation is still stimulated by ATP. The ATP-dependent breakdown of amino group-blocked proteins is inhibited by hemin (84) and vanadate (91). It was suggested that ATP has two roles in protein breakdown, one requiring ubiquitin and another that is independent of ubiquitin (84). It is possible that some α-amino-blocked proteins are degraded by non-ubiquitin-dependent (but ATP-dependent) proteolytic systems. It is also possible, however, that N^α-acetylated proteins are subject in cells to deacetylation, or even to a single endoproteolytic cleavage, which would expose a new amino-terminal α-amino group and thus become subject to the action of the ubiquitin system. These alternative possibilities remain to be investigated.

While a free α-amino group of proteins appears to be an important recognition determinant, it seems reasonable to assume that it is not the only one. Solubilized brain hexokinase is rapidly degraded by the ubiquitin proteolytic system from reticulocytes, while mitochondria-bound hexokinase is not (92). It remains to be seen whether the dissociation of hexokinase from mitochondria exposes its amino-terminus or whether some other structural alteration takes place. Reduction and thiol-alkylation of bovine serum albumin increase its rate of degradation by the ubiquitin system (93). It has been suggested that this is not due to denaturation, since reduced albumin (without alkylation) is

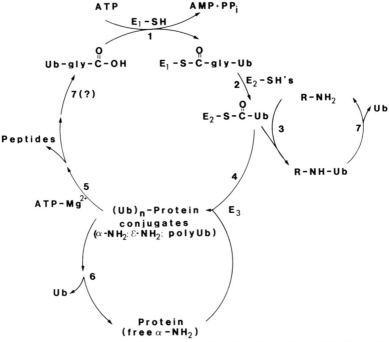

FIG. 2. Proposed sequence of reactions in the ubiquitin pathway. Ub, Ubiquitin.

degraded at a much slower rate. No correlation was found between the charge, hydrophobicity, or aggregation of the various derivatives of serum albumin and their susceptibility to degradation (93). It is not clear from this study which specific structural alteration (produced by reduction–alkylation) makes this protein more susceptible to the action of the ubiquitin system.

VIII. Proposed Sequence of Events in the Ubiquitin Proteolytic Pathway

Figure 2 summarizes our present knowledge of the intermediary reactions in the ubiquitin proteolytic pathway and our working hypothesis on the sequence of events in this process. The following steps are indicated. (1) Activation of the carboxyl-terminal glycine residue of ubiquitin and its transfer to the thiol site of the ubiquitin-activating enzyme (E_1). (2) Transfer of activated ubiquitin from E_1 to a variety of ubiquitin-carrier proteins (E_2s). (3) Some species of E_2 can transfer ubiquitin directly to small amines or to certain basic proteins

(60). Such types of reactions may be involved in the formation of the monoubiquitin derivatives of histones. (4) Some type(s) of E_2 are donors of ubiquitin for the formation of multiubiquitin conjugates of proteins, which are substrates for degradation. For this reaction, the action of E_3 is required, and a free α-amino group of the protein substrate is necessary. The conjugates formed may contain ubiquitin linked to the α-amino group, to ε-amino groups, or linked in polyubiquitin branched chains. (5) A major fate of proteins conjugated to multiple molecules of ubiquitin is their degradation to small peptides by an enzyme system which requires Mg-ATP. The role of ATP in this process is not known, and its intermediary reactions remain to be elucidated. (6) An alternative pathway for the degradation of ubiquitin–protein conjugates is the release of free ubiquitin and undegraded protein by the action of ubiquitin–protein lyases. These enzymes may have a corrective function or some other roles. (7) Ubiquitin linked to small amines or thiol compounds is released by the action of a specific hydrolase. This enzyme may also carry out the release of ubiquitin from peptide products of the breakdown of ubiquitin-conjugated proteins.

IX. Involvement of tRNA in Ubiquitin-Mediated Protein Breakdown

Recently, it was found that the ubiquitin-mediated and ATP-dependent degradation of labeled serum albumin in reticulocyte extracts is strongly and specifically inhibited by ribonucleases (RNases) (94). Deoxyribonuclease I (EC 3.1.21.1) had no inhibitory effect.

To determine whether the inhibitory effect of the RNases is due to their enzymatic activities rather than to some unusual effects of the protein molecules, enzymatic activity was inhibited prior to addition of the enzyme to the proteolytic system. RNase A (EC 3.1.27.5) was incubated with human placental RNase inhibitor, and micrococcal nuclease (EC 3.1.31.1) was incubated with thymidine 3',5'-bisphosphate (pTp), a specific inhibitor of the enzyme (95). Incubation of RNases with their specific inhibitors completely abolished their ability to inhibit protein breakdown. Likewise, omission of Ca^{2+} ions, essential for the activity of micrococcal nuclease (95), prevented inhibition of proteolytic activity.

To test directly the notion that RNA is required for the ubiquitin- and ATP-dependent proteolytic system, phenol-extracted total RNA from crude reticulocyte fraction II was added to an RNA-depleted proteolytic system. Added RNA completely restored proteolytic activity. To determine which RNA species could restore proteolytic activity, total RNA from fraction II was separated by gel electrophoresis. Only the tRNA-sized species could restore the activity of the nuclease-treated system; equivalent amounts of 7-SL, 5.8-S, and 5-S RNAs had no stimulatory activity.

Since the active component comigrated with tRNA, it was interesting to determine whether any individual tRNA species might be sufficient to reconstitute proteolytic activity. Certain patients with autoimmune diseases, such as systemic lupus erythematosus and polymyositis, produce antibodies directed against subsets of tRNAs (96). Serum from three such patients was used to isolate pure tRNA species for addition to the nuclease-treated proteolytic system. Serum MN precipitated a single RNA, identified by RNA-sequence analysis as tRNA^{His}. Serum SU precipitated five major species, one of the most prominent spots being tRNA^{His}. Serum LL also precipitated several species, none of them being tRNA^{His}. When tRNAs isolated by immunoprecipitation were added to the nuclease-treated proteolytic system, tRNA^{His} (precipitated by serum MN) was sufficient to restore over 80% of the proteolytic activity. The tRNAs precipitated by serum SU (which includes tRNA^{His}) also restored proteolytic activity, but tRNAs precipitated by serum LL (lacking tRNA^{His}) had no effect.

The tRNA requirement of the ubiquitin- and ATP-dependent proteolytic system shows an apparent substrate specificity: although the degradation of bovine serum albumin, α-lactalbumin, and soybean trypsin inhibitor is RNase-sensitive, the degradation of lysozyme, casein, β-lactoglobulin, and oxidized ribonuclease is not inhibited by RNases. Furthermore, tRNA is necessary for ubiquitin conjugation to RNase-sensitive proteolytic substrates. Specific ubiquitin–substrate conjugates of these proteins are not formed in reticulocyte fraction II treated with RNase. Upon addition of tRNA, these conjugates are reformed (96a).

Further studies are required to elucidate the exact role of tRNA in the ubiquitin- and ATP-dependent proteolytic system. Does tRNA play a regulatory role in proteolysis, or does it participate in some reaction of the ubiquitin pathway? Stimulation of rates of protein breakdown has been correlated with decreased levels of aminoacyl-tRNA in *E. coli* (97) and cultured mammalian cells (98), but there is no evidence that the ubiquitin system is involved in these phenomena. Furthermore, it is not known whether decreased aminoacylation is required for the stimulatory effect of tRNA in the reticulocyte cell-free system. An alternate explanation is that tRNA is directly involved in the mechanism of the degradation of specific proteins. The finding that tRNA increases steady-state levels of ubiquitin–protein conjugates is compatible with the notion that either it is required for ubiquitin conjugation to certain proteins, or that it inhibits the activity of a specific competing isopeptidase. It will be interesting to find out whether proteins that require tRNA for degradation share a common structural feature not present in proteins that do not require tRNA. This may give a clue to another aspect of the recognition of protein structure by the ubiquitin system. It was noted that the degradation of reduced-carboxymethylated bovine serum albumin, α-lactalbumin, and soybean trypsin inhibitor is inhibited by RNases to the same extent as that of their native counterparts. It is therefore conceivable that the structural feature common to all tRNA-requiring substrates resides in the primary structure of these proteins, since most secondary and tertiary structures are destroyed by reduction–carboxymethylation.

X. Evidence for Ubiquitin-Dependent Proteolysis in Various Cells

The ubiquitin-mediated proteolytic pathway was characterized and analyzed in rabbit reticulocytes. Since reticulocytes are terminally differentiating cells, it was important to know whether the ubiquitin-mediated proteolytic pathway is involved in energy-dependent degradation of cellular proteins in other eukaryotic cells as well. To date, attempts to establish ubiquitin-dependent cell-free proteolytic systems from cells other than reticulocytes have not been successful. This might be due to technical problems, such as the inactivation of ubiquitin by a lysosomal protease in liver extracts (99). However, experiments with intact cells of various types indicate the widespread occurrence and major physiological functions of the ubiquitin system, as described below.

A. Conjugation of Ubiquitin to Abnormal Proteins Is Proportional to Their Rates of Degradation

Several groups of investigators have used microinjection techniques to study intracellular protein degradation. Microinjected proteins are degraded at different and specific rates and show correlations between protein half-lives and molecular size or charge (74, 100, 101). Furthermore, the degradation of some microinjected proteins is under physiological control. The degradation of glutamine synthetase is subject to regulation by glutamine, as is the case with the endogenous enzyme (102, 103). The degradation of microinjected RNase A is under the regulation of serum (104), as is the degradation of cellular proteins (105). Proteins microinjected into cultured cells are degraded mainly in the cytosol (106). The degradation of microinjected proteins in hepatoma cells is energy-dependent and cannot be inhibited by lysosomotropic agents (83). Taken together, these results indicate that the degradation of most microinjected proteins is carried out by a soluble, nonlysosomal, and energy-dependent proteolytic system.

To study the role of the ubiquitin pathway in the breakdown of abnormal proteins, Chin et al. (107) injected HeLa cells with either ^{125}I-ubiquitin or ^{125}I-hemoglobin. Following incubation with phenylhydrazine (which oxidizes hemoglobin and thus denatures it), each radioactive protein was converted to a species of higher molecular weight. The molecular sizes of the derivatives were in agreement with the notion that they consist of increasing numbers of ubiquitin molecules conjugated to a single molecule of globin. Furthermore, the ratios of radioactivity of ubiquitin to that of globin in the various conjugates increased in proportion to increasing molecular weight. Experiments in which ^{125}I-ubiquitin and ^{131}I-hemoglobin were simultaneously microinjected into HeLa cells yielded similar results. If ubiquitin serves to mark globin chains for degradation, it is expected that the intracellular concentration of ubiquitin–globin conjugates should be proportional to the rate of hemoglobin degradation. This was indeed the case. When HeLa cells injected with ^{125}I-hemoglobin were incubated with increasing concentrations of phenylhydrazine, there was a linear relationship between the intracellular concentration of ubiquitin–globin conjugates (as determined by polyacrylamide-gel electrophoresis) and the rate of protein degradation (as determined by the release of acid-soluble ^{125}I label to the medium).

Immunochemical analysis of the turnover of ubiquitin–protein conjugates in intact cells was consistent with the hypothesis that covalent attachment of ubiquitin to the proteolytic substrate is an intermediate step in its degradation (108). A polyclonal antibody that reacts with ubiquitin–protein conjugates was used to isolate such conjugates from cells. If ubiquitin–protein conjugates are intermediates in protein degradation, their levels should reflect a balance between rates of conjugate formation and breakdown. Rates of conjugate formation may be affected, in turn, by the availability of rapidly degradable cellular proteins. To test this prediction, the rapidly degraded amino-acid analog containing abnormal proteins and the slowly turning-over normal proteins in reticulocytes and Ehrlich ascites tumor cells were compared. In reticulocytes, the major product of protein synthesis, hemoglobin, is normally stable, but abnormal globin chains containing lysine or valine analogs are degraded rapidly. Ehrlich ascites tumor cells selectively and rapidly degrade abnormal proteins and short-lived normal proteins (109). Reticulocytes were labeled with [^{3}H]leucine in the presence of the lysine analog, 4-thialysine, or the valine analog, t-α-amino-β-chlorobutyric acid, and immunoprecipitated with affinity-purified anti-ubiquitin immunoglobulins. In cells labeled without the addition of analogs, about 0.5% of total pulse-labeled protein was immunoprecipitated. A 10-fold increase in labeled immunoreactive protein was observed under conditions of abnormal protein formation. In a similar experiment with Ehrlich ascites tumor cells, the ubiquitination pattern resembled that observed in reticulocytes, except that the levels of ubiquitin conjugates in Ehrlich tumor cells were generally higher and the effect of the analogs was less pronounced.

To ascertain that immunoprecipitated label represents abnormal proteins conjugated to ubiquitin, cells were labeled with [^{3}H]tryptophan instead of leucine. Since ubiquitin does not contain tryptophan (13), labeled tryptophan incorporated into immunoreactive proteins represents exclusively the nonubiquitin portion of the ubiquitin–protein conjugate. The results obtained in this experiment were identical to the results obtained with leucine labeling, indicating that under the conditions employed, most of the immunoprecipitated label is derived from the nonubiquitin moiety of ubiquitin–protein conjugates (108).

Next, the kinetics of decay of ubiquitin–protein conjugates was examined. It is expected that the decay of ubiquitin–protein conjugates will reflect the decay of the pool of degradable proteins from which they are derived. Reticulocytes and Ehrlich ascites tumor cells were labeled with [^{3}H]leucine in the presence of the lysine analog; further protein synthesis was then blocked by cycloheximide and the incubation continued. Proteins synthesized in the presence of the analog are degraded at fast but heterogeneous rates, and following the "chase" incubation, a population of relatively stable labeled proteins remained. The decay of labeled ubiquitin–protein conjugates proceeded in parallel with the degradable portion of the labeled protein pool. Thus, after 1 hour, about 85% of the initial amount of labeled conjugates was degraded, compared to only 40% decay in total labeled proteins. Although these studies bear only on the case of abnormal proteins, they do not exclude the participation of the ubiquitin pathway in some other types of protein degradation. It was noted (108) that the extent of decay of ubiquitin conjugates of normal proteins markedly exceeds that of total labeled proteins, which indicates that some of those conjugates might be derived from a rapidly turning-over protein pool.

B. A Mammalian Cell-Cycle Mutant with Thermolabile Ubiquitin-Activating Enzyme Is Also Temperature-Sensitive for Intracellular Protein Breakdown

Identification of cell mutants defective in the ubiquitin pathway can help to define the cellular roles of this system. One such mutant has already been described and partially characterized (110, 111). The temperature-sensitive cell cycle mutant, designated ts85, was isolated from the mouse mammary carcinoma cell line FM3A (112). At the nonpermissive temperature (39°C), the mutant cells are arrested at the G_2 phase of the cell cycle and their chromatin fails to condense (112, 113). A marked decrease in the phosphorylation of H1 histone was observed in the mutant cells at 39°C, which was taken as supportive evidence for the role of H1 phosphorylation in chromatin condensation (112, 114). However, no thermolability of a histone kinase could be observed in cell-free extracts of mutant cells, which may indicate some other primary defect. The same investigators also noted that the ubiquitin–histone conjugate uH2A rapidly disappears in ts85 cells incubated at 39°C (115). Based on indirect evidence, it was suggested that the decrease in uH2A at the nonpermissive temperature is due to reduced synthesis, and not to accelerated degradation (116). Subsequently, the defect in ts85 cells was identified by direct methods (110). Conjugation of ubiquitin to proteins in crude extracts of ts85 cells was greatly reduced at 39°C, as compared to extracts of wild-type cells. Similar thermolability of the ubiquitin conjugation system was observed when its three components were purified by affinity chromatography. Seeking to identify which component of the ligase system is thermolabile in ts85 cells, the ubiquitin-activating enzyme (E_1) was purified to apparent homogeneity by covalent ubiquitin-affinity chromatography (58). The formation of ubiquitin-E_1 thioester was determined. The levels of the thioesters formed upon incubation at 30°C of purified E_1 from ts85 and FM3A cells were comparable. Incubation at 40°C, however, rapidly and specifically inactivates the formation of the ubiquitin thioester of E_1 from ts85 cells (110). This indicates that the mutation in E_1 is responsible for the thermolability of the ubiquitin conjugation system in ts85 cells.

If conjugation of ubiquitin to intracellular proteins is required for their degradation, the temperature-sensitive defect in the ubiquitin-activating enzyme of ts85 cells should result in defective protein degradation at the nonpermissive temperature. Cells were pulse-labeled with [^{35}S]methionine in the presence of analogs of valine and lysine. The degradation of the labeled proteins was followed during a "cold-chase" incubation (111). At the permissive temperature, all three cell lines examined (the mutant, the wild type, and a temperature-resistant revertant ts85R-MN3) degraded their proteins at approximately equal rates, with more than 70% of the labeled proteins degraded within 4 hours. In striking contrast, less than 15% of the labeled proteins were degraded during the same time in ts85 cells at the nonpermissive temperature (111). Degradation of abnormal proteins in the wild-type and the revertant cells did not display the temperature sensitivity seen with ts85 cells. The degradation of truncated puromycyl peptides is also specifically temperature sensitive in ts85 cells, but not in the wild-type or revertant cells. In another experiment, the degradation of normal proteins was examined under thermoinactivation conditions. Cells were pulse labeled for a short time (4 minutes) with [^{35}S]methionine, and the release of ^{35}S counts into acid-soluble products was monitored. Degradation of prelabeled proteins was strongly suppressed in ts85, but not in FM3A cells, at the nonpermissive temperature (111). Thus, not only abnormal proteins, but also normal, relatively short-lived proteins, fail to be degraded efficiently in ts85 cells at the nonpermissive temperature.

Immunochemical analysis of ubiquitin–protein conjugates at the permissive and nonpermissive temperatures, using polyclonal antibodies directed against the ubiquitin moiety of the conjugate (108), showed that at the nonpermissive temperature ts85 cells fail to ubiquitinate abnormal proteins. At the permissive temperature, the level of ubiquitin–protein conjugates in ts85 and FM3A cells treated with amino-acid analogs is approximately 4% of total pulse-labeled proteins. While similar results were demonstrated with the wild-type cells at the nonpermissive temperature, the formation of ubiquitin–protein conjugates was strongly suppressed in ts85 cells at the nonpermissive temperature (111).

In the course of these studies it was noted that at the nonpermissive temperature (39.5°C) there was a strong induction of heat-shock proteins in ts85 cells, but not in the wild-type cells (110, 111). This is consistent with the notion that the ubiquitin system has a role in the heat-shock response (see Section II). It was suggested that abnormal or damaged proteins, which would be expected to accumulate in ts85 cells at the nonpermissive temperature, may induce the heat-shock response by some direct or indirect mechanism (110).

Taken together, microinjection experiments, the immunochemical analysis of ubiquitin–protein conjugates, and experiments with the E_1-defective mutant cells provide strong evidence that covalent attachment of ubiquitin to the proteolytic substrate is required for its degradation in the eukaryotic nucleated cells studied. It should be noted, however, that all these studies bear directly only on the case of short-lived normal and abnormal proteins. The ubiquitin pathway mutant can also be used to study the mode of degradation of long-lived and other specific cellular proteins.

XI. Possible Roles of Ubiquitin in Histone Modification

In eukaryotic cell nuclei, the major ubiquitin–protein conjugate is ubiquitin–H2A semihistone (uH2A; see Section I). The function(s) of chromatin-associated ubiquitin–histone conjugates is not known. One possibility is that these conjugates are proteolytic intermediates in histone degradation. The turnover of nucleosomal core histone is generally very slow; half-life estimates range from several days to several months (117, 118). It is difficult to determine absolute rates of turnover for these stable proteins, because of reutilization of label released from proteins with shorter half-lives. It was possible, however, to determine relative rates of turnover among related proteins such as the histone variants under steady-state conditions where the relative amounts of proteins remain constant. In exponentially growing murine erythroleukemia cells, the H2A.1 variant has the highest relative rate of turnover, while H2A.2 has the lowest (119). A similar finding was observed for variants H2B.2 and H2B.1, respectively. These differences in turnover rates increased even further after induction of these cells to terminal differentiation. If ubiquitin–histone conjugates are intermediates in the breakdown of histones, it is expected that the level of these conjugates would be different for the variants differing in their rate of turnover and that this difference should increase upon induction of cells to differentiation. Quantification of ubiquitin conjugates of the different histone variants did not reveal any difference in their relative levels. The authors concluded, therefore, that it is unlikely that these conjugates are intermediates in the proteolysis of histones (119). The ubiquitin moiety of uH2A turns over more rapidly than its H2A moiety, the level of which was almost stable during the 24-hour period of the experiment (120). These observations also make it unlikely that the ubiquitin–histone adducts are intermediates of histone breakdown. It should be noted that, in the reticulocyte system (51), conjugation with several ubiquitin molecules is necessary to render most proteins susceptible to proteolytic attack, while all the ubiquitin–histone adducts described so far have a single ubiquitin modification.

A possible function in chromosome condensation during mitosis was advanced by Matsui and colleages (121), who observed that uH2A disappears from the nucleosomes shortly before metaphase and reappears in the G_1 phase. The reappearance of conjugated histone in the G_1 phase did not require protein synthesis. It was suggested that the removal of uH2A from metaphase chromosomes plays a role in triggering mitotic chromosome condensation. Disappearance of ubiquitinated histones during mitosis was also observed by Wu et al. (120), who suggested that there is a dynamic equilibrium between ubiquitin, H2A, and uH2A, and that the equilibrium is shifted at mitosis, perhaps by inhibiting the formation of uH2A. They further observed that all ubiquitinated histones behave similarly. Mueller and colleagues (122) recently described similar events in the slime mold *Physarum polycephalum*. In this organism, all nuclei divide synchronously, which allows very precise studies of cell-cycle-related events. In the early prophase, all ubiquitinated histones (uH2A variants and uH2B) are present, but they all disappear in the metaphase, which lasts 7 minutes. When the nuclei enter anaphase, which lasts 3 minutes, all ubiquitin–histone conjugates reappear. It was concluded that the cleavage of ubiquitin from histone conjugates is a very late event in chromatin condensation, and that ubiquitination is an early event in their decondensation. It should be noted, however, that removal of ubiquitin from core chromosomal histones might be an obligatory, but not sufficient condition for chromatin to condense. The $ts85$ mutants (see Section X,B) lose all their ubiquitinated histones at the nonpermissive temperature, but their chromatin remains largely dispersed (116).

Another possible function of histone ubiquitination is the regulation of gene expression. Early experiments by Busch and co-workers indicated that increased transcription of ribosomal genes is accompanied by a decrease of nucleolar uH2A (reviewed in 2, 23). By contrast, another study from the same group showed that the level of uH2A is increased in actively transcribing avian erythroid cells as compared to inactivate mature chicken erythrocytes (123). More recent work by Varshavsky and co-workers (124–126) suggested that ubiquitin attachment to histone molecules of nucleosomes of specific genes is associated with increased transcription.

One of the major difficulties in designing experiments which will address the problem of the role of ubiquitin-modified histones is that it is not known whether H2A molecules in all nucleosomes have an equal probability of being ubiquitinated, or whether only a specific subset of nucleosomes is modified. Major changes in ubiquitin modification occurring in a small specific subset of nucleosomes can be overlooked when changes in total chromatin are determined. To approach this problem, Levinger and Varshavsky (29) developed a two-dimensional electrophoretic system for fractionation of nucleosomes following micrococcal nuclease digestion. The system resolves core mononucleosomes in which either one or both H2A molecules are substituted by uH2A; the single replacement is most abundant (29). They used the system to compare the structure of mononucleosomes from different regions of the *Drosophila melanogaster* genome (124). DNA of resolved mononucleosomes was hybridized to different specific probes following denaturation *in situ*. They found that approximately 50% of the nucleosomes of the transcribed *Copia* and heat-shock 70 (hsp 70) genes in nonshocked cells contain uH2A semihistone. In contrast, less than 4% of the nucleosomes of the tandemly repeated, nontranscribed 1.688 satellite DNA contain uH2A. They concluded that nucleosomes containing uH2A semihistone reside largely in transcribed genes. It should be noted that hsp 70 genes are transcribed at a low rate in nonshocked cells, but the rate of transcription is dramatically accelerated upon heat-shock. Upon induction of heat-shock, there is total loss of recognizable nucleosomal organization of the heat-shocked genes. Virtually no hsp 70-specific mononucleosomes are detected in even mild staphylococcal nuclease digests of chromatin from shocked cells. The authors suggested that, while moderate levels of transcription are associated with a recognizable nucleosomal organization that is modified by ubiquitin, much higher transcriptional rates result in loss of the organized nucleosomal structure. It should be noted that in these studies, the state of ubiquitination of different genes was compared, and it is not certain that satellite DNA is a suitable control for the genes studied. In order to draw more definite conclusions as to the role of histone modification in gene expression, it would be more appropriate to compare the same gene at different states of activity. Another problem is that the identification of ubiquitinated nucleosomes relies solely on their electrophoretic mobility, and it is possible that some other alteration in nucleosome structure produces a similar electrophoretic migration.

Barsoum and Varshavsky (126) further examined the distribution of variant nucleosomes within a transcriptionally active gene. Digestion of genomic and cDNA clones of dihydrofolate reductase (EC 1.5.1.3) with restriction endonucleases (EC 3.1.21.3–.5) yielded fragments that could be used as specific hybridization probes for different regions of the gene. Two-dimensional hybridization mapping of nucleosomes with a full-length cDNA probe showed a 2-fold increase in monoubiquitinated nucleosomes, as compared to bulk nucleosomes. By contrast, when a 5'-end-specific probe was used (which hybridizes to the first exon of the 6-exon dihydrofolate reductase gene), a striking increase was revealed in the content of nucleosomes in which both H2A molecules are substituted by uH2A. These disubstituted nucleosomes are a very minor species in bulk chromatin. When a probe specific for the second exon was used (220 nucleotides downstream from the first region), there was an abrupt fall in the abundance of diubiquitinated nucleosomes. The abundance of mono-uH2A-substituted nucleosomes is also high at the 5' end of the gene, and decreases toward the 3' terminal region. These results were taken as evidence for the role of histone ubiquitination in gene expression, since several other alterations had been previously detected at the 5' termini of active genes. It should again be noted that more proof seems to be needed that the altered electrophoretic migration of these variant nucleosomes is due to histone ubiquitination. Such criticism has recently been raised concerning the electrophoretic migration of nucleosomes that package the active IgK chain gene (126a).

In summary, both proteolytic and nonproteolytic functions of ubiquitin conjugation to histones are possible, but there is no conclusive evidence yet for either possibility. Studies on the turnover of bulk cellular ubiquitinated histones (119, 120) appear to rule out a function in the degradation of most histones. A likely possibility is that the conjugation of a single ubiquitin to histones serves to modify chromatin structure, but the functions of the modification are unknown. Further investigation is needed to settle the problem of whether the structural alterations have a role in the packaging of mitotic chromatin or in exposure to factors involved in the transcription of specific genes.

XII. Concluding Remarks

In the past few years, considerable progress has been made in the elucidation of the mode of action and cellular roles of the ubiquitin pathway. The main enzymatic steps in the formation of ubiquitin–protein conjugates have been delineated, and a broad outline of the major routes of the degradation of ubiquitin-conjugated proteins has been described. Powerful tools are now available to study the biological functions of the ubiquitin system, including specific antibodies, a temperature-sensitive mutant, microinjection techniques, and cloned genes. Still, many major problems remain unsolved, and the unknown greatly exceeds what we presently know of the ubiquitin system.

What determines the specificity of the ubiquitin conjugation system for commitment of a certain protein for degradation? It appears reasonable to assume that a free α-amino group is only one of several features of protein structure recognized by the ubiquitin ligation system. What are the roles of E_3 or of the different species of E_2 in the recognition of protein structures? How does the system distinguish between ubiquitin ligation leading to protein breakdown and that involved in protein modification? What determines whether a particular protein is conjugated with a single ubiquitin or with many ubiquitin molecules? While we have no answers to these questions at present, the available information suggests that there are several different ubiquitin–protein ligation systems, and that these may act on different types of cellular proteins. Examples are the E_3-independent conjugation reaction, which acts on certain basic proteins (60); tRNA-dependent conjugation reactions, specific to certain other proteins; and the α-amino-independent ubiquitin ligation and degradation of *Dictyostelium* calmodulin (86). Only in one of these cases had the responsible enzyme been traced to particular species of E_2 (60). It remains to be seen whether the different substrate-specific conjugation systems are carried out by different species of E_2, E_3, or by additional factors. Among other specific questions which still await elucidation, the intermediary reactions in the degradation of ubiquitin-conjugated proteins, the role of ATP in conjugate breakdown, and that of tRNA in the ubiquitin system are noteworthy.

Concerning the cellular functions of the ubiquitin proteolytic pathway, the available evidence is limited to its involvement in the breakdown of abnormal and rapidly turning-over normal proteins. It is now possible to do more sophisticated experiments to examine the role of the ubiquitin system in the turnover of specific cellular proteins. In addition, the roles of the ubiquitin system in a variety of basic cellular processes such as cell-cycle-related events, gene expression, and the heat-shock response are of considerable interest. In some of these cases, possible nonproteolytic functions of ubiquitin conjugation have to be taken into account. It is quite possible that histone ubiquitination is but one example of ubiquitin function in protein modification, rather than breakdown. Using immunochemical (108) and microinjection (127) probes, it was noted that a considerable fraction of cellular ubiquitin–protein conjugates is stable; these may represent modification products, rather than degradation intermediates. The conjugation of a single, or a few ubiquitin molecules may modulate enzyme activity, produce alterations in structural proteins, or change the activity of regulatory proteins. Evidently, much more progress is required to elucidate the functions of ubiquitin in protein modification and breakdown (see Addendum, p. 301).

Acknowledgments

We thank Dr. Irwin A. Rose for helpful comments on the manuscript. A.H. acknowledges support from U.S. Public Health Service Grant AM-25614 and a grant from the United States–Israel Binational Science Foundation. A.C. was supported by a Research Career Development Award from the Israel Cancer Research Fund. A portion of the work of A.H. at the Institute for Cancer Research, Philadelphia, was supported by a grant of the American Cancer Society to Irwin A. Rose.

References

1. R. Schoenheimer, "The Dynamic State of Body Constituents," pp. 25–46. Harvard Univ. Press, Cambridge, Massachusetts, 1942.
2. A. Hershko and A. Ciechanover, *ARB* **51**, 335 (1982).
3. A. Hershko, *Cell* **34**, 11 (1983).
4. A. Ciechanover, D. Finley and A. Varshavsky, *J. Cell Biochem.* **24**, 27 (1984).
5. A. L. Goldberg and J. F. Dice, *ARB* **43**, 835 (1974).
6. A. L. Goldberg and A. C. St. John, *ARB* **45**, 747 (1976).
7. F. J. Ballard, *Essays Biochem.* **13**, 1 (1977).
8. E. A. Khairallah, J. S. Bond and J. W. C. Bird (eds.), "Intracellular Protein Catabolism." Liss, New York, 1985.
9. G. Goldstein, M. Scheid, U. Hammerling, E. A. Boyse, D. H. Schlesinger and H. D. Niall, *PNAS* **72**, 11 (1975).
10. M. Scheid, G. Goldstein, U. Hammerling and E. A. Boyse, *Ann. N.Y. Acad. Sci.* **249**, 531 (1975).
11. T. L. K. Low, G. B. Thurman, M. McAdoo, J. McClure, J. L. Rossio, P. H. Naylor and A. L. Goldstein, *JBC* **254**, 981 (1979).
12. T. L. K. Low and A. L. Goldstein, *JBC* **254**, 987 (1979).
13. D. H. Schlesinger, G. Goldstein and H. D. Niall, *Bchem* **14**, 2214 (1975).
14. D. H. Schlesinger and G. Goldstein, *Nature* **255**, 423 (1975).
15. D. C. Watson, W. B. Levy and G. H. Dixon, *Nature* **276**, 196 (1978).
16. E. Dworkin-Rastl, A. Shrutkowski and M. B. Dworkin, *Cell* **39**, 321 (1984).
17. J. G. Gavilanes, G. G. de Buitrago, R. Perez-Castells and R. Rodriguez, *JBC* **257**, 10267 (1982).
18. E. Özkaynak, D. Finley and A. Varshavsky, *Nature* **312**, 663 (1984).
19. S. Vijay-Kumar, C. E. Bugg, K. D. Wilkinson and W. J. Cook, *PNAS* **82**, 3582 (1985).
20. A. Ciechanover, Y. Hod and A. Hershko, *BBRC* **81**, 1100 (1978).
21. R. E. Lenkinsky, D. M. Chen, J. D. Glickson and G. Goldstein, *BBA* **494**, 126 (1977).
22. I. L. Goldknopf and H. Busch, *Cell Nucl.* **6**, 149 (1978).
23. H. Busch and I. L. Goldknopf, *MCBchem* **40**, 173 (1981).
24. H. Busch, *Methods Enzymol.* **106**, 238 (1984).
25. M. H. P. West and W. M. Bonner, *Bchem* **19**, 3238 (1980).
26. M. H. P. West and W. M. Bonner, *NARes* **8**, 4671 (1980).

27. I. L. Goldknopf, M. F. French, R. Musso and H. Busch, *PNAS* **74**, 5492 (1977).
28. H. G. Martinson, R. True, J. B. E. Burch and G. Kunkel, *PNAS* **76**, 1030 (1979).
29. L. Levinger and A. Varshavsky, *PNAS* **77**, 3244 (1980).
30. A. M. Kleinschmidt and H. G. Martinson, *NARes* **9**, 2423 (1981).
31. W. M. Bonner and J. D. Stedman, *PNAS* **76**, 2190 (1979).
32. U. Bond and M. J. Schlesinger, *MCBiol.* **5**, 949 (1985).
33. O. Wiborg, M. S. Pedersen, A. Wind, L. E. Berglund, K. A. Marcker and J. Vuust, *EMBO J.* **4**, 779 (1985).
34. M. Jzquierdo, C. Arribas, J. Galcerari, J. Burke and V. M. Cabrera, *BBA* **783**, 114 (1984).
35. P. K. Lund, B. M. Moats-Staats, J. G. Simmons, E. Hoyt, A. J. D'Ercole and J. J. Van Wyk, *JBC* **260**, 7609 (1985).
36. M. Ashburner and J. J. Bonner, *Cell* **17**, 241 (1979).
37. M. J. Schlesinger, G. Aliperti and P. K. Kelley, *TIBS* **7**, 222 (1982).
38. F. C. Neidhardt, R. A. VanBogelen and V. Vaughn, *Annu. Rev. Genet.* **18**, 295 (1984).
39. P. M. Kelley and M. J. Schlesinger, *Cell* **15**, 1277 (1978).
40. T. A. Phillips, R. A. VanBogelen and F. C. Neidhardt, *J. Bact.* **159**, 283 (1984).
41. T. A. Baker, A. L. Grossman and C. A. Gross, *PNAS* **81**, 6779 (1984).
42. S. A. Goff, L. P. Casson and A. L. Goldberg, *PNAS* **81**, 6647 (1984).
43. S. A. Goff and A. L. Goldberg, *Cell* **41**, 587 (1985).
44. M. V. Simpson, *JBC* **201**, 143 (1953).
45. A. Hershko and G. M. Tomkins, *JBC* **246**, 710 (1971).
46. J. D. Etlinger and A. L. Goldberg, *PNAS* **74**, 54 (1977).
47. A. Ciechanover, S. Elias, H. Heller, S. Ferber and A. Hershko, *JBC* **255**, 7525 (1980).
48. K. D. Wilkinson, M. K. Urban and A. L. Haas, *JBC* **255**, 7529 (1980).
49. A. Ciechanover, H. Heller, S. Elias, A. L. Haas and A. Hershko, *PNAS* **77**, 1365 (1980).
50. A. Hershko, A. Ciechanover, H. Heller, A. L. Haas and I. A. Rose, *PNAS* **77**, 1783 (1980).
51. A. Hershko, E. Leshinsky, D. Ganoth and H. Heller, *PNAS* **81**, 1619 (1984).
52. A. Ciechanover, H. Heller, R. Katz-Etzion and A. Hershko, *PNAS* **78**, 761 (1981).
53. A. L. Haas, J. V. B. Warms, A. Hershko and I. A. Rose, *JBC* **257**, 2543 (1982).
54. A. L. Haas and I. A. Rose, *JBC* **257**, 10329 (1982).
55. A. L. Haas, J. V. B. Warms and I. A. Rose, *Bchem* **22**, 4388 (1983).
56. A. Hershko, A. Ciechanover and I. A. Rose, *JBC* **256**, 1525 (1981).
57. K. D. Wilkinson and T. K. Audhya, *JBC* **256**, 9235 (1981).
58. A. Ciechanover, S. Elias, H. Heller and A. Hershko, *JBC* **257**, 2537 (1982).
59. A. Hershko, H. Heller, S. Elias and A. Ciechanover, *JBC* **258**, 8206 (1983).
60. C. M. Pickart and I. A. Rose, *JBC* **260**, 1573 (1985).
61. S. Speiser and J. D. Etlinger, *PNAS* **80**, 3577 (1983).
62. E. Eytan and A. Hershko, *BBRC* **122**, 116 (1984).
63. R. Katznelson and R. G. Kulka, *EJB* **146**, 437 (1985).
64. R. Hough and M. Rechsteiner, *PNAS* **81**, 90 (1984).
65. S. Rapoport, W. Dubiel and M. Müller, *FEBS Lett.* **180**, 249 (1985).
66. K. H. S. Swamy and A. L. Goldberg, *Nature* **292**, 652 (1981).
67. F. S. Larimore, L. Waxman and A. L. Goldberg, *JBC* **257**, 4187 (1982).
68. M. Desautels and A. L. Goldberg, *JBC* **257**, 11673 (1982).
69. S. Watabe and T. Kimura, *JBC* **260**, 5511 (1985).

70. M. F. Charette, G. W. Henderson and A. Markovitz, *PNAS* **78**, 4728 (1981).
71. C. H. Chung and A. L. Goldberg, *PNAS* **78**, 4931 (1981).
72. L. Waxman and A. L. Goldberg, *PNAS* **79**, 4883 (1982).
73. A. P. Seddon, L. Li and A. Meister, *JBC* **259**, 8091 (1984).
74. M. Rechsteiner, N. Carlson, D. Chin, R. Hough, S. Rogers, D. Roof and K. Rote, in "Protein Transport and Secretion" (D. L. Oxender, ed.), pp. 391–402. Liss, New York, 1984.
75. A. Hershko and H. Heller, *BBRC* **128**, 1079 (1985).
76. I. A. Rose and J. V. B. Warms, *Bchem* **22**, 4234 (1983).
77. C. M. Pickart and I. A. Rose, *JBC* **260**, 7903 (1985).
78. M. W. Andersen, N. R. Ballal, I. L. Goldknopf and H. Busch, *Bchem* **20**, 1100 (1981).
79. M. W. Andersen, I. L. Goldknopf and H. Busch, *FEBS Lett.* **132**, 210 (1981).
80. S. I. Matsui, A. A. Sandberg, S. Negoro, B. K. Seon and G. Goldstein, *PNAS* **79**, 1535 (1982).
81. F. Kanda, S. I. Matsui, D. E. Sykes and A. A. Sandberg, *BBRC* **122**, 1296 (1984).
82. A. Hershko, H. Heller, E. Eytan, G. Kaklij and I. A. Rose, *PNAS* **81**, 7021 (1984).
83. R. Katznelson and R. G. Kulka, *JBC* **258**, 9597 (1983).
84. K. Tanaka, L. Waxman and A. L. Goldberg, *J. Cell Biol.* **96**, 1580 (1983).
85. F. Wold, *ARB* **50**, 783 (1981).
86. L. Gregori, D. Marriott, C. M. West and V. Chau, *JBC* **260**, 5232 (1985).
87. D. R. Marshak, M. Clarke, D. M. Roberts and D. M. Watterson, *Bchem* **23**, 2891 (1984).
88. H. Jörnvall, *J. Theor. Biol.* **55**, 1 (1975).
89. J. L. Brown, *JBC* **254**, 1447 (1979).
90. K. B. Hendil, *J. Cell. Physiol.* **105**, 449 (1980).
91. K. Tanaka, L. Waxman and A. L. Goldberg, *JBC* **259**, 2804 (1984).
92. M. Magnani, V. Stocchi, M. Dacha and G. Fornaini, *MCBchem* **61**, 83 (1984).
93. A. C. Evans and K. D. Wilkinson, *Bchem* **24**, 2915 (1985).
94. A. Ciechanover, S. L. Wolin, J. A. Steitz and H. F. Lodish, *PNAS* **82**, 1341 (1985).
95. P. Cuatrecasas, S. Fuchs and C. B. Anfinsen, *JBC* **242**, 1541 (1967).
96. J. A. Hardin, D. R. Rahn, C. Shen, M. R. Lerner, S. L. Wolin, M. D. Rosa and J. A. Steitz, *J. Clin. Invest.* **70**, 141 (1982).
96a. S. Ferber and A. Ciechanover, *JBC* **261**, 3128 (1986).
97. A. C. St. John, K. Conklin, E. Rosenthal and A. L. Goldberg, *JBC* **253**, 3945 (1978).
98. O. A. Scornik, *JBC* **258**, 882 (1983).
99. A. L. Haas, K. E. Murphy and P. M. Bright, *JBC* **260**, 4694 (1985).
100. N. T. Neff, L. Bourret, P. Miao and F. J. Dice, *J. Cell Biol.* **91**, 184 (1981).
101. M. Rechsteiner, D. Chin, R. Hough, T. McGarry, S. Rogers, K. Rote and L. Wu, in "Cell Fusion," Ciba Found. Symp. 103 (D. Evered and J. Whelan, eds.), pp. 181–201. Pitman, London, 1984.
102. G. Arad, A. Freikopf and R. G. Kulka, *Cell* **8**, 95 (1976).
103. A. Cassel-Freikopf and R. G. Kulka, *FEBS Lett.* **128**, 63 (1981).
104. J. M. Backer, L. Bourret and F. J. Dice, *PNAS* **80**, 2166 (1984).
105. A. Hershko, P. Mamont, R. Shields and G. M. Tomkins, *Nature NB* **232**, 206 (1971).
106. S. Bigelow, R. Hough and M. Rechsteiner, *Cell* **25**, 83 (1981).
107. D. T. Chin, L. Kuehl and M. Rechsteiner, *PNAS* **79**, 5857 (1982).
108. A. Hershko, E. Eytan, A. Ciechanover and A. L. Haas, *JBC* **257**, 13964 (1982).
109. W. D. Yushok and R. Frech, *Proc. Am. Assoc. Cancer Res.* **15**, 94 (1974).
110. D. Finley, A. Ciechanover and A. Varshavsky, *Cell* **37**, 43 (1984).

111. A. Ciechanover, D. Finley and A. Varshavsky, *Cell* **37**, 57 (1984).
112. Y. Matsumoto, H. Yasuda, S. Mita, T. Marunouchi and M. Yamada, *Nature* **284**, 181 (1980).
113. S. Mita, H. Yasuda, T. Marunouchi, S. Ishiko and M. Yamada, *Exp. Cell Res.* **126**, 407 (1980).
114. H. Yasuda, Y. Matsumoto, S. Mita, T. Marunouchi and M. Yamada, *Bchem* **20**, 4414 (1981).
115. T. Marunouchi, H. Yasuda, Y. Matsumoto and M. Yamada, *BBRC* **95**, 126 (1980).
116. Y. Matsumoto, H. Yasuda, T. Marunouchi and M. Yamada, *FEBS Lett.* **151**, 139 (1983).
117. S. L. Commerford, A. L. Carsten and E. P. Cronkite, *PNAS* **79**, 1163 (1982).
118. L. P. Djondjurov, N. Y. Yancheva and E. C. Ivanova, *Bchem* **22**, 4905 (1983).
119. C. W. Grove and A. Zweidler, *Bchem* **23**, 4436 (1984).
120. R. S. Wu, K. W. Kohn and W. M. Bonner, *JBC* **256**, 5916 (1981).
121. S. I. Matsui, B. K. Seon and A. A. Sandberg, *PNAS* **76**, 6386 (1979).
122. R. D. Mueller, H. Yasuda, C. L. Hatch, W. M. Bonner and E. M. Bradbury, *JBC* **260**, 5147 (1985).
123. I. L. Goldknopf, G. Wilson, N. R. Ballal and H. Busch, *JBC* **255**, 10555 (1980).
124. L. Levinger and A. Varshavsky, *Cell* **28**, 375 (1982).
125. A. Varshavsky, L. Levinger, D. Sundin, J. Barsoum, E. Özkaynak, P. Swerdlow and D. Finley, *CSHSQB* **47**, 511 (1983).
126. J. Barsoum and A. Varshavsky, *JBC* **260**, 7688 (1985).
126a. S-Y. Huang, M. B. Barnard, M. Xu, S. I. Matsui, S. E. Rose and W. T. Garrard, *PNAS* **83**, 3738 (1986).
127. J. Atidia and R. G. Kulka, *FEBS Lett.* **142**, 72 (1982).

DNA Polymerase-α: Enzymology, Function, Fidelity, and Mutagenesis

> LAWRENCE A. LOEB,*
> PHILIP K. LIU,* AND
> MICHAEL FRY†
>
> * The Joseph Gottstein Memorial
> Cancer Research Laboratory
> University of Washington
> Department of Pathology SM-30
> Seattle, Washington 98195
> † Unit of Biochemistry
> Faculty of Medicine
> Technion-Israel Institute of
> Technology
> Haifa, 31096, Israel

This review is focused on the structure and catalytic properties of DNA polymerase-α,[1] the major DNA polymerase in eukaryotic cells, because of its central role in DNA synthesis and because of new, exciting experimental results. Until recently, progress in understanding the biochemistry of the enzyme was slow and the pathways to knowledge mimicked mainly studies in bacterial systems. Conditional mutants in DNA replication, so instrumental in isolating and defining the role of DNA replication proteins in bacteria, only recently became available from animal cells. In recent years, progress has been made in ascertaining the structure of this enzyme; mutant cell lines that contain an altered enzyme have been established, and the gene coding for this enzyme has been assigned to a specific chromosome. Moreover, protocols to clone the gene for DNA polymerase-α are developing. The purpose of this review is to assemble knowledge on the structure and catalytic properties of the α polymerase and use it as a framework for applying and evaluating the results of new molecular approaches to this problem.

---

[1] The proposal, to name these as DNA polymerase-α, -β, and -γ, replacing the original and often confused I (or II), II (or I), and III, originated with Baltimore, Bollum, Gallo, Korn, and Weissbach; it was published in ref. 18, and by Gillespie, Saxinger, and Gallo in tabular form in this series, Vol. 15, p. 88, 1975 (Eds.).

I. Identification of DNA Polymerase-α

The accurate synthesis of the genetic information in living cells is catalyzed by a class of enzymes officially named DNA-directed DNA polymerase (EC 2.7.7.7), usually simply designated DNA polymerases and abbreviated Pol. These enzymes occur in all prokaryotes and eukaryotes examined, and are also encoded by the genomes of many viruses. Despite considerable variation in the structure of DNA polymerases from different sources, diverse association with auxiliary proteins, and dissimilarity in catalytic behavior and *in vivo* roles, DNA polymerases from every source have the same basic catalytic characteristics.

The generic requirements for catalysis by a DNA polymerase were initially defined (*1*, *2*), as follows. (1) DNA polymerases utilize deoxyribonucleoside triphosphates (dNTPs) as substrates for the formation of DNA by sequentially incorporating deoxyribonucleoside monophosphate residues (dNMPs) by cleavage of the α–β phosphodiester bond with release of inorganic pyrophosphate. A template DNA and a metal ion activator are required. The order of polymerization of the deoxynucleotide substrates in the product DNA is dictated primarily by the base sequence of the template DNA. Thus, DNA polymerases and other such polymerases are unique in that the specificity of catalysis is determined by another molecule. (2) In addition, DNA polymerases require a short segment of DNA or RNA complementary to the template that has a free 3′-hydroxyl group in the terminal deoxyribose or ribose moiety. This oligonucleotide acts as a primer, and polymerization of the first nucleotide residue proceeds from that 3′-hydroxyl position. (3) All DNA polymerases catalyze DNA synthesis exclusively in the 5′ → 3′ direction.

The major DNA polymerase activity in eukaryotic cells has been designated DNA polymerase-α.[1] This activity was initially shown (*3*) to exist in calf thymus, and soon was identified, partially purified, and characterized in diverse organisms, tissues, and cells. In animal tissues, the amount of DNA polymerase activity can be correlated with the proliferative activity of the cells (*4–12*). For this reason, initial efforts at purification utilized rapidly dividing cells exhibiting one predominant DNA polymerase activity. The first extensive purification of the major animal cell DNA polymerase (*13*) from calf thymus yielded a high-molecular-weight DNA polymerase sedimenting at approximately 12 S (*13*). DNA polymerase purified shortly thereafter from rapidly dividing sea urchin embryos (*14*) was mostly localized in the cell nucleus, and utilized double-stranded DNA templates more efficiently than denatured DNA. That the initially described Pol-α activity was associated with replication of genomic DNA was clear from studies showing that its activity increased dramatically as cell division was induced in quiescent human lymphocytes by the mitogenic plant lectin phytohemagglutinin (4, 5, 8), during liver regeneration after partial hepatectomy (6, 15), and when quiescent cells in tissue culture were stimulated to divide (7, 8–12). Common properties of polymerases of the α type eventually led to their classification as DNA polymerases with molecular size in excess of 100 kDa, an acidic isoelectric point, inhibition by salt (>50 mM), and by reagents that react irreversibly with sulfhydryl groups (18).

About a decade after the initial isolation of DNA polymerase-α (Pol-α) from calf thymus, a distinctly different polymerase (later termed DNA polymerase-β),[1] was isolated from several types of cells (19–21). This polymerase was distinguished from Pol-α by virtue of its uniquely smaller molecular size, basic isoelectric point, and resistance to chemicals that react with thiol groups (18). Subsequently, a third type of animal cell polymerase, later designated DNA polymerase-γ, was isolated (22–26). Polymerase-γ differs from α- and β-polymerases by its synthetic template requirements, preference for Mn^{2+} as a metal activator, and a requirement for >100 mM salt for maximum DNA synthesis *in vitro,* as well as by its sensitivity to agents that react with protein sulfhydryl groups (18, 22–26).

Following their discoveries, each of these animal DNA polymerases was identified in a wide variety of organisms, tissues, and cell types. In 1975, Weissbach *et al.* (18) proposed a set of criteria for the classification of known animal cell DNA polymerases into the three main groups by which they are known today: α, β, and γ.[1] Whereas these three types of polymerase are all devoid of associated exonuclease activity, a fourth type of DNA polymerase intimately associated with a 3' → 5' deoxyribonuclease activity was isolated from rabbit bone marrow (27). That activity, denoted DNA polymerase-δ, was later shown to exist also in calf thymus (28). However, it remains to be seen whether the δ-polymerase is sequestered in specific tissues or cells, or whether it is as omnipresent as are the α-, β-, and γ-polymerases.

As more was learned about the physical, catalytic, and biological properties of DNA polymerase-α in higher eukaryotes, it became increasingly apparent that it plays a major role both in the replication of genomic DNA and in its repair synthesis. Pol-β and -γ appear to have relatively more restricted functions in repair synthesis of nuclear DNA and in the replication of mitochondrial DNA, respectively (29).

II. Biochemical Characteristics of DNA Polymerase-α

Until recently there has been no generally accepted method for the purification of the catalytic core component of Pol-α and the establishment of a definitive "holoenzyme" complex in which each of the subunits is present in constant stoichiometric ratio. Thus, we review the structure and catalytic properties of the enzyme only briefly, as these must be regarded as tentative.

A. Structure of Polymerase-α

Despite intense efforts to define the precise molecular structure of Pol-α, we still do not have a firm notion of the detailed multisubunit composition of this cardinal enzyme. Perhaps the major single obstacle is the difficulty encountered in attempts to purify it in its native form. The α-polymerases display a high degree of heterogeneity among different species and, more importantly, it frequently appears in a number of physically and catalytically distinct molecular forms within a single cell type. Purification of Pol-α is further impeded by its relatively low concentration even within dividing cells and by its instability under conditions of conventional cell fractionation and enzyme purification.

Pol-α appears to be exceptionally sensitive to proteolytic cleavage. The unexpected observation that polymerase fragments are catalytically active has been a major hindrance to obfuscation in defining the size of the catalytic "core" subunit (30–34). It is interesting to speculate that this proteolytic degradation to catalytically active subunits may have important biologic implications. Second, it is now recognized that Pol-α is composed of several polypeptide subunits and that it may be associated with additional auxiliary proteins. However, the precise number, size, and function of the various subunits have not yet been established.

Most of the accumulated knowledge of the enzymology of DNA polymerase-α comes from functional *in vitro* assays for the enzyme itself and for activities associated with it. Assays for nucleotide polymerization *in vitro* permitted a detailed description of the properties of the "catalytic unit" of α-polymerase. Likewise, a specific assay for the production of an RNA-primer allowed the definition and characterizations of a "DNA-primase" activity intimately associated with Pol-α (35).[2] The structure and function of Pol-α, its subunits, and its auxiliary proteins are now being actively investigated by the develop-

---

[2] "DNA primase" is a generic name for an RNA polymerase activity synthesizing oligoribonucleotides that "prime" the initiation of DNA synthesis (2).

ment of new functional assays (36, 38), immunochemical techniques (39–44), selection for specific mutations in DNA replication proteins (45, 46), and by the application of DNA transfection (47, 48) and molecular cloning methodology (43).

Until the recent introduction of immunochemical isolation techniques (41, 42, 44), the structure of polymerase-α was deduced from analysis of purified enzyme preparations obtained by the conventional techniques of proteins purification. Several research groups employing laborious and prolonged purification procedures obtained preparations of α-polymerase of near or apparent homogeneity (15, 17, 58). These groups used proliferating tissues such as calf thymus (49, 52), regenerating liver (15, 53), bone marrow (54), embryonic tissues (17, 31, 55), or dividing cultured cells (50, 57) as sources for the enzyme. These choices were based on the observation that dividing cells show an elevated activity of Pol-α and that it is involved in DNA replication. Purification from cellular extracts was conducted by the usual steps of salting-out, ion-exchange and DNA cellulose column chromatographies, gel filtration, and velocity sedimentation. These procedures yielded purified polymerase preparations at a low recovery, thus necessitating large quantities of cells, which in turn required larger columns and even longer purification procedures.

Early reports on the molecular size of native Pol-α varied widely with values ranging between 70 and 1000 kDa (59). From more recent experimental results, it can now be assumed that the lower range of molecular size (perhaps <180 kDa) was due to isolation of partially degraded but catalytically active enzyme molecules (reviewed in ref. 60). The higher range of observed sizes (perhaps >500 kDa) may also be artifactual, resulting in part from aggregation in media of low ionic strength (61, 62). Furthermore, the presumed asymmetric nature of the Pol-α molecule often led to different apparent sizes deduced from gel filtration and velocity sedimentation analyses (31, 53, 63, 64). Lastly, association of Pol-α with auxiliary proteins that are not formally subunits of the enzyme molecule may also increase its observed size (38, 58).

The introduction of either monospecific polyclonal (65, 76) or monoclonal (39) antibodies against Pol-α eventually led to their use as powerful tools for enzyme purification. Monospecific immunoglobulins enabled the isolation of pure Pol-α by procedures requiring only one or a few steps using either immunoprecipitation or immunoaffinity column chromatography (41–44). "Activity gel electrophoresis" provided another technique for the identification of the catalytic subunit (32, 67, 68). This procedure permits a direct detection of the catalytic polypeptide of a polymerase after resolution by denaturing gel electrophoresis renaturing after removal of sodium dodecyl sulfate from the gel, and assaying for polymerization activity *in situ*. Finally, an increasing concern about the detrimental effect of proteolysis during purification has led to inclusion of multiple protease-inhibitors in each and every buffer used during the purification (see, for instance, *31* and *42*). With these techniques, a consensus on the size of the catalytic subunit of DNA polymerase-α is emerging. However, the number and size of its other subunits are still uncertain.

Studies on extensively purified DNA polymerase-α isolated by conventional chromatography yielded a plethora of sizes for catalytically active polypeptides. However, recent analyses of the enzyme purified by immunoaffinity increasingly indicate that it contains an exceptionally large catalytic polypeptide of 180–200 kDa. Thus it seems likely that the smaller size previously obtained for the enzymatically active polypeptides represents fragments resulting from proteolysis. However, the possibility that such fragments exist *in vivo* and that controlled proteolysis is a regulatory mechanism in the activation and inactivation of Pol-α cannot yet be excluded. Table I summarizes data on the sizes of the catalytic polypeptides of Pol-αs isolated from different types of eukaryotic cells. Note that there is close agreement on a size of 160–180 kDa for the catalytic core for all the Pol-αs isolated by the least disruptive immunoaffinity procedures. Also note that in many cases, polypeptides that lack DNA-polymerase activity copurify with the catalytic polypeptide. Whether these 40–85 kDa polypeptides represent subunits of the "holoenzyme" and what their functions may be is not clear at present.

B. Catalytic Mechanism for DNA Synthesis

DNA polymerase-α, like all known prokaryotic and eukaryotic DNA polymerases, replicates polynucleotide templates by the sequential addition of deoxyribonucleoside monophosphate residues to the 3′-hydroxyl terminus of a primer stem (*2*). Its catalytic properties are defined, therefore, by its interaction with template, primer, dNTP substrates, and divalent cation activators. The kinetic analysis of Pol-α action is very difficult because of the large variety of templates and primers that the enzyme can utilize, because of the multiple substrates, and because of the effects of protein cofactors and small molecules. Furthermore, much of the kinetic data that have been collected on DNA polymerase-α were obtained either for enzymes that contained only a part of the full complement of subunits and/or for polymerases that were partially degraded. In most cases, therefore, findings

TABLE I
SIZE AND SUBUNIT COMPOSITION OF HIGHLY PURIFIED PREPARATIONS OF POLYMERASE α[a]

Source of polymerase	Molecular form[b]	Specific activity[c] (fold purification)	Size of native enzyme (kDa)	Size of catalytic subunit (kDa)	Size of other subunit (kDa)	Number of other subunits	Remarks[d]	References
Calf thymus	A1, A2 (8.0–8.15)	(2000–3000)	200 ± 15	150–170 (1)	50–70 (1)	1		49, 64
	B (5.2 S)	(2000–3000)	103 ± 8	[e]				49, 64
	C (7.3 S)	0.15×10^5 (2000–3000)	165 ± 10	150–170 (1)				49, 64
	D (6.8 S)	0.1×10^5 (3000)	144 ± 10	[e]				70
	9 S	0.6×10^5 (12,000)	~500	148 or 158	48, 55, 59			16
	5.7 S	1.0×10^5 (20,000)	~250	123 and 134				71
[f]	10 S	1.0×10^5 (25,000)	200–250	140, 145, 150	45–50	3–4		72
[f]	10 S			180, 150 (240)	43–50		Immun.	73
[f]	10.2 S	0.35×10^5 (5000)	404	170, 180	60, 65, 70	3	Immun.	42
				118 (>140) + 54–64 (5)				10
	7.4 S		200–250	125, 150				74, 75
	>11.3 S		>500	125		11		74, 75
[f]	9.3 S	1.05×10^5 (5000)	223	160 (185)	68	1	Immun.	44
Rat liver			404–413		52–64	7		51
		0.45×10^5 (×3000)	155–250	156	54–64	4		53, 62

(continued)

TABLE I (Continued)

Source of polymerase	Molecular form[b]	Specific activity[c] (fold purification)	Size of native enzyme (kDa)	Size of catalytic subunit (kDa)	Size of other subunit (kDa)	Number of other subunits	Remarks[d]	References
Chick embryo		0.45×10^5 (49,000)	200	130–155 (3–4) (3–4)	51–59	4		55
Drosophila[f]								
Drosophila[f]		0.5×10^5		166, 185	85, 60		Immun.	76
Drosophila[f]		($\times 7500$)	280 (ref. 4)	182	60, 50			77
Xenopus laevis	α_2[g]	2.9×10^5 ($\times 41{,}600$)	180	120	55			78
Rabbit marrow	α_1	$0.3–2.0 \times 10^5$	215	135	54–66	4		54
Monkey BSC-1 cell				~190			Immun.	66
Monkey CV-1 cell		2.0×10^5 ($\times 53{,}000$)		(40, 70, 115) 118–176 (9)	30–62	4		79
Mouse myeloma				120 (76)				80
Human KB cell[f]				125–180 (4)	77, 49–60	6	Immun.	41
Mouse ascites		0.3×10^5		125–178 (3)	48–70	4		56
Human HeLa cell	α_2[h]	2.6×10^5	600	140 (1)	24–69	4		58
	α_3	Homogeneous	220	140 (1)	65	1		58

[a] Data are presented in the order of their description in Section II,B. Modified from Fry and Loeb (29).
[b] Molecular form is specified when one of several distinct molecular forms of α-polymerase was purified.
[c] Specific activity is expressed in units/mg purified enzyme protein. "Unit of activity" is according to each author's definition.
[d] Purification by immunoaffinity is specified as Immun. All other enzyme preparations were obtained by conventional protein purification procedures.
[e] Forms B and D are products of partial proteolytic digestion of form C (refs. 49, 64, 71). In parentheses: number of polypeptides with catalytic activity.
[f] An enzyme preparation that contains DNA primase activity.
[g] Form α_2 is devoid of DNA primase activity.
[h] An enzyme preparation that contains cofactors C1 and C2 (ref. 58).

on kinetic parameters of purified DNA Pol-α pertain mostly to the catalytic core of the enzyme. The possibility that the catalytic properties of an intact multimeric Pol-α that interacts with other cofactors may differ from those of the isolated core should therefore be kept in mind.

C. Interaction with Template and Primer Stem

Although the relative efficiency with which natural and synthetic polynucleotide templates are copied by α-polymerase depends on the source of the enzyme and is often different for its different molecular forms (*49, 69, 72, 81, 93*), all polymerases of the α class copy most efficiently templates composed of deoxyribonucleotides. Of these, a template that contains a large number of 3′-hydroxy termini, usually "activated" DNA, is the most effective. Although it was originally claimed that only Pol-α and Pol-β can copy the ribo-strand of oligo(dT) · poly(rA), more recent reports demonstrate that Pol-α from a variety of cells can utilize a primed poly(rA) template (*30, 81–83*). So far, there is no strong evidence that any Pol-α can copy a natural RNA template, although based on catalytic similarity to *E. coli* DNA polymerase I (EC 2.7.7.7) and to reverse transcriptase (EC 2.7.7.49), both of which can copy a synthetic ribopolymer (*85*), it would not be surprising to find that Pol-α can do so at a very low efficiency.

A catalytic hallmark of the α-polymerases is their preferential interaction with single-stranded DNA. Korn and associates (*33, 34, 83, 86–89*) made a detailed analysis of the interaction of nearly homogeneous Pol-α from human KB cells with the DNA template and with its other reactants. Analysis of the competition between different templates for Pol-α as well as direct velocity sedimentation analysis of enzyme–template complex formation indicated that whereas the polymerase did not bind to supercoiled or relaxed DNA duplexes and did not interact with the 3′-hydroxyl end of single-stranded DNA, it did associate with single-stranded templates (*33, 86, 87*). Furthermore, both human (*33, 89*) and mouse (*90*) Pol-αs bind preferentially to dT-rich sequences in homo- and heteropolymeric DNA. There is evidence that interaction of KB cell Pol-α with single-stranded DNA is followed by an allosteric activation of a second template-binding site within the enzyme molecule (*83, 90*). According to the proposed model, Pol-α identifies single-stranded regions on the template, binds to them, and activates a second site on the enzyme that binds to the primer stem. This model is based solely on kinetic evidence and needs to be confirmed by physical identification of template and other binding sites on the polymerase.

DNA Pol-α interacts *in vitro* with and extends both deoxyribonucleotide and ribonucleotide primer stems. Further, α-polymerases from various cell types, in contrast to β- and γ-polymerases, elongate synthetic and natural RNA primers *in vitro* (54, 57, 91, 92, 94–97). The failure of β- and γ-polymerases to utilize RNA primers and the ability of the α-polymerases to extend these primers constituted an early indication of the possible involvement of Pol-α in RNA-primed discontinuous synthesis of DNA.

The inability of mitrochondrial DNA polymerase to utilize RNA primers is not in accord with the recent finding that mitochondrial DNA Pol-γ can be found associated with primase activity (98). The ability of Pol-α to add deoxynucleotides onto ribonucleotide termini is well documented (2). By use of "hooked" polymers (single-stranded polynucleotides self-complementary at the 3' end) terminated with either ribo- or deoxyribonucleotides, it was demonstrated that α-polymerase extends both types of primers with similar K_m and V_{max} values. The minimum length of primer required by the KB cell Pol-α to initiate DNA polymerization is an octamer. This minimum size was invariant with changing reaction temperature (over a limited range) or with base composition of the primer (88). Furthermore, a single mismatch at the end of the primer stem prevents binding of the polymerase to the primer terminus (88). Although H, OH, or PO_4 groups at the 2' position in the terminal residue of the primer did not affect binding of the Pol-α, a 3'-phosphate in this residue diminished polymerase–primer association (88). With calf-thymus Pol-α (the 9-S form), a minimum primer length of four nucleotides is required for initiating the copying of M13 bacteriophage DNA, and a hexamer of either deoxy- or ribonucleotides served as a very efficient primer (99). A terminally mismatched primer is extended by calf thymus α-polymerase at a lower efficiency than a perfectly matched primer (100). The rate-limiting step in elongating a terminally mismatched primer is the addition of the first nucleotide after the mismatch (100). The low rate of Pol-α binding to a mismatched primer terminus and its inefficient elongation may have mechanistic significance with respect to the fidelity of DNA synthesis.

D. Nucleotide Substrates, Metal Activators, and Initial Order of Reactants

The K_m for the deoxynucleotide substrates for DNA Pol-α is usually estimated to be between 1 and 20 μM. Considering the scarcity of deoxynucleotide triphosphates in eukaryotic cells and even in nuclei at the time of DNA replication (101, 102), and the relatively high K_m of these enzymes for their substrates, a mechanism should exist for the efficient utilization of dNTPs by polymerase. One proposed mechanism is the existence of a DNA-replicating complex that channels the triphosphates onto the DNA polymerase (*103, 106*).

In addition to the four major dNTP substrates, Pol-α can also utilize dUTP as a substrate for DNA polymerization *in vitro* (*107, 108*). However, dUMP is not present in DNA made *in vitro* in isolated nuclei (*109*). Hence, either dUTP is hydrolyzed prior to its incorporation (*109*), or it is excised from the DNA product by uracil *N*-glycosylase (*109a*). These processes must be exceptionally efficient, since deoxyuridine is not found in detectable amounts in eukaryotic DNA from normal cells.

Studies on the competitive inhibition of incorporation between the antibiotic aphidicolin and dCTP and dTTP by Pol-α, but not of the incorporation of dATP and dGTP (*110, 111*), can be interpreted to indicate that Pol-α contains at least two substrate binding sites, one for purine and the other for pyrimidine nucleotides. Even though the concept of multiple nucleotide binding sites is appealing and can account for many kinetic phenomena observed with DNA polymerases (*86, 112*), it is unlikely to be correct considering the universality in the mechanism for catalysis by these enzymes and the considerable evidence for only one substrate binding site on *E. coli* DNA polymerase-I (*2*).

Catalysis by Pol-α requires a metal activator; activation is exerted by magnesium ions whereas many of the other divalent metal ions studied are much less efficient (*113*). As initially demonstrated by electron-spin resonance for the single metal-ion binding site of *E. coli* Pol-I, the major function of divalent metal ions in catalysis by Pol-α is likely to involve coordination between an incoming nucleotide substrate and the active catalytic site on the polymerase molecule (*114*). However, recent studies with crystals of the Klenow fragment of *E. coli* Pol-I indicate the presence of a tightly bound metal ion only at the $3' \rightarrow 5'$ exonucleolytic site on *E. coli* Pol-I (*115*). Conceivably, crystallization at a high concentration of ammonium sulfate displaces other metal ions.

Other aspects of Pol-α interaction with magnesium were investigated by Fisher and Korn (*88*). Magnesium ions enhance the binding of human Pol-α to single-stranded DNA and to polyprimidines, but not to polypurines. In addition to its effect on template binding, magnesium may be involved in the binding of the primer stem by Pol-α. Kinetic analysis demonstrated competitive inhibition by magnesium of binding of the enzyme to a $(dT)_{15}$ primer (*88*).

The requirement for a primer of minimum length of eight nucleotides as well as the lack of structural similarity between competing metal and primer led to the proposal that normal binding of an octameric primer stem by Pol-α occurs through a magnesium–primer complex that involves the coordination of four magnesium ions with the seven phosphate residues in the phosphodiester backbone of the primer. Furthermore, multiple magnesium binding sites in α-polymerase were suggested by kinetic analysis of the effect of metal ion on purine–pyrimidine "hooked-template" primers. It was postulated, therefore, that KB cell Pol-α contains four magnesium–primer-binding subsites (88). Since octameric primers possess only seven phosphodiester bonds and yet serve as optimal length initiators for KB cell Pol-α, it was suggested that one of the four magnesium ions might be available for polymerase translocation along the template, subsequent to the insertion of the next dNMP residue. The effect of magnesium on the binding of Pol-α to template and primer have been shown by direct sedimentation analysis to be independent of the presence or absence of dNTP substrates. Hence, it is argued that magnesium ions are required for the interaction of Pol-α and nucleic acid independently of catalysis. The low intracellular concentration of magnesium and the very high affinity of dNTPs for that metal raised the possibility that the dNTPs bring magnesium to the enzyme and, in turn, the metal ions are cycled during the polymerization (88).

Kinetic studies suggest an order to the reaction of Pol-α with its template, the primer, and the dNTP substrates. Early kinetic analyses of substrate incorporation suggested that mouse Pol-α first reacts with an activated DNA template followed by interactions with the nucleotide substrates (116). Steady-state kinetics of template and substrate binding by Pol-α suggested a kinetic model to rationalize initial velocity data and termination probabilities (117). The proposed model included sequential interaction of Pol-α with the primer–template, followed by nucleotide substrate. The model also distinguished between the initial polymerization step and elongation. As proposed, the rate-limiting step is the transition between initiation and elongation; this step is facilitated at higher levels of primer template (117).

A detailed steady-state kinetic analysis of KB cell Pol-α action established the order of the initial events of the interaction of this human cell polymerase with its reactants (88). This analysis was used in conjunction with velocity sedimentation analysis of polymerase–substrate binding, and with the use of a 2',3'-dideoxy blocked primer stem, which cannot be elongated and thus permits analysis of enzyme–substrate complex formation without DNA polymerization catalysis. The results of these combined approaches yielded a model that specifies binding of Pol-α to the template, followed by primer and then the dNTP substrates.

E. Elongation of DNA

1. Processivity

The number of phosphodiester bonds formed by Pol-α per template binding event has been investigated in several laboratories. A nonprocessive enzyme will bind to the templates and catalyze the formation of a single phosphodiester bond, and then detach from the template; a quasiprocessive polymerase will polymerize several nucleotides per template binding events; a fully processive enzyme will copy an entire template molecule before dissociating. Considering that the rate of collisions between two macromolecules is infrequent in solution and is limited by rates of diffusion, it must be assumed that rapid synthesis of DNA *in vivo* depends on a completely processive mechanism. Thus, if a purified Pol-α is not processive, it is very probable that ancillary cellular factors render it so *in situ*.

Several general methodologies have been devised for the measurement of the processivity of polymerases. (Detailed descriptions of these methods are contained in references *118–121*.) With the pioneering and relatively insensitive method of template challenge, Pol-α appeared to be entirely nonprocessive (*118, 122*). With the introduction of more sensitive methods for assessment of processivity, Pol-α was found to be quasiprocessive. Under various measurement conditions, and with Pol-αs from different sources, it polymerizes 11 ± 5 (*83*), 14.2 ± 3.1 (*91*), 5.6 ± 1.3 (*123*), and 19.5 ± 4.5 (*124*) nucleotides per template binding event. It should be noted that small molecules such as spermine (*123*) and ATP (*125*) increase the processivity of Pol-α somewhat. Also, different molecular forms of the polymerase often display different processivities (*93*), and addition of bacterial DNA-binding protein enhances the processivity of *Drosophila* Pol-α about 20-fold (*91*). It appears, therefore, that the overall processivity of the Pol-α holoenzyme *in vivo* in the presence of affector small molecules may very well be higher than that of the purified enzyme, but unlikely to be completely processive. Considering the importance of processivity to DNA synthesis *in vivo*, and the lack of high processivity of all eukaryotic DNA polymerases that have so far been described, with the exception of Pol-γ, a direct search for factors in animal cells that enhance processivity might be very fruitful (*126*).

2. BARRIERS TO DNA ELONGATION

Using a general method for the determination of sites of replication arrest, Weaver and DePamphilis (127) reported that during *in vitro* synthesis of Pol-α from calf thymus, monkey, and human cells, each of these enzymes decreased its rate of synthesis at the same positions using φX DNA as a template. About one-third of the stop signals possessed computer-predicted "hairpin" secondary structures and the Pol-α usually halted proximally to its stalk. However, in most cases, the advancement of Pol-α was stopped up to 25 bases upstream from any predicted hairpin structure. Moreover, results obtained in that study did not support the possibility that arrest sites are formed by long-range interactions between remote regions on the template (127). With singly- or multiply-primed single-stranded phage M13 DNA serving as a template for homogeneous Pol-α from *Drosophila melanogaster*, DNA elongation terminated after several hundred nucleotides, but addition of *Escherichia coli* single-stranded binding protein (SSB) increased the length of the products by about 2-fold (91). These results were interpreted as an indication of the ability of SSB to relax secondary structures in the template that may act as barriers for the advancement of Pol-α.

It is likely that DNA propagation barriers are a property of the template rather than of the polymerase since similar stoppage patterns are observed with *E. coli* DNA polymerase III and *Drosophila* Pol-α on a mitochondrial DNA template inserted into M13 phage DNA (128). A search for common structural features of the stop signals revealed that although some of those sites had the potential to form hairpin structures, not all of them did (128), and those that did not might be only kinetic pause sites. In studies with homogeneous and intact *Drosophila* Pol-α, the enzyme dissociated repeatedly at specific sites on singly primed φX174 DNA (129).

That these sites are kinetic pause sites rather than obstructions to polymerization could be demonstrated by the production of longer DNA molecules with extended incubation times and by the appearance of fully replicated molecules when Pol-α was added in excess to the template (129). Similarly, calf thymus 9-S Pol-α paused at several discrete sites along single-stranded M13mp7 template DNA, and the addition of *Escherichia coli* SSB permitted the polymerase to proceed through some of the pausing loci (99). In another study, computer-predicted secondary structures in M13mp8 single-stranded DNA could be correlated with pausing sites for calf thymus Pol-α (130). This analysis indicated that all replication barriers could be correlated with secondary structures predicted by computer calculations that allowed for long-range base-pairing, G · T mispairs, and "looping-out" of bases. Interestingly, calf thymus 5.7-S and 9-S Pol-α stopped at two bases before the hairpin structure; from there on, they polymerized a few more nucleotides in a fully distributive manner (130). The difference in correlations between experimentally determined stoppage by DNA polymerase and theoretically determined pause sites or secondary structures in the templates could be due to differences in stringency of the rules for secondary structure formation put forward by the different authors.

Finally, a possible permissive role for ATP in copying over pause sites has been suggested (56, 131), although a lack of effect of ATP on advancement of Pol-α has also been reported (130). The extent to which secondary structures in DNA form *in vivo*, and their potential as a barrier for the movement of the replicative complex in the cell, is completely unknown at present. However, *in vitro* studies may provide an assay for associated factors that enable the enzyme to proceed through replicative obstructions in the template.

3. GAP FILLING AND LONG-STRETCH DNA SYNTHESIS

Two catalytic parameters that distinguish between different DNA polymerases are preference for gaps of discrete size in duplex DNA and ability to fill such gaps to completion (as measured by the ability of DNA ligase to seal gaps). These parameters may reflect *in vivo* activities of DNA polymerases in short- and long-patch DNA repair, and thus could serve as *in vitro* criteria for involvement of these enzymes in different modes of DNA repair.

Unfortunately, the disparity of results obtained with Pol-αs from different cell types does not allow a simple unifying concept as to its involvement in different types of DNA repair. Human KB cell Pol-α is most reactive with gaps of 30 to 60 nucleotides, but cannot fill them to completion (33, 83, 132). A similar preference for gaps of 40 residues that could not be filled to completion was observed with monkey CV-1 cell Pol-α (133). However, calf thymus and HeLa Pol-α can fill such gaps (133). Moreover, human lymphoblastic Pol-α could fill single-nucleotide gaps (134). In another study, HeLa Pol-α failed to utilize gaps smaller than 15 nucleotides, whereas gaps of 20 to 63 residues were filled partially, leaving an unsealable gap of 15 bases remaining (84). These differences in gap utilization by Pol-αs from different sources may perhaps be explained by differences in the subunit composition of the enzymes used in the various studies. In accord with this presumption is the finding that form A1 of calf thymus Pol-α utilized gaps of 65 nucleotides more efficiently than form C, which lacked a 50 to 70 kDa subunit (93).

The ability of Pol-α to extend primers opposite long stretches of single-stranded template DNA has also not yielded uniform results in different investigations. The enzyme from calf thymus elongated RNA primers on linear φX174 DNA template for only 230 to 390 residues (97). Addition of Pol-β to the reaction system led to completion of the 4500-long RF structure. Similarly, Pol-α from chick and mouse cells synthesized *in vitro* relatively short DNA chains, about the size of *in vivo* Okazaki fragments, which could then be further extended by Pol-β (135). However, many studies report conditions in which Pol-α synthesizes long stretches of DNA on single-stranded DNA templates. The Pol-αs from *D. melanogaster* (91), *Xenopus laevis* (131), Ehrlich ascites mouse cells (37, 56), and calf thymus (99) can extend RNA or DNA primers by up to several thousand nucleotides.

III. Auxiliary Activities Associated with DNA Polymerase-α

The structural complexity of DNA polymerase-α (Table I) suggests that the catalytic core of the holoenzyme is associated with a number of other polypeptides that may play roles in the intricate replication process. Much effort has been put into the identification of functions other than DNA polymerization that reside in the Pol-α "holoenzyme." It is clear that most of these associated components are not present in extensively purified enzyme preparations. However, they might function in close association with Pol-α in the course of DNA replication *in vivo*.

A. Exonuclease and DNA Primase

All bacterial DNA polymerases studied have an $3' \rightarrow 5'$-exonuclease activity that hydrolyzes noncomplementary nucleotides immediately after their incorporation and thus could make a substantial contribution to the fidelity of DNA synthesis by these enzymes. It was a logical expectation that a similar activity would be an integral part of the eukaryotic DNA polymerases. In fact, in lower eukaryotes, exonucleolytic activities copurify with an α-like DNA polymerase. Exonucleases were identified in partially purified α-like polymerases from *Ustilago maydis* (136), *Euglena gracilis* (137), *Tetrahymena pyriformis* (138), and *Saccharomyces cerevisiae* [polymerase-B (II) but not polymerase-A (I), ref. 139]. Although early reports indicated that in some cases, Pol-α from higher eukaryotes is associated with an exonucleolytic activity, later studies using more highly purified enzyme preparations demonstrated that, in the large majority of cases, exonucleolytic activity could not be detected in purified Pol-α from sea urchin (*14*), *Drosophila melanogaster* (*31*), regenerating rat liver (*62*), calf thymus (*52*), and human KB cells (*140*), to mention but a few examples. However, at least two important exceptions should be noted: first, one of the two highly purified mouse myeloma Pol-α forms exhibited exonucleolytic activity that hydrolyzed DNA processively in the $5' \rightarrow 3'$ direction, and distributively in the $3' \rightarrow 5'$ direction (*57*). A second, more extensively documented case is that of Pol-δ isolated from rabbit bone marrow cells and from calf thymus, which displayed properties generally similar to those of Pol-α, but which, even in the most purified enzyme fractions, still contained an associated $3' \rightarrow 5'$ exonuclease activity (*27, 28*).

Discontinuous synthesis of nascent strands of DNA in both prokaryotes and eukaryotes is initiated by the formation of short oligoribonucleotides that are subsequently elongated by DNA polymerase (*141–145, 150*). In prokaryotes, the enzymes responsible for this activity can readily be isolated as distinct entities and are called "DNA primase." The possible association of Pol-α with RNA primers was first suggested by the finding that the α enzyme was the only class of eukaryotic DNA polymerase able to extend RNA primers (*91, 92, 94–97*). The first clue that DNA primase may be associated with Pol-α was the finding that the polymerase can copy unprimed synthetic homopolymers such as poly(dC) or poly(dT) in an ATP-requiring reaction (*146*).

In the last few years, several laboratories reported on the purification of DNA primase from a variety of animal cells. The tightness of the association between DNA primase and α-polymerase varies extensively among different cell types. In many cases, DNA primase activity is detected in a distinct molecular form of Pol-α whereas other form(s) of the polymerase are devoid of primase activity (*79, 147–149, 151*). In most cases, DNA primase could not be separated from the catalytic core of Pol-α (*79, 152–154*), and it was even considered to reside in the single catalytic polypeptide of calf thymus Pol-α (*156*). Despite the tight association between Pol-α and primase, the two activities could often be distinguished by their different heat labilities (*41*), or by differential sensitivity to N-ethylmaleimide or aphidicolin (*154*).

Finally, in some cells, DNA primase is strongly bound to the Pol-α catalytic core, but under favorable *in vitro* conditions can be dissociated from it. For example, in 3T3 cells that were arrested at the G_1–S boundary, primase can be separated from Pol-α by heparin-Sepharose chromatography (155). Primase from *Drosophila melanogaster* constitutes a subunit of the homogeneous primase–Pol-α complex, consisting of three polypeptide chains of 182, 60, and 50 kDa that are present in stoichiometric amounts (65, 77). Glycerol-gradient velocity-sedimentation in the presence of 2.8 M urea resolved the two activities. The polymerase activity was associated with the α catalytic subunit, while primase activity was detected in a mixture of the β and γ polypeptides (129). However, in the case of DNA primase from human lymphocyte or mouse hybridoma cells, the primase could be completely freed of Pol-α activity and consisted of two polypeptides of 46 and 56 kDa in 1 : 1 stoichiometry (157, 158). Clearly, this case of isolation of primase free of Pol-α stands in contrast to the tight association between the two described for other cell types. Interestingly, primase from the yeast *Saccharomyces cerevisae* is also readily dissociable from Pol-I the "analogue" of Pol-α, and is devoid of polymerase activity (159, 161).

In addition to the above studies, DNA primase has recently been isolated from mitochondria of human KB cells, and it appears to be associated with the latter's Pol-γ (98). If this mitochondrial primase is proved to be identical with the primase associated with Pol-α, it will strongly suggest that primase is an independent enzyme that binds to any polymerase that requires *de novo* initiation of DNA synthesis using RNA primers.

B. Primer Recognition and Ap₄A-Binding Proteins

Animal cell α-polymerase may associate with proteins that serve as primer recognition factors by decreasing the K_m of the enzyme for the 3′-hydroxyl primer terminus (38, 58, 162). A protein that binds ssDNA, called C1, has been isolated from HeLa cells (163). This protein, which was associated with α-polymerase in the crude cell extract, was separated from the enzyme by DNA-cellulose chromatography. Reconstitution of α-polymerase with partially purified C1 increased the rate of nucleotide polymerization with a denatured DNA template 15- to 30-fold (163). The product DNA synthesized by the C1 · Pol-α complex was covalently linked to the DNA template, suggesting that priming occurred by a backfolded hairpin structure.

Subsequently, three molecular entities containing Pol-α activity were separated from HeLa cells and purified (58). One of these, an enzyme complex of 600 kDa, which was designated polymerase-$α_2$, was associated with two protein cofactors (called C1 and C2), which were, respectively, a tetrameric 24 kDa polypeptide and a monomeric chain of 51 kDa (58). The C1C2 cofactors preferentially enhanced the utilization of denatured DNA by Pol-α_2 and could be dissociated from the enzyme by exposure to the nonionic detergent Triton (58). Analogous C1C2 factors have been purified from CV-1 cultured monkey cells (38). The single observed effect exerted by C1C2 on polymerase is the considerable decrease in the K_m of the enzyme for the 3'-hydroxyl terminus of the primer stem (162). Essentially all monkey cell Pol-α is complexed with C1C2 in crude extracts, and the purified core enzyme can reassociate with the dissociated factors. It seems, therefore, that C1C2 cofactors serve as accessory subunits that enable Pol-α to replicate long stretches of DNA that are relatively sparsely primed, as is the case in the advancing "replicating fork" (141, 162).

A less well-defined protein factor that may be analogous to C1C2 has been purified from calf thymus (165). This 70 kDa protein increases the rate of dNTP incorporation by Pol-α in copying ϕX174 ssDNA singly primed with a complementary DNA restriction fragment, and has been postulated to increase the affinity of the Pol-α for the primer stem (165).

The unusual purine nucleotide P^1,P^4-bis(adenosine 5')tetraphosphate diadenosine 5',5'''-P^1,P^4-tetraphosphate (Ap$_4$A) is a naturally occurring product and results from the back reaction of the amino-acid activation step in protein synthesis (166, 167). Changes in the intracellular level of Ap$_4$A have been implicated in the positive control of DNA replication during cell cycling (168–172). Moreover, Ap$_4$A has been found to bind to specific proteins that may be auxiliary proteins of Pol-α (51, 173, 174). Thus, one can speculate that Ap$_4$A might be a participant in the initiation of DNA replication during the cell cycle.

It was originally reported (168) that the intracellular concentration of Ap$_4$A in several cultured cell lines ranges from 0.1×10^{-4} to 20×10^{-4} of the ATP concentration and parallels the rate of cell division, which contrasted to the constancy of ATP concentrations. The opposite parallelism was also demonstrated: drug-induced arrest of cell entry into S phase led to a decline to 1/50th to 1/100th of the intracellular concentration of Ap$_4$A (168). This correlation was supported by the finding (169) that mitogenic stimulation of cultured 3T3 and BHK cells led to a 1000-fold increase in the intracellular level of Ap$_4$A. Others also noted a tight positive correlation between the intracellular concentration of Ap$_4$A and cell proliferative activity (170). Lastly, exposure of permeabilized hamster cells obtained at the G_1 phase of the cell cycle to Ap$_4$A led within minutes to enhanced incorporation of labeled thymidine into DNA, and this corresponded to the rate of thymidine incorporation using permeabilized S phase cells (171, 172).

The effect of Ap_4A on DNA synthesis was not mimicked by other adenosine nucleotides nor by pyrophosphate, and was abolished by inhibitors of DNA replication. That Ap_4A stimulates replication rather than repair synthesis is supported by the electron-microscope observation of an increased number of replication "eyes" in DNA of quiescent cells exposed to Ap_4A (*171, 172*).

A possible link between Pol-α and the effects of Ap_4A on DNA replication in animal cells has been explored. Equilibrium dialysis of nearly homogeneous calf thymus Pol-α and Ap_4A demonstrated a strong binding between the two (*51*). The binding was specific to Ap_4A as was shown by lack of competition in experiments using various nucleotide analogues. With the calf thymus DNA-polymerase complex containing seven proteins, a polypeptide of 57 kDa was the only one that bound specifically to Ap_4A.

Strong affinity between purified Pol-α and Ap_4A was also found for the human enzyme (*173*). Ap_4A bound to 660 and 145 kDa forms of the HeLa enzyme at molar ratios of 0.5 and 0.008, respectively. In the latter studies, the preferential association of Ap_4A with the 660 kDa polymerase was due to its linkage to an Ap_4A-binding protein that could be separated from the polymerase by hydrophobic column chromatography (*173, 174*). The Ap_4A-binding protein from HeLa cells was isolated as a dimer of two 47-kDa polypeptides (*174*). In experiments on calf thymus, about half the Ap_4A-binding activity copurified with a 450-kDa Pol-α, whereas the remaining binding activity was resolved as a 100-kDa polymerase-free protein. Both the polymerase-bound and free Ap_4A-binding protein were composed of a 54-kDa polypeptide, which differed only in that the former displayed Ap_4A phosphohydrolase activity (*175*). Tryptophanyl-tRNA synthetase copurified with HeLa Pol-α (*174*). Both it and the Ap_4A-binding protein can be dissociated from the polymerase by hydrophobic chromatography (*174*). It is not clear whether the association of Pol-α with the tRNA activating enzyme is adventitious or reflects an unknown function of significance *in vivo*.

The simplest logical mechanism for both the stimulatory effect of Ap_4A on Pol-α and the increased initiation of DNA replication could involve the utilization of this dinucleotide as a primer for DNA replication. Ap_4A can serve as an efficient primer for the *in vitro* synthesis of poly(dA) on a poly(dT) template by Pol-α (*176*). Furthermore, Ap_4A can prime the *in vitro* copying by α-polymerase of a synthetic double-stranded octadecamer containing a part of the origin of replication of SV40 DNA; the primer Ap_4A was covalently linked to the product DNA (*177*). Whether the priming capability of Ap_4A in an *in vitro* α-polymerase-directed DNA synthesis reflects its role *in vivo* is a matter of speculation. It still remains to be demonstrated whether the priming Ap_4A molecule is the same one that binds to the α-polymerase-associated protein.

The correlations between the intracellular level of Ap_4A and the initiation of DNA replication, as well as between the binding of Ap_4A to an α-polymerase-associated protein and increased polymerase activity are highly intriguing, as they potentially offer a mechanistic model for the regulation of α-polymerase activity, and perhaps for DNA replication in animal cells. Some major questions concerning the Ap_4A system, however, remain unresolved. First, it is not clear how Ap_4A is able to transform quiescent cells into cells that start actively to replicate their genome several minutes after exposure to the nucleotide (*171, 172*).

A second question concerns the association between tryptophanyl-tRNA synthetase and Pol-α. It is not clear whether the former is a fortuitously copurified protein or an accessory protein of the Pol-α. Although the ability of Ap_4A to serve as a primer in Pol-α-directed *in vitro* synthesis of DNA has been demonstrated (*176, 177*), its possible role in priming DNA replication *in vivo* is still a matter of speculation. It should nonetheless be noted that Ap_4A inhibits the activity of DNA primase associated with calf thymus Pol-α and does not act as a primer on single-stranded phage M13mp7 DNA (*99*).

In summary, the tantalizing observations on a possible functional association among Ap_4A production and level in cells, DNA replication, and the activity of Pol-α do not yet provide a comprehensive mechanistic model for the involvement of Ap_4A in the regulation of Pol-α activity and perhaps DNA replication *in vivo*.

C. Template Binding Proteins

Proteins that bind to DNA and modulate the activity of Pol-α with the template have been isolated from several cell types. The properties of these DNA-binding proteins, their mechanism of action, and their roles *in vivo* are diverse, although in numerous cases not thoroughly documented. Despite this diversity in observed properties, it is likely that many of these template-binding proteins have helix-destabilizing activity. Moreover, based on the assays used for detection, each of these auxillary proteins can stimulate the activity of Pol-α and appear to do so primarily by binding to the template. Table II summarizes the major characteristics of DNA binding proteins that affect the activity of Pol-α.

TABLE II
Template Interactive Stimulatory Proteins of DNA α-Polymerase

Protein	Source	Size (kDa)	Activity	Enzyme Selectivity	Mechanism of Action	Reference
UP1	Calf thymus	24	Stimulation	High	Helix destabilization	180, 181
—	Calf thymus	18–20	Stimulation	High		182
HD25	Rat liver	25	Stimulation	Low	Helix destabilization	183, 184
HD1	Mouse cells				Enhancement of processivity	117, 185, 186
Factor D	Mouse liver	10–20	Template-selective stimulation	Low	Formation of factor–template complex	187, 188
—	Mouse ascites	30–35	Stimulation with denatured DNA	High	Formation of ternary complex with polymerase and template	190
—	Calf thymus				Reproduction of nonproductive binding of polymerase to template	189

IV. Roles of DNA Polymerase-α in Replication and Repair

Shortly after DNA polymerase-α was identified as the major DNA polymerizing enzyme in dividing animal cells, it was realized that the level of its activity is proportional to the proliferative activity of the cells. Despite early contradictory results, it has also been substantiated in recent years that Pol-α is localized within the cell nucleus and that it is associated with actively replicating DNA. Less extensively documented observations indicate that it is associated with the nuclear matrix. Experiments with selective inhibitors indicated that blocking the activity of Pol-α arrests DNA replication. All these observations bolstered early predictions that Pol-α is the major enzyme responsible for the replication of genomic DNA.

In contrast to the compelling evidence for the central role of Pol-α in DNA replication, it has been demonstrated that neither Pol-β nor the mitochondrial and nuclear Pol-γ appears to have a significant role in DNA replication. Pol-β was long considered to be the major, and perhaps the only, DNA polymerase that participates in repair synthesis of DNA. Only in recent years has it been shown, mainly by use of cells with a mutated Pol-α and through the employment of selective inhibitors of Pol-α, that α-polymerases also play a major role in aspects of the repair synthesis of DNA. We summarize here briefly the evidence on the role of Pol-α in DNA replication and DNA repair. (For a more comprehensive review of these subjects see ref. 29.)

A. DNA Replication

Different approaches have been employed to elucidate the function of Pol-α in DNA replication. One of the first questions was whether it is localized within the cell nucleus, the site of DNA replication. Employment of diverse cell fractionation procedures yielded contrasting results. Whereas under certain fractionation conditions Pol-α was associated exclusively or mainly with the cell nucleus (*191–201*), separation of nucleus from cytoplasm under other conditions resulted in localization of the polymerase in the cytoplasm (*63, 164, 202–205*). The controversy that ensued was recently more definitively resolved by the application of monoclonal antibodies for the *in situ* intracellular detection of Pol-α, which showed that the enzyme lies exclusively within the cell nucleus throughout the cell cycle except for mitosis, during which the nuclear membrane dissolves (*206–208*). It appears, therefore, that it is only under certain conditions involved in cell fractionation that Pol-α appears in the cytoplasm. Furthermore, it is selectively bound to the nuclear matrix, a subnuclear skeletal structure in which DNA replication is believed to take place (*209*).

The amount of Pol-α detected immunologically within the nucleus (*206, 208, 210*) and its activity in the nuclear matrix (*210, 211*) appear to be correlated with the DNA replication rate. These observations are in good agreement with many earlier reports on the close correlation between the intracellular activity of Pol-α and the replicative activity of cells. Several lines of evidence indicate that Pol-α is unique among all classes of DNA polymerase in that its activity parallels cell division. Postmitotic or quiescent cells normally contain very low levels of Pol-α. Differentiated cells with the least mitotic activity (nerve cells) contain only Pol-β and Pol-γ (*212, 213*). Mature liver, which contains a low proportion of dividing cells, does not contain significant Pol-α activity (*199, 214–216*). Similar observations have been made for postmitotic sperm cells (*207, 217*), cardiac and skeletal muscle, kidney, and brain (*218*).

Early studies showed that liver regeneration after partial hepatectomy results in an increase in the activity of Pol-α, whereas the level of Pol-β remains unchanged (6, 219). Since actinomycin D and cycloheximide abolished the increase in Pol-α activity, it was implied that stimulation of cell division involved *de novo* synthesis of the enzyme (220–222). There is a similar selective increase in the intracellular activity of Pol-α in lymphocytes exposed to mitogens (4, 8, 222–224), as well as quiescent cells in culture stimulated to divide (7, 225–229). That this increase in Pol-α activity in dividing cells is indeed due to *de novo* synthesis of the enzyme was firmly established with the demonstration that in dividing cells an increased amount of enzyme is immunolabeled by anti-Pol-α monoclonal antibodies (206–208).

Observations on the kinetics of DNA replication and the appearance of Pol-α in synchronized cycling cultured cells have yielded divergent results as to the exact stage of cell cycling at which the enzyme activity is at maximum. However, it has been shown consistently that among the three types of animal polymerases, it is only Pol-α whose level increases in synchrony with cell division (9, 227–230).

Strong support for the central role of Pol-α in DNA replication is provided by cell mutants with a modified enzyme. A mutant hamster V79 cell line resistent to aphidicolin and designated as Aph^r-4-2 (see mutants in Poly-α, Section VII) has been isolated (231). Pol-α purified from the mutant cells is resistant to aphidicolin *in vitro* (46). A correlation between the genetic change in Pol-α and DNA replication *in vivo* is also shown by the extended division time and the prolongation of S phase in the mutant cells. This mutant cell also provides considerable support for the contention that Pol-α plays a cardinal role in repair of several types of damage to DNA (Section IV,B).

Cultured FM3A cell mutants in which DNA replication is temperature-sensitive have been isolated (232); Pol-α isolated from one of these mutants is thermolabile (233). If, as suggested, a single mutation is responsible for the thermosensitivity of DNA replication, then the central role of Pol-α in DNA replication is clearly indicated.

The last line of evidence that sustains the central place of Pol-α in DNA replication comes from studies with selective inhibitors of DNA polymerases. The most advantageously used selective inhibitor is aphidicolin, which is highly selective for the α-polymerases (234). In animal cells permeable to it, aphidicolin blocks Pol-α without affecting the activities of Pol-β and Pol-γ. At the same time, aphidicolin also selectively and reversibly arrests the replication of genomic DNA without inhibiting synthesis of mitochondrial DNA (235–237). Similarly, butylanilinouracil (238) and butylphenylguanine (239) inhibit both purified HeLa Pol-α and the *in vivo* replication of DNA at comparable concentrations. Finally, a very convincing argument in favor of the cardinal place of Pol-α in DNA replication is the recent demonstration that DNA replication in lysolecithin-permeabilized human cells in culture was inhibited specifically and in a dose-responsive manner by monoclonal antibodies against Pol-α (*240*). The monospecific antibodies were localized in the cell nucleus as is Pol-α, and whereas an inhibition of 80% of DNA synthesis was uniformly attained in all the cells, RNA synthesis remained unaffected by the antibodies (*240*). Selective inhibitors of Pol-α are listed in Table III.

B. DNA Repair Synthesis

Early studies on the function of Pol-α *in vivo* focused exclusively on its role in the replicative synthesis of nuclear DNA. At that time, its involvement in DNA repair was not actively investigated (*241*), and DNA repair was considered to be mediated by DNA Pol-β. However, in the late 1970s, evidence began to accumulate indicating that selective inhibitors of Pol-α diminished the rate and/or extent of DNA repair synthesis. More recently, mutant hamster cells with an alteration in Pol-α have been shown to be defective in DNA repair. These combined studies indicated that Pol-α plays a major role in DNA repair.

Several experimental approaches have been employed to elucidate specific roles for Pol-α in repair synthesis. Early attempts to correlate intracellular levels of DNA polymerases with repair activity met with only limited success because of both the variety and complexity of repair processes in animal cells, and because, unlike DNA replication, the induction of DNA repair does not necessarily involve significant elevation in DNA polymerase activity (*242–247*). A different type of approach, involving an attempt to correlate hereditary deficiencies in repair with altered levels of polymerase(s), did not provide insights into the relative contribution of any of the DNA polymerases to DNA repair processes (*248–249*). Two alternative experimental approaches were, however, much more fruitful in indicating the possible functions of Pol-α in repair mechanisms.

First, the use of selective inhibitors has been instrumental in demonstrating the participation of Pol-α in the repair of damaged DNA. Although some experiments indicated that inhibition of Pol-α by aphidicolin had no effect on repair synthesis in UV-damaged cells (*250–253*), later developments established that the repair of UV-induced lesions in DNA and of other types of damage were strongly blocked by aphidicolin (*254–257*). The discrepancy between early and subse-

TABLE III
SELECTIVE INHIBITORS OF DNA POLYMERASE-α

Inhibitor	Mechanism of action	Specificity to DNA polymerase-α[a]	Competitors
Sulfhydryl group reactive reagents	Interaction with thiol group on a protein	Low	Dithiothreitol, 2-mercaptoethanol
Aphidicolin	Competing with dCTP for polymerase-binding site	High (in vivo and in vitro)	dCTP, dTTP
6-(Arylamino)-uracils (238); N^2-arylaminopurines (239)	Guanine analogue	High	dGTP
Arabinose deoxynucleotides	Deoxynucleoside analogues	Low	dNTP
Phosphonacetic acid	Interaction with the pyrophosphate binding site of polymerases	Moderate (in vitro)	PP_i
Acyclovir[b]	Guanosine analogues inhibitory to herpes DNA polymerase	Moderate	dGTP
Novobiocin	Inhibition of DNA topoisomerase	?	?
Hemin	Induction of dissociation of DNA polymerase-α and template	?	Template
Modified DNA	Crosslinkage of DNA by mitomycin C, X-ray irradiation polyvinyladenosine-hybridized DNA	None	?
2',3'-Dideoxynucleotides	Unavailability of 3'-OH for chain elongation	None; potent inhibitor of DNA polymerases-β and -γ	Deoxynucleotides

[a] Compared with polymerases-β and -γ from homologous or heterologous animal cells.
[b] Acycloguanosine; 9-[(2-hydroxyethoxy)methyl]guanine; 2'-nor-2'-deoxyguanosine; 2'-DNG (239a).

quent observations can be explained by cell-type and growth condition, to differences in the contribution of Pol-α to repair of damage by different genotoxic agents, and to differences in the sizes of repair patches after damage by different agents. In many cases, results on the correlation between diminished repair synthesis and inhibition of

Pol-α by aphidicolin were sustained by analogous findings with other selective inhibitors, including MalNEt, phosphonacetic acid, araC, and d_2TTP (255, 258–260, 263–265). In summary, although a detailed discussion of particular findings is beyond the scope of this chapter, it is clear at present that Pol-α participates in the repair of different types of damage to DNA. Whether it alone is responsible for repair synthesis or cooperates with a non-α-polymerase (probably the β enzyme) in repair may depend primarily on the extent of damage to DNA (261–263), on the type of damaging agent and lesion formed (264–265), and the type and growth conditions of cell tested (258, 266–268).

A second type of evidence in support of the role of Pol-α in repair synthesis is provided by the aphidicolin-resistant hamster mutant cell Aph^r-4-2, which possesses a Pol-α whose affinity for dCTP is increased 10-fold (46). This mutant cell exhibits MNNG- and UV-induced as well as spontaneous hypermutability at the 6-thioguanine, diphtheria toxin, and ouabain loci (269–271). It has been argued that, if indeed aphidicolin resistance is due to a single mutation at the Pol-α gene, hypermutability of the Aph^r-4-2 cell strain after DNA damage strongly argues for the central role of the α enzyme in DNA repair synthesis.

V. Chromosomal Localization of the Gene for DNA Polymerase-α

Monoclonal antibodies against Pol-α have not only provided a new protocol for the rapid purification of this enzyme with the likelihood of establishing the structure of the intact catalytic core component and identifying the complement of associated proteins, but have also been successfully used to pinpoint the chromosomal locus for the gene coding for Pol-α. Korn and associates (39) have established a "catalogue" of hybridoma cell lines, each producing monoclonal antibodies against Pol-α isolated from human KB cells. One of these monoclonal antibodies, SJK132-20 IgG, binds to and inactivates Pol-α from most mammalian species. This antibody, when bound covalently to Sepharose, has been used successfully for the purification of Pol-α from calf thymus (42) (M. Reyland and L. A. Loeb, unpublished).

Another of this series of monoclonal antibodies, SJK237-71 IgG, binds and inactivates Pol-α from KB cells, yet does not bind to nor inactivate Pol-α from mouse cells or from Chinese hamster ovary fibroblasts (272). The SKJ237-71 IgG has been used by Wang et al.

(272) to detect human KB Pol-α in a series of human mouse hybrid cell lines containing different combinations of human chromosomes. The unexpected finding is that the presence of human DNA polymerase-α in these hybrids correlates exactly with the presence of the human X-chromosome. This absolute concordance in these human–rodent cell hybrids indicates that the X-chromosome contains either the gene for Pol-α or a locus controlling its expression. Furthermore, by fine-structure mapping using human cells containing partial deletions of the X-chromosome, these investigators determined that the ability to express human Pol-α in human–rodent cell hybrids is associated with the presence of the P21.2-22.1 segment of the X-chromosome. This segment is adjacent to the locus associated with Duchenne muscular dystrophy.

The X-chromosomal location of Pol-α has been confirmed and extended by Hanaoka et al. (273). In their studies with human mouse hybrids they utilized the monoclonal antibody-sensitive mutant of FM3A cells that possesses a heat-labile Pol-α. In one of the heterokaryotes formed, the only human chromosome present was the X-chromosome, and the hybrid cell was able to grow at the nonpermissive temperature. Furthermore, its Pol-α was inactivated by SKJ287-38 antibody, which is specific to the human enzyme.

VI. Role of DNA Polymerase-α in the Fidelity of DNA Synthesis

In a eukaryotic cell with some 3×10^9 nucleotides in its genome, DNA is copied during each division cycle with few if any errors. Studies on the fidelity of purified Pol-α may be critical for understanding the overall accuracy of DNA replication in animal cells. Clearly, Pol-α affects the fidelity of DNA synthesis; it copies DNA templates with an accuracy much greater than that predicted from the energies of base-pairing alone (274, 275). Yet the fidelity of catalysis by the purified enzyme is not sufficiently high to copy DNA with an accuracy similar to what occurs in vivo. Thus, other cellular factors, by mechanisms still to be defined, must function in concert with DNA polymerase-α to achieve the high fidelity that occurs during DNA replication in animal cells.

The fidelity of DNA replication in animal cells is likely to be in the range of 10^{-9} to 10^{-11} errors per base-pair replicated (276). This estimate is based on the size of the human genome, on the theoretical argument that a mutation rate greater than one per genome per cell generation in germ line cells might be incompatible with species stability (276), and on experimental measurements of spontaneous mutation frequencies (277).

Evidence from bacteria suggests that this overall accuracy of DNA replication results from a multistep sequential process. Studies with bacterial DNA polymerases and accessory replicating proteins permit an estimate of the contribution of each of the individual steps (274, 278, 279). First, thermodynamic physical calculations indicate that the difference in free energy between correct and incorrect base-pairings in an aqueous environment is 1–3 kcal/mol, which would predict an error rate of 10^{-2} (275, 280). Actual measurements of nonenzymatic template-directed assemblage of activated nucleotides demonstrate an error rate of 1 in 200 (281, 282), and confirm these predictions. Second, as we will consider in detail, animal DNA polymerases copy DNA with an error rate of 10^{-3} to 10^{-5} (283) and thus enhance the fidelity of base-pairing by two to three orders of magnitude. This occurs in the absence of any exonucleolytic activity. Third, the contribution of the associated $3' \rightarrow 5'$ proofreading exonucleolytic activity of prokaryotic DNA polymerases further enhances fidelity by two to three orders of magnitude (274). Fourth, in bacterial cells, errors in DNA replication can be excised by a mismatch correction system that operates after DNA synthesis and can enhance accuracy by an additional two or three orders of magnitude (284, 285). In considering the even greater requirements for accuracy in eukaryotic cells as a result of their large genomes (276), it seems likely that a multicomponent system should be operative. Furthermore, considering that Pol-α is a cardinal element in the DNA replication complex, studies on the fidelity of this enzyme and its associated proteins may be the key to understanding the overall mechanisms for accuracy of DNA duplications in animal cells.

A. Fidelity of the Catalytic Core Component

Until recently, all studies on the fidelity of Pol-α have been limited to measurements of single base substitutions using extensively purified preparations of Pol-α core enzyme (283). In this context, we use the term "core enzyme" to refer to the catalytic polypeptide, which might be partially degraded, plus several other polypeptides whose function is unknown in most cases. These preparations have invariably been free of any detectable exonucleolytic activity. The lack of detectible $3' \rightarrow 5'$ exonuclease was most rigorously demonstrated by Chang and Bollum with calf thymus DNA Pol-α (286). Their studies utilized homopolymer templates containing a single noncomplementary nucleotide at the $3'$ terminus of the primer and demonstrated that this single noncomplementary terminal residue is not excised and can serve as an effective primer terminus for the addition of subsequent nucleotides. Moreover, the lack of generation of complementary deoxynucleoside monophosphates during polymerization suggested that any form of proofreading requiring the excision of segments containing complementary nucleotides is unlikely to occur with the catalytic core component.

One final and technically difficult experiment remains to be carried out with Pol-α to eliminate any scheme of proofreading involving hydrolysis of noncomplementary bases. This would be to measure directly the rate of generation of noncomplementary nucleoside monophosphates during catalysis, as has been done in the case of *E. coli* Pol-I (*287*). Even lacking this, it seems likely that proofreading is not carried out by the catalytic core component of DNA Pol-α and that the errors observed result from misinsertions in the absence of excision. Since the error rates of Pol-α—varying from 1 in 5000 to 1 in 40,000 using polynucleotide and natural DNA templates—are much lower than the error frequency of base-pairings in the absence of a polymerase (*274, 275, 281*), the enhancement in accuracy implies that Pol-α participates in base-selection by some mechanism that does not involve exonucleolytic excision.

1. Use of Polynucleotide Templates

Measurements of misincorporation by the catalytic core component of Pol-α using synthetic polynucleotide templates have been carried out in many laboratories using preparations of DNA polymerases from diverse sources (*278, 288–292*). In these studies, the error rate is defined by the frequency of incorporation of a nucleotide not complementary to any nucleotide in the template. High reproducibility can be obtained by the use of a double label assay: a noncomplementary nucleotide labeled with a radioisotope of very high specific activity and a complementary nucleotide labeled with a different radioisotope at very low specific activity (*293*). By definition, the sensitivity of these assays is limited by the chemical purity of both the polynucleotide templates and the (labeled) nucleotide substrates. In this assay, the enzyme, template, and nucleotide substrates must be of exceptional purity. Any contaminating DNA would serve as an alternate substrate and direct the incorporation of substrates noncomplementary to the added polynucleotide.

In more recent studies, the alternating copolymer poly[d(A-T)] has been used as a template-primer because of the exceptional purity by which the template can be synthesized *in vitro* using *E. coli* Pol-I

(287). Nevertheless, the chemical purity of other reaction components precludes measurements of error rates less than 1 in 10^5. In order to measure fidelity in crude extracts, Fry et al. (294) designed a procedure in which any copied DNA would be hydrolyzed by a series of restriction enzymes that failed to hydrolyze the products whose synthesis was directed by synthetic polynucleotide templates. By this method, it was possible to measure the fidelity of DNA polymerization in preparations of chromatin that contained large amounts of endogenous DNA.

The error rates of Pol-α with poly[d(A-T)] as a template and dGTP as the noncomplementary nucleotide have been reported to vary from 1 in 4000 to 1 in 20,000 residues of dGMP incorporated per total nucleotides polymerized. Nearest-neighbor analysis of the product of the reaction synthesized with human placenta Pol-α (290) indicated that at least 95% of the dGMPs were incorporated in phosphodiester linkage and in place of the "correct" dAMP. In selected situations, greater accuracy has been observed for DNA polymerase-α, but these results were from pool bias experiments (295, 296) in which reactions were carried out in the presence of high concentrations of one of the noncomplementary nucleotides and the results were extrapolated to error rates than would be obtained at equal concentrations of complementary and noncomplementary nucleotides. Even though there is considerable support for such an extrapolation, one study (296) failed to observe competition between complementary and noncomplementary nucleotides with Pol-α.

The core component of Pol-α commonly exhibits molecular heterogeneity after purification. It is likely that this heterogeneity is due to proteolytic degradation during isolation and purification. Using the purification procedure of Holmes et al. (49, 297), Loeb et al. (298) observed no differences in fidelity between the two major species from calf thymus, forms C and D. In contrast, Brosius et al. (299) extensively purified three forms of Pol-α and found a 15-fold difference in fidelity among the different forms, with the smallest species exhibiting the highest error rate. Conceivably, proteolysis of Pol-α results in smaller catalytically active species that have higher error rates. These in vitro observations could be of profound physiologic importance. Krauss and Linn (300) observed that when cells in culture grow to confluency, there is a decrease in the fidelity of purified DNA polymerases. Two linked speculative hypotheses can be formulated. (1) As cells cease DNA replication, Pol-α is degraded by cellular proteases so that it can no longer function as an integral part of a multienzyme DNA replication complex. (2) The degraded enzyme in confluent cells is catalytically active and is still capable of functioning in DNA repair, a process that may tolerate errors at greater frequency.

2. Single-Base Substitution Using Natural DNA Templates

Ideally, one desires to measure the fidelity of DNA synthesis using a natural DNA template containing all four bases in which one can quantitate frequencies of misincorporations that occur *in vivo*. Weymouth and Loeb (*301*) designed a genetic reversion assay to measure the frequency of errors by a DNA polymerase in copying a single-stranded φX174 DNA containing an amber mutation. Each time the amber mutation is copied, there is the possibility of inserting an incorrect nucleotide, which could revert the amber mutation to a wild-type phenotype. The copied DNA is used to infect *E. coli* spheroplasts, which are plated with indicator bacteria, either permissive or nonpermissive for the amber mutation. The reversion frequency of the amber mutation thus reflects the fidelity of the DNA polymerase. In the initial system, a purified restriction endonuclease fragment was used as a primer and synthesis by the polymerase had to proceed 83 nucleotides in order to reach the amber site (*302*). By using chemically synthesized oligonucleotides instead of restriction fragments, it has been possible to position the 3'-terminus only a few nucleotides from the amber codon (*303*). This modification has permitted studies on the effects of primer position (*304*) and has facilitated measurements of fidelity in nonextensively purified preparation of eukaryotic DNA polymerases (*303*).

The *am*3 φX mutation offers an exceptionally high sensitivity. It is located in a stretch of the φX genome which contains the sequence for two required genes that are coded for by two overlapping reading frames separated by a single nucleotide. All three nucleotide substitutions at position 587 will produce both active gene products and viable phage. With *am*3 φX174 DNA, one can detect misincorporations at a frequency approaching 10^{-6} in reactions when all four dNTPs are present at equal concentrations (*302*). From measurements of reversion frequency of DNA copied in reactions containing increased concentrations of noncomplementary nucleotides (pool bias experiments), the results can be extrapolated to error rates as low as 10^{-8} (*305*).

In order to obtain a limited spectrum of single-base substitutions, other investigators (*306*, *307*) have utilized a different φX mutant, amber 16, in which substitutions at all three positions in the amber locus produce viable phage. By analyzing plaque morphology and growth at different temperatures, a set of criteria for designating types of substitutions prior to sequencing DNA of progeny phage has been established.

The frequency of single-base substitutions using the catalytic core component of Pol-α from a variety of sources has been quantitated (283). In order to quantitate the percentage of template molecules copied, the partially double-stranded reaction product was isolated on neutral sucrose gradients prior to transfection into spheroplasts. In order to control for penetrance (301), the expression of the minus strand, transfection experiments were also carried out with model templates containing minus strands of varying lengths. The results indicate that at equal nucleotide concentration, Pol-α inserts one noncomplementary nucleotide opposite deoxyadenosine at position 587 for each 30,000 complementary (deoxy)thymidines incorporated at this position. Even though deoxycytidine was the noncomplementary nucleotide most frequently incorporated at this position, misincorporation of deoxyadenosine and deoxyguanosine also occurs, but at a very low frequency.

Similar error rates for the core component or Pol-α were obtained by Grosse et al. (307) using amber 16 and the 9-S DNA-polymerase from calf thymus. By utilizing the endogenous primase activity associated with the 9-S polymerase, these investigators were also able to measure the fidelity of Pol-α in DNA synthetic reactions primed with oligoribonucleotides. By increasing the concentration of each deoxynucleotide in individual reaction mixtures and by measuring plaque morphology and temperature sensitivity of the progeny phage, they estimated that the frequencies of the commonest mismatches at equal nucleotide concentrations were dGPT:dT (1/13,500), dGTP:dG (1/22,000), and dATP:dG (1/35,700). It should be noted that other mispairings, if they occurred, were below the level of detection, estimated to be 5×10^{-6} (307).

3. MUTATIONAL SPECTRA

Kunkel (308) has established a forward mutation assay to analyze the mutation spectrum of DNA polymerases *in vitro*. It measures errors produced during DNA synthesis within a single-stranded gap containing the *lacZ* gene cloned in M13mp2 DNA. The single-stranded region contains the C-terminal coding sequence of the *lacI* gene, the *lac* promotor and operator, and the DNA sequence coding for the first 145 amino acids of the N-terminal end of the *lacZ* region (the α region). The *E. coli* host used for transfection contains a chromosomal deletion of the *lacZ* gene, but also harbors an F' episome to provide the remaining coding sequences of the *lacZ* region. The two partial proteins produced within the M13mp2-infected host cell reconstitute enzyme activity by intracistronic α-complementation. Mutants with a loss of α-complementation activity are detected, using an indicator dye, as light blue or colorless plaques. This assay is not as sensitive as a reversion assay. However, since β-galactosidase production is not required for phage viability, this assay is capable of detecting frameshifts, deletions, additions, and complex errors in addition to single-base substitutions.

Using this system, Kunkel (309) analyzed the mutational spectrum of extensively purified DNA polymerase-α from five different sources and compared the results to those obtained with preparations of Pol-β and Pol-γ. With Pol-α, the majority of errors are single-base substitutions (309); the frequency with which they are observed is similar to that reported previously in studies with the φX system (283). Interestingly, with all five enzyme preparations, a significant number of single-base frameshifts that could not have been observed in the φX system were detected. For each of the classes of mutants observed, Pol-α was more accurate than Pol-β. However, in contrast to the results with the φX system (283), Pol-γ was more accurate than Pol-α (309).

B. Fidelity of DNA Polymerase-α Holoenzyme Complexes

Given that the catalytic core component of Pol-α is highly error-prone and that replication is an exceptionally accurate process *in vivo*, it becomes important to determine whether putative eukaryotic DNA replication complexes are more accurate than purified preparations of Pol-α. If complexes are more accurate, what is the mechanistic contribution of each of the subunits to this enhanced accuracy? Kaguni *et al.* (77) isolated a four-subunit complex from *Drosophila* that contains a high-molecular-weight Pol-α, a primase, and two other subunits of unknown function. This complex is not highly processive and lacks any $3' \rightarrow 5'$ exonucleolytic activity, yet exhibits high fidelity in copying *am*3 φX174 DNA. From experiments carried out at equal nucleotide concentrations and from nucleotide pool bias experiments, it is estimated that the fidelity of catalysis by the replicating complex is 1 in 200,000 (310). These results have been independently confirmed (Reyland and Loeb, unpublished observations) and extended to a DNA replication complex isolated from calf thymus by purification on affinity columns containing monoclonal antibody prepared against Pol-α from KB cells (Table IV). The antibody-purified Pol-α from calf thymus is also devoid of any associated exonucleolytic activity. There are two simple explanations for the enhanced fidelity of the com-

TABLE IV
Fidelity of DNA Polymerase-α with Natural DNA Templates[a]

DNA polymerase-α	Template	Mismatched Nucleotide · template base pair	Error rate	Reference
Calf thymus	am3 φX	dCTP, dATP, or dGTP · A	1/24–31,000	283
Mouse myeloma	am3 φX	dCTP, dATP, or dGTP · A	1/47,500	283
Calf thymus	am16 φX	dGTP · T	1/13,200	307
Calf thymus	am16 φX	dATP · G	1/35,700	307
Calf thymus	M13mp2	All substitutions	1/19,200	309
HeLa cells	M13mp2	All substitutions	1/33,000	309
Drosophila polymerase–primase complex	am3 φX	dCTP · A	~1/250,000	310
Calf thymus polymerase complex	am3 φX	dCTP · A	~1/200,000	b

[a] For methods, see the indicated references. The error rate in the M13mp2 systems (*309*) was estimated from the observed mutation frequency, dividing by 250 (the number of nucleotides copied *in vitro*) and thus is the average frequency of all mutations that yield mutant plaques. The estimate of 1/250,000 for the *Drosophila* polymerase–primase complex is obtained from the values given in ref. *310*, assuming a penetrance of 40%.

[b] The values for calf thymus complex are from unpublished results of Reyland and Loeb, using calf thymus DNA polymerase-α purified by immunoaffinity columns.

plexes. The polymerase activity in both complexes is of very high molecular weight and less likely to have suffered proteolytic degradation during isolation. Alternatively, associated proteins work in concert with the polymerase to enhance fidelity by a mechanism still to be defined.

C. Mechanisms for Enhancement in Fidelity by DNA Polymerase-α

These studies with natural DNA templates allow one to conclude that the frequency for the most common single-base substitution errors by the catalytic core component of Pol-α is approximately 1 in 30,000, and that certain substitutions occur at a frequency of less than 5×10^{-6} (*283, 307, 309*). Since this accuracy is greater than that predicted on the basis of base-pairings in the absence of DNA polymerase, these results suggest that the core component of Pol-α enhances the fidelity of DNA synthesis. The fact that this accuracy is greater than that exhibited by Pol-β using the same assay systems adds further weight to the argument that Pol-α enhances the fidelity of DNA synthesis.

After extensive purification by column chromatography, Pol-α is devoid of detectable exonucleolytic activity (*288*). Kinetic experiments indicate that if any proofreading activity is present, it does not contribute to enhanced accuracy (*311*). The addition of deoxynucleoside monophosphates, pyrophosphate, and increased amounts of the next complementary nucleotide, each results in increased misincorporation by *E. coli* Pol-I presumably by diminution of proofreading (*312*). Yet these same factors do not enhance misincorporation by Pol-α (*311*). Thus, the involvement of Pol-α in enhancing fidelity does not involve exonucleolytic proofreading.

A number of kinetic schemes have been invoked to explain the fidelity of DNA polymerases in general (*313–316*), and some of these are relevant to Pol-α. The K_m discrimination model postulates an enhancement in the free energy of base pairings due to the hydrophobic environment at the substrate binding site on the polymerase (*317*). The recent X-ray crystallographic data on *E. coli* Pol-I predicts that the DNA template is located in a hydrophobic cleft in the protein (*115*). The idea is that DNA polymerases exclude water and as a result amplify differences between "correct" and "incorrect" base-pairing. The key prediction of the model is that error rates should be proportional to differences in K_m values between "correct" and "incorrect" nucleotides. With Pol-α, the 6-fold difference in K_m values between 2-aminopurinenucleoside triphosphate and dATP is indeed proportional to the difference in error rates. However, verification of this model will require rigorous studies involving correct and incorrect nucleotide substrates. Also, the model does not account for the differences in mutation spectra observed by Kunkel (*309*) using DNA polymerases-α, -β, and -γ.

In the "energy-relay" model, enhanced fidelity occurs in the absence of proofreading; the energy released by pyrophosphate bond cleavage is used by the polymerase to monitor the insertion of the following nucleotide (*318*). The key prediction is that incorporation of the first nucleotide is more error-prone than the incorporation of subsequent nucleotides. However, two laboratories have tested the key prediction by comparing the error rate for incorporating the first nucleotide to that of incorporating subsequent nucleotides using Pol-α or -β (*304, 307*). No differences were observed as a function of the distance of the 3' terminus of the primer from the site for mutagenesis. Nevertheless, an energy-relay mechanism could account for the enhanced fidelity observed using the DNA replicating complexes (see Table IV).

The relationship of fidelity to processivity could be central to mechanism for accuracy. In comparing eukaryotic DNA polymerases, the order of increased accuracy in the M13mp2 system ($\gamma > \alpha > \beta$)

correlated with increased processivity (309). Moreover, substitution of Mg^{2+} by an Mn^{2+} as a metal activator of Pol-β changes the processivity from 1 to 4–6 nucleotides per association event, and results in increased accuracy (309). Also, it should be noted that increased processivity as a concomitant of increased fidelity is consistent with the energy-relay model. However, the increased fidelity of the DNA polymerase complexes purified from *Drosophila* (310) and from calf thymus (Reyland and Loeb, unpublished results) does not appear to be correlated with a corresponding increase in processivity. Experiments are needed in which processivity is systematically varied while the effects on fidelity are carefully monitored.

The addition of pyrophosphate to reactions catalyzed by Pol-α does not result in decreased accuracy as would be expected from a mechanism that involves kinetic proofreading (314, 315). In fact, with Pol-α, pyrophosphate causes the opposite effect, a 5-fold increase in accuracy (311). The latter finding is compatible with a proposed model for enhanced fidelity that involves rejection of incorrect NTPs immediately prior to incorporation by pyrophosphate exchange or by pyrophosphorolysis (319).

A conformational model for enhancement of fidelity by Pol-α involves a change in the conformation of the nucleotide binding site on the polymerase with each nucleotide addition step. A number of phenomenological considerations provide support for such a model. Different classes of eukaryotic DNA polymerases have different error rates and different mutation spectra when copying the same DNA segment (283, 309). Even though differences in mutation spectra cannot easily be accounted for by mechanisms of fidelity that involve nonspecific increases in the free energy differences between correct and incorrect base-pairings, these differences in mutation spectra do not provide positive evidence for a conformation mechanism of enhanced fidelity.

Other mechanisms, such as selection by the enzyme for different nucleotide conformations in solution, might also produce characteristic types of misincorporations at different template sites. In the case of *E. coli* DNA Pol-I, there is strong kinetic and binding evidence (320) for a single nucleotide-binding site on the polymerase at which all four deoxynucleotide substrates compete. A single nucleotide binding site is likely to be found on all DNA polymerases, even though in the case of Pol α there is circumstantial evidence to indicate multiple nucleotide binding sites during catalysis (43, 88). If there are less than four nucleotide binding sites, Pol-α must have a mechanism to select and accommodate each of the four nucleotide substrates or to recognize different types of base-pairs. Lastly, NMR studies with Pol-I indicate that this polymerase, in the absence of the DNA template, immobilizes the nucleotide substrate in a conformation it would occupy in β-DNA (*321, 322*), suggesting that the enzyme has multiple contact points with the nucleotide substrate. The high level of discrimination for deoxy- over the corresponding ribonucleotide substrates by all DNA polymerases that have been studied provides an additional argument against a loose nonspecific nucleotide binding site on DNA polymerase.

A conformation model for enhanced fidelity by Pol-α would predict that the difference in V_{max} between the correct and incorrect nucleotide substrates at each nucleotide addition step is proportional to the error rate. It is likely that such measurements can now be obtained using singly primed DNA templates and measuring rates of elongation by the addition of single nucleotides. Measurements of rates of incorporation using nucleotide analogues containing methyl groups at different positions should help to define the extent and sites for coordination at the substrate site. Nevertheless, definite evidence for a detailed mechanism for increased accuracy by Pol-α is unlikely to be defined by kinetic studies alone. It may be necessary to await cloning of the enzyme, the introduction of amino acid changes at defined sites of the enzyme, and a knowledge of the three-dimensional structure of the polymerase. These goals are presently obtainable.

D. Enhancement of Fidelity *in Vivo*

The mechanism by which associated proteins increase the fidelity of catalysis by Pol-α is entirely unknown. There is the likelihood of multiple steps, considering that this enhancement is estimated to be about 10^6. When one considers the aesthetic simplicity of an excision type of proofreading step for achieving accuracy with prokaryotic DNA polymerases, and the fact that in one very well documented case, *E. coli* polymerase-III, the exonuclease step is provided by a separate subunit of the polymerase (*323*), it is hard to conclude that a similar proofreading activity does not also operate in animal cells. The exonucleolytic function might simply be lost during extensive purification or might be associated only with the polymerase in a ternary complex involving a DNA template. DNA polymerases obtained from simple eukaryotes frequently contain an exonuclease; a loss in the fidelity of Pol-α during purification has been reported to parallel removal of exonuclease activity (*324*). However, even if proofreading occurs with complexes containing Pol-α it is unlikely to be a major contributor to accuracy. Energy considerations suggest that extensive proofreading is cost-inefficient, and would require a major fraction of the cell's total ATP (*325*).

Recent observations support a postreplicative mechanism for error correction in animal cells (*326*). The suggestion is that a subgroup of methylated cytidines designate the parental DNA strands so that all misincorporations are specifically excised from the newly synthesized daughter strands. By analogy to prokaryotes, error-correction might proceed by excision of a mismatched base by a specific endonuclease followed by the formation of a gap, with resynthesis by DNA polymerase and closure by ligase. The enhancement of accuracy by such a system could be very high; in principle, it would be equivalent to the accuracy of the polymerase used for resynthesis multiplied by the number of nucleotides replaced. However, for correction to be energy-inexpensive, the accuracy in recognizing and excising only a mismatched base would also have to be exceptionally high.

VII. Mutants in DNA Polymerase-α

Our understanding of the details of DNA replication in prokaryotes has benefited enormously from the isolated bacterial strains containing conditional mutants in genes encoding DNA replicating proteins. The study of these mutants has allowed one to assign physiological functions to catalytic activities measured *in vitro*, and has provided a new method for the purification of normal replication proteins based on their ability to complement the altered function in extracts derived from mutant cells. Until recently, mutant animal cells that contained altered DNA replicating proteins were not available. Also, there are no diseases known that result from an alteration in an enzyme believed to be involved in DNA replication. The availability of aphidicolin, a highly specific inhibitor of Pol-α, and its ability to pass through the membrane of cells in tissue culture provided the first method for selecting mutations in a DNA replicating enzyme of animal cells.

A mutant cell line containing an altered Pol-α was obtained from a cell line of *Drosophila melanogaster* (*45*). After exposure to ethyl methanesulfonate, mutants were selected on the basis of growth in increasing concentrations of aphidicolin. Using stepwise selection, two types of DNA polymerase mutants were isolated. Mutant aph-13 overproduced Pol-α by 8-fold in one of two measurements using unfractionated cell extracts. The change in polymerase activity as a function of the concentration of aphidicolin or dCTP was the same with partially purified enzymes from the parental strain and from the over-producing strain. Thus, the enzyme does not appear to be altered, and it is argued that the increased amount of Pol-α per cell accounts for resistance to aphidicolin. However, such resistance based on overproduction of DNA polymerase-α alone seems unlikely if the latter is a stoichiometric component of a DNA replication complex. Pol-α from one of the other mutants, aph-10, exhibited a higher apparent K_i for aphidicolin (100 nM) than that demonstrated for the wild-type enzyme (12 nM). After partial purification of Pol-α, the molecular weight (150,000) and sensitivity to MalNEt were the same for the mutant and parent. This altered Pol-α has not been further purified, nor have the properties of the mutant been described in detail.

The *Drosophila* cell mutants resistant to aphidicolin demonstrate that the target of this antibiotic is indeed Pol-α. However, a detailed analysis of the functions of this enzyme has come from the isolation and characterization of a mammalian Pol-α mutant isolated by selecting viable colonies from 3×10^7 mutagenized Chinese-hamster lung-fibroblast V79 cells in the presence of aphidicolin (231). Chang et al. (231) incubated 2.5×10^6 V79 cells in 100 μM bromodeoxyuridine (BrdU) for 48 hours (approximately 3–4 cell divisions) and then irradiated them with visible light, which reduced cell survival by 10^{-4}. The surviving cell population was grown to 3×10^7 cells over 7–10 days and then mutagenized a second time with UV irradiation (10–20 J/m²). The mutagenized cells were then cultured in the presence of the selective agent (1 μM aphidicolin) for 3 weeks. The selection scheme was based on the assumption that the gene for Pol-α is autosomal and that two mutational events might be required to create an aphidicolin-resistant polymerase mutant. Mutagenesis by BrdU and visible light was used to induce functional hetero- and hemizygosity in the parental *wt* cells, and UV irradiation was used to produce a point mutation or small deletion in the second copy of the polymerase gene. This procedure for mutagenesis yielded the first successful isolation of a mutant Pol-α from animal cells, although the gene for Pol-α was subsequently localized to the human X-chromosome (272). This mutant, originally designated Aph[r]-4, was later recloned and designated Aph[r]-4-2 (231); it contains an aphidicolin-resistant Pol-α (46).

The phenotypic characteristics of Aph[r]-4-2 are summarized in Table V. The mutant exhibits a stable resistance to aphidicolin; at 0.8 μM aphidicolin, survival of the mutant is 80%, while at the same concentration, survival of parental cells is less than 10^{-6}. The normal sensitivity of the mutant cells to arabinocytidine (arabinosylcytosine, cytarabine, araC) provides a rapid method to distinguish polymerase mutants from other mutants that contain altered nucleotide pools. The latter are frequently resistant to both aphidicolin and arabinosylcyto-

TABLE V
Phenotypes of Aph^r-4-2, a Polymerase-α Mutant Resistant to Aphidicolin[a]

Phenotypes	Parental wt cells (743X)	Mutant Aph^r-4-2
Colony-forming ability		
Aphidicolin (0.8 μM)	<10^{-6}	0.8
araC (1.0 μM)	<10^{-3}	<10^{-3}
BrdU (100 μM)	<10^{-6}	<10^{-6}
Cell doubling time (hours at 37°C)	12	31
Length of S phase (hours)	6	15
UV irradiation (J/m^2 for 37% survival)	16	8
UV repair synthesis ([^{3}H]thymidine incorp. dpm/μg DNA)	90	90
UV mutagenesis (ouabain-resistant clones per 10^6 cells after UV dose = 37% survival rate)	25	130
MNNG (μg/ml for 37% survival)	>>0.8	0.8
Spontaneous mutation rate (× 10^8)		
Ouabain resistance	3	42
Diphtheria toxin resistance	1	59
6-Thioguanine resistance	2	21
Partially purified DNA polymerase-α		
Percentage inhibition by 1.2 μM aphidicolin	80%	<10%
Michaelis constant (μM)		
dCTP	10 ± 4	1.0 ± 0.4
dTTP	8 ± 2	1.5 ± 0.5
dATP	4	5
Percentage inhibition by SJK 132–20 IgG	90%	90%

[a] The data in this table are taken from references 46, 231, 269–271, and unpublished results from the author's laboratories.

sine (araC) (327, 328). Of primary importance is the fact that Pol-α purified from this mutant is resistant to aphidicolin (46); 1.2 μM aphidicolin inhibits the ability of polymerase-α purified from the parent by 80%, while the activity of the mutant enzyme is inhibited by less than 10%. Pol-α purified from the mutant and parent cells is equally inhibited with a monoclonal antibody prepared against KB cell Pol-α (Reyland and Liu, unpublished results). Moreover, the purified mutant enzyme shows a marked reduction in the K_m for dCTP and dTTP, while the K_m for dATP is unchanged. This is in agreement with the known competition between aphidicolin and dCTP or dTTP (329), and suggests a mechanism for the aphidicolin-resistant phenotype based on enhanced binding for pyrimidine nucleotides by the mutant polymerase. A mutant DNA polymerase from herpes simplex virus resistant to aphidicolin also exhibits a reduction in the K_m for dCTP (330).

The differences in K_m values for the pyrimidine nucleotide substrates between the mutant and the parent polymerase suggest that the mutant enzyme is error-prone, and that the mutant cells have a mutator phenotype (46, 270). In 100 μM BrdU for 48 hours, the frequency of chromosomal aberrations in the mutant is 9-fold that in wild-type cells (231). The mutant cells exhibit a 5- to 40-fold increase in spontaneous mutation rates (46, 269–271) at three different loci: Na/K-ATPase (ouabain resistance), hypoxanthine phosphoribosyltransferase (6-thioguanine resistance), and elongation factor 2 (diphtheria toxin resistance). If the alterations in K_m values observed *in vitro* are the mechanism for enhanced mutagenesis, one would predict that many of the resultant mutants should consist of single-base substitutions, predominantly substitutions by pyrimidine nucleotides. A comparison of the fidelity of DNA synthesis as a function of relative nucleotide concentration by the parent and mutant Pol-α is needed to address this problem.

Even though changes in DNA replicative and repair processes in the mutant cells are likely to result from an alteration in Pol-α, mutations in other genes might have occurred during the extensive treatment required for mutagenesis and subsequent growth of the altered cells. Bearing this caveat in mind, it is instructive to consider the properties of this mutant with respect to the function(s) of Pol-α. The slow growth of the mutant and the prolongation of S phase (Liu, unpublished results) are in accord with a primary role for Pol-α in DNA replication. The hypersensitivity of the mutant to UV irradiation, MNNG, and 2-(N-acetoxy)acetamidofluorene supports the concept that Pol-α is also involved in DNA repair-synthesis. This sensitivity to mutagens is not likely to result from a deficiency in excision repair, since unscheduled DNA synthesis is the same in mutant and wild-type cells (271). The involvement of Pol-α in DNA repair-synthesis is further suggested by the enhancement of mutation frequency per lethal hit after the exposure of mutant cells to either UV irradiation (269) or MNNG (271). These results point to a mechanism in which unrepaired DNA damage is more frequently bypassed by the mutant DNA polymerase, or one in which the mutant DNA polymerase is involved in repair-synthesis and is error-prone. The fact that this mutant is as sensitive to BrdU (271) argues against the notion that previous BrdU-visible light treatment has preselected clones resistant to BrdU. The pleiotrophic phenotype exhibited by Aph^r-4-2 appears to be tightly associated; reversion to aphidicolin sensitivity is invariably accompanied by increased growth rate, UV resistance, and decreased mutator activity (269–271).

A number of temperature-sensitive mutants of cultured animal cells have been isolated and characterized (331). One of these mutants, clone FT20, isolated from a mouse mammary carcinoma cell line, FM3A, is temperature sensitive in DNA replication and contains a temperature-sensitive Pol-α (273). Growth at the restrictive temperature (39°C) immediately halts DNA replication, and cells accumulate at the G_1/S boundary or in the S phase of the cell cycle. A prior incubation of FM3A cell extracts at 39°C for 20 minutes results in an 80% loss in Pol-α activity, whereas the loss of this activity from the parent strain is less than 15%. Since measurements on the reversion frequency of the mutant suggest that the mutation is on a single gene, an analysis of DNA repair and mutagenesis in the mutant will be highly informative concerning the role of Pol-α in these processes.

The transfer of a mutant phenotype by DNA from a mutant cell provides a rigorous test that a specific mutation is responsible for a physiological function. Gene transfer, in addition, is a useful initial step for the cloning of that gene. Reconstruction experiments allow one to test the feasibility of this approach. For example, in the experiment shown in Fig. 1, 50 Aph^r-4-2 mutant cells were mixed with 2 $\times$ 10^6 V79 parent cells, and mutant clones could be recovered at concentrations of aphidicolin that abolished survival of all parent cells, indicating that the "window" for selection might be sufficient to allow the recovery of wild-type (Aph^s) transfected with DNA from cells resistant to aphidicolin.

Direct transfection experiments have been carried out using purified DNA from Aph^r-4-2 and wild-type V79 recipients (47). Aphidicolin-resistant clones were recovered at a frequency of 6 $\times$ 10^{-8}/μg of mutant DNA. No aphidicolin-resistant clones were recovered in the absence of added DNA, or after sonication of Aph^r-4-2 DNA to <5 kilobases or after digestion of the DNA with the restriction enzyme EcoRI. Cotransfection of mutant and pSV2-gpt DNA produced stable mutants resistant to both aphidicolin and mycophenolic acid at a frequency of 15 per 10^6 recipient cells. This frequency is a least 10^8-fold higher than that expected from the product of the spontaneous mutation frequencies for resistance to each of the inhibitors. Most significantly, the demonstration that Pol-α purified from one of the transfectants was resistant to aphidicolin provides strong evidence that Aph^r-4-2 contains a mutation in the gene for Pol-α. The further characterization of the phenotype of the transfected cells should establish unambiguously a role for Pol-α in DNA repair and mutagenesis.

VIII. Prospects for Cloning the Gene for DNA Polymerase-α

In addition to DNA transfection, other methods involving recombinant DNA are now being applied to cloning the gene for Pol-α.

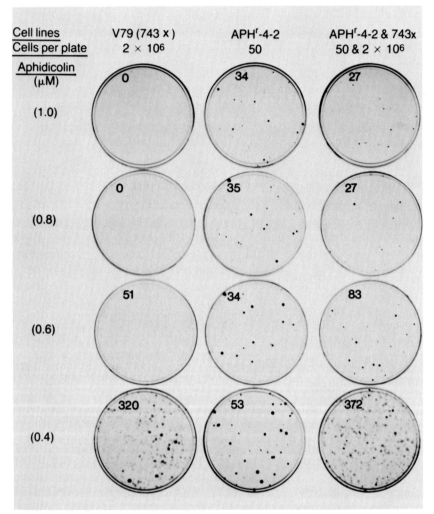

FIG. 1. Reconstruction experiments: Wild-type V79 cells (743X) and Aph^r-4-2 cells were seeded at indicated density per 100-mm plate and incubated in the presence of increasing concentrations of aphidicolin at 37°C. Cells were fed with fresh medium and aphidicolin every 2 days. After 7–14 days incubation, the plates were stained with crystal violet and colonies (≥50 cells/colony) were scored. The number on each plate represents the average colony of triplicate plates. Wild-type Aph^s cells [V79 (743X)] could not form colonies ($<2 \times 10^{-7}$) at concentrations of aphidicolin greater than 0.6 μM. The mutant cells, in contrast, formed colonies with high efficiency (70%) at 1.0 μm aphidicolin. When a mixture of 50 mutant and 2×10^6 wild-type cells was incubated in the presence of 0.4 to 1.0 μM aphidicolin, the number of viable colonies approximated the sum of the numbers of colonies obtained when the mutant and wild-type cells were incubated separately.

These include the screening of cDNA libraries contained in expression vectors, such as lambda gt11, with monoclonal and polyclonal antipolymerase antibodies (332); the sequencing of polymerase polypeptides, and the synthesis of corresponding oligonucleotides for use as hybridization probes with which to screen cDNA and genomic DNA libraries (333); and the production of genomic libraries of mutant DNA in cosmid vectors (334). Major impediments in the cloning of this gene include a large mRNA (greater than 5 kilobases) that is difficult, but not impossible, to isolate; the size of the genomic DNA that encodes the mRNA, which may be too large to be contained intact in a lambda or cosmid cloning vector; and the possible need for posttranscriptional modifications to produce a catalytically active enzyme. Furthermore, without the production of a catalytically active enzyme or in the absence of complementation of a conditional mutant, it will be difficult to establish that a cloned gene is indeed the gene for Pol-α.

Cloning the gene for Pol-α may be the most efficient means to establish the structure and amino-acid sequence of this enzyme. The identification of DNA fragments that contain sequences coding for the Pol-α polypeptides should make it possible to isolate polymerase mRNA. The availability of a cDNA encoding Pol-α should permit the following experiments: quantitation of the synthesis of mRNA during the cell cycle, after induction of DNA replication or after exposure to DNA-damaging agents, and *in vitro* synthesis of large amounts of Pol-α protein for the determination of protein structure and for studying the effects of site-specific modifications on catalysis. These experimental approaches should permit one to study how DNA polymerase—and ultimately DNA replication—is regulated during the cell cycle.

Acknowledgments

Research in the Joseph Gottstein Memorial Cancer Research Laboratory is supported by the U.S. National Institutes of Health (R35-CA-39903, AG-07151, and R23-CA-42702) and the U.S. Environmental Protection Agency (R809-623-010). Research in M.F.'s laboratory is supported by the Israel–U.S. Binational Science Foundation, Jerusalem, Israel. We thank colleagues who generously and critically read the manuscript: E. Snow, M. Horowitz, M. Reyland, and R. J. Monnat; and M. Whiting, who typed and edited the many drafts.

References

1. A. Kornberg, I. R. Lehman, M. J. Bessman and E. S. Simms, *BBA* **21**, 197 (1956).
2. A. Kornberg, "DNA Replication." Freeman, San Francisco, California, 1080. (Suppl. 1982).
3. F. J. Bollum and V. R. Potter, *JACS* **79**, 3603 (1957).
4. L. A. Loeb, S. S. Agarwal and A. M. Woodside, *PNAS* **61**, 827 (1968).
5. L. A. Loeb and S. S. Agarwal, *Exp. Cell Res.* **66**, 299 (1971).

6. W. E. Lynch, J. Short and I. Lieberman, *Cancer Res.* **36**, 901 (1976).
7. H. M. Keir, R. K. Craig and A. G. McLennon, *Biochem. Soc. Symp.* **42**, 37 (1977).
8. U. Bertazzoni, M. Stefani, G. P. Noy, E. Giulotto, F. Nuzzo, A. Falaschi and S. Spadari, *PNAS* **73**, 785 (1976).
9. S. Spadari, G. Villani and N. Hardt, *Exp. Cell Res.* **113**, 57 (1978).
10. M. S. Coleman, J. J. Hutton and F. J. Bollum, *Blood* **44**, 19 (1974).
11. R. D. Barr, P. Sarin, G. Sarna and S. Perry, *Eur. J. Cancer* **12**, 705 (1976).
12. G. D. Roodman, J. J. Hutton and F. J. Bollum, *BBA* **425**, 478 (1976).
13. M. Yoneda and F. J. Bollum, *JBC* **240**, 3385 (1965).
14. L. A. Loeb, *JBC* **244**, 1672 (1969).
15. O. Fichot, M. Pascal, M. Mechali and A.-M. de Recondo, *BBA* **561**, 29 (1979).
16. F. Grosse and G. Krauss, *Bchem* **20**, 5470 (1981).
17. G. Villani, B. Sauer and I. R. Lehman, *JBC* **255**, 9479 (1980).
18. A. Weissbach, D. Baltimore, F. J. Bollum, R. C. Gallo and D. Korn, *Science* **190**, 401 (1975).
19. A. Weissbach, A. Schlabach, B. Fridlender and A. Bolden, *Nature NB* **231**, 167 (1977).
20. E. F. Baril, O. E. Brown, M. D. Jenkins and J. Laszlo, *Bchem* **10**, 1981 (1971).
21. L. M. S. Chang and F. J. Bollum, *JBC* **246**, 5835 (1971).
22. B. Fridlender, M. Fry, A. Bolden and A. Weissbach, *PNAS* **69**, 452 (1972).
23. A. Bolden, M. Fry, R. Muller, R. Citarella and A. Weissbach, *ABB* **153**, 26 (1972).
24. M. Yamaguchi, K. Tanabe, Y. N. Taguchi, M. Nishizawa, T. Takahashi and A. Matsukage, *JBC* **255**, 9942 (1980).
25. S. Yoshida, T. Ando and T. Kondo, *BBRC* **60**, 1193 (1974).
26. A. Matsukage, E. W. Bohn and S. H. Wilson, *PNAS* **71**, 578 (1974).
27. J. J. Byrnes, K. M. Downey, V. L. Black and A. G. So, *Bchem* **15**, 2817 (1976).
28. M. Y. W. T. Lee, C.-K. Tan, K. M. Downey and A. G. So, *Bchem* **23**, 1906 (1984).
29. M. Fry and L. A. Loeb, "Animal Cell DNA Polymerases." CRC Press, Boca Raton, Florida, 1986.
30. C. L. Brakel and A. B. Blumenthal, *EJB* **88**, 351 (1978).
31. G. R. Banks, J. A. Boezi and I. R. Lehman, *JBC* **254**, 9886 (1979).
32. U. Hubscher, A. Spanos, W. Albert, F. Grummt and G. R. Banks, *PNAS* **78**, 6771 (1981).
33. D. Korn, P. A. Fisher, J. Battey and T. S.-F. Wang, *CSHSQB* **43**, 613 (1978).
34. D. Korn, P. A. Fisher and T. S.-F. Wang, *in* "New Approaches in Eukaryotic DNA Replication" (A.-M. de Recondo, ed.). Plenum, New York, 1983.
35. R. C. Conaway and I. R. Lehman, *PNAS* **79**, 2523 (1982).
36. J. K. Burke, J. Plummer, J. A. Huberman and M. J. Evans, *BBA* **609**, 205 (1980).
37. E. A. Faust and C. D. Rankin, *NARes* **10**, 4181 (1982).
38. C. G. Pritchard and M. L. DePamphilis, *JBC* **258**, 9801 (1983).
39. S. Tanaka, S.-Z. Hu, T. S.-F. Wang and D. Korn, *JBC* **257**, 8386 (1982).
40. A. Matsukage, M. Yamaguchi, K. Tanabe, M. Mishizawa, T. Takahashi, M. Seto and T. Takahashi, *Gann* **73**, 850 (1982).
41. T. S.-F. Wang, S.-Z. Hu and D. Korn, *JBC* **259**, 1854 (1984).
42. A. F. Wahl, S. P. Kowalski, L. W. Harwell, E. M. Lord and R. A. Bambara, *Bchem* **23**, 1895 (1984).
43. S. Detera-Wadleigh, E. Karawya and S. H. Wilson, *BBRC* **122**, 420 (1984).
44. E. Karawya, J. Swack, W. Albert, J. Fedorko, J. D. Minna and S. H. Wilson, *PNAS* **81**, 7777 (1984).
45. A. Sugino and K. Nakayamara, *PNAS* **77**, 7049 (1980).

46. P. K. Liu, C.-C. Chang, J. E. Trosko, D. K. Dube, G. M. Martin and L. A. Loeb, *PNAS* **80**, 797 (1983).
47. P. K. Liu and L. A. Loeb, *Science* **226**, 833 (1984).
48. L. A. Loeb, P. K. Liu, M. E. Reyland, W. R. Pendergrass and K. P. Gopinathan, *FP* **44**(3), 855 (1985).
49. A. M. Holmes, I. P. Hesselwood, R. F. Wickremasinghe and I. R. Johnston, *Biochem. Soc. Symp.* **42**, 17 (1977).
50. P. A. Fisher and D. Korn, *JBC* **252**, 6528 (1977).
51. F. Grummt, G. Waltl, H.-M. Jantzen, K. Hamprecht, U. Hubscher and C. C. Kuenzle, *PNAS* **76**, 6081 (1979).
52. F. Grosse and G. Krauss, *Bchem* **20**, 5470 (1981).
53. M. Mechali and A.-M. de Recondo, in "New Approaches in Eukaryotic DNA Replication" (A.-M. de Recondo, ed.). Plenum, New York, 1983.
54. L. P. Goscin and J. J. Byrnes, *NARes* **10**, 6023 (1982).
55. M. Yamaguchi, K. Tanabe, T. Takahashi and A. Matsukage, *JBC* **257**, 4484 (1982).
56. E. A. Faust, G. Gloor, M.-F. Macintyre and R. Nagy, *BBA* **781**, 216 (1984).
57. Y.-C. Chen, E. W. Bohn, S. R. Planck and S. H. Wilson, *JBC* **254**, 11678 (1979).
58. P. Lamothe, B. Baril, A. Chi, L. Lee and E. Baril, *PNAS* **78**, 4723 (1981).
59. A. Weissbach, *ARB* **46**, 25 (1977).
60. U. Hubscher, *Experientia* **39**, 1 (1983).
61. J. G. Salisbury, P. J. O'Connor and R. Saffhill, *BBA* **517**, 181 (1978).
62. M. Mechali, J. Abadidebat and A.-M. de Recondo, *JBC* **255**, 2114 (1980).
63. R. K. Craig and H. M. Keir, *BJ* **145**, 215 (1975).
64. A. M. Holmes and I. R. Johnston, *FEBS Lett.* **29**, 1 (1973).
65. G. R. Banks, J. A. Boezi and I. R. Lehman, *EJB* **88**, 351 (1978).
66. E. Karawya, J. A. Swack and S. H. Wilson, *Anal. Biochem.* **135**, 318 (1983).
67. A. Spanos, S. G. Sedgwick, G. T. Yarranton, U. Hubscher and G. R. Banks, *NARes* **9**, 1825 (1981).
68. A. Blank, J. R. Silber, M. P. Thelan and C. A. Dekker, *Anal. Biochem.* **135**, 423 (1983).
69. K. McKune and A. M. Holmes, *NARes* **6**, 3341 (1979).
70. I. P. Hesslewood, A. M. Holmes, W. F. Wakeling and I. R. Johnson, *EJB* **84**, 123 (1978).
71. F. Grosse and G. Krauss, *NARes* **8**, 5703 (1980).
72. S. Masaski, O. Koiwai and S. Yoshida, *JBC* **257**, 7172 (1982).
73. S. Masaki, K. Tanabe and S. Yoshida, *NARes* **12**, 4455 (1984).
74. U. Hubscher, P. Gerschweiler and G. K. McMaster, *EMBO J.* **1**, 1513 (1982).
75. U. Hubscher and H.-P. Ottinger, in "Mechanisms of DNA Replication and Recombination" (N. Cozzarelli, ed.). Liss, New York, 1983.
76. P. Sauer and I. R. Lehman, *JBC* **257**, 12394 (1982).
77. L. S. Kaguni, J.-M. Rossignol, R. C. Connaway and J. R. Lehman, *PNAS* **80**, 2221 (1983).
78. R. M. Benbow, D. J. Stowers, K. T. Hiriyanna, V. C. Dunne and E. C. Ford, in press (1985).
79. M. Yamaguchi, E. A. Hendrickson and M. L. De Pamphilis, *JBC* **260**, 6254 (1985).
80. E. M. Karawya and S. H. Wilson, *JBC* **257**, 13129 (1982).
81. S. Yoshida, S. Masaki and O. Koiway, *BBA* **654**, 104 (1981).
82. F. M. Racine and P. M. Morris, *NARes* **5**, 3945 (1978).
83. P. A. Fisher, T. S.-F. Wang and D. Korn, *JBC* **254**, 6128 (1979).
84. D. W. Mosbaugh and S. Linn, *JBC* **259**, 10247 (1984).

85. L. A. Loeb, K. D. Tartof and E. C. Travaglini, *Nature NB* **242**, 66 (1973).
86. P. A. Fisher and D. Korn, *JBC* **254**, 11033 (1979).
87. P. A. Fisher and D. Korn, *JBC* **254**, 11040 (1979).
88. P. A. Fisher and D. Korn, *Bchem* **20**, 4560 (1981).
89. P. A. Fisher, J. T. Chen and D. Korn, *JBC* **256**, 113 (1981).
90. S. N. Wilson, A. Matsukage, E. W. Bohn, Y. C. Chen and M. Sivarjan, *NARes* **4**, 3981 (1977).
91. G. Villani, P. G. Fay, R. A. Bambara and I. R. Lehman, *JBC* **256**, 8202 (1981).
92. L. M. S. Chang and F. J. Bollum, *BBRC* **46**, 1354 (1972).
93. J. W. Hochensmith and R. A. Bambara, *Bchem* **20**, 227 (1981).
94. W. Keller, *PNAS* **69**, 1560 (1972).
95. S. Spadari and A. Weissbach, *PNAS* **72**, 503 (1975).
96. G. Brun and A. Weissbach, *PNAS* **75**, 5931 (1978).
97. J.-E. Ikeda, M. Longiaru, M. S. Horwitz and J. Hurwitz, *PNAS* **77**, 5827 (1980).
98. T. W. Wong and D. A. Clayton, *JBC* **260**, 11530 (1985).
99. F. Gross and G. Krauss, *EJB* **141**, 109 (1984).
100. B. Reckman, F. Grosse and G. Krauss, *NARes* **11**, 7251 (1983).
101. L. Thelanda and P. Reichard, *ARB* **48**, 133 (1979).
102. S. K. Das, T. A. Kunkel and L. A. Loeb, in "Genetic Consequences of Nucleotide Pool Imbalances" (F. J. de Seeres and W. Sheridan, eds.). Plenum, New York, (1985).
103. G. P. V. Reddy and A. B. Pardee, *PNAS* **77**, 3312 (1980).
104. G. P. V. Reddy and A. B. Pardee, *JBC* **257**, 12526 (1982).
105. H. Noguchi, G. P. V. Reddy and A. B. Pardee, *Cell* **32**, 443 (1983).
106. G. P. V. Reddy and A. B. Pardee, *Nature* **303**, 86 (1983).
107. D. K. Dube, T. A. Kunkel, G. Seal and L. A. Loeb, *BBA* **561**, 369 (1979).
108. S. Yoshida and S. Masaki, *BBA* **561**, 396 (1979).
109. E. Wist, *BBA* **565**, 98 (1979).
109a. T. Lindahl, *This Series* **22**, 135, 145 (1979).
110. S. Spadari, F. Sala and G. Pedrali-Noy, *TIBS* **7**, 29 (1982).
111. K. Ono, Y. Iwata and H. Nakane, *Bior.ed. Pharmacother.* **37**, 27 (1983).
112. E. C. Travaglini, A. S. Mildvan and L. A. Loeb, *JBC* **250**, 8647 (1975).
113. M. A. Sirover and L. A. Loeb, *JBC* **252**, 3605 (1977).
114. J. P. Slater, I. Tamir, L. A. Loeb and A. S. Mildvan, *JBC* **247**, 6784 (1972).
115. D. L. Ollis, P. Brick, R. Hamlin, N. G. Xuong and T. A. Steitz, *Nature* **313**, 762 (1985).
116. K. Tanabe, Y. N. Taguchi, A. Matsukage and T. Takahashi, *J. Biochem. (Tokyo)* **88**, 35 (1980).
117. S. D. Detera, S. P. Becerra, J. A. Swack and S. H. Wilson, *JBC* **256**, 6933 (1981).
118. L. M. S. Chang, *JMB* **93**, 219 (1975).
119. D. K. Dube and L. A. Loeb, *Bchem* **15**, 3605 (1976).
120. R. A. Bambara, D. Uyemura and T. Choi, *JBC* **253**, 413 (1978).
121. S. K. Das and R. K. Fujimara, *JBC* **254**, 1227 (1979).
122. K. McKune and A. M. Holmes, *BBRC* **90**, 864 (1979).
123. V. S. Mikhailov and I. M. Androsova, *BBA* **783**, 6 (1984).
124. D. Grossberger and W. Clough, *Bchem* **20**, 4049 (1981).
125. J. V. Wierowski, K. G. Lawton, J. W. Hockensmith and R. A. Bambara, *JBC* **258**, 6250 (1983).
126. H. D. Riedel, H. Konig and R. Knippers, *BBA* **783**, 158 (1985).
127. D. T. Weaver and M. L. DePamphilis, *JBC* **257**, 2075 (1982).

128. L. S. Kaguni and D. A. Clayton, PNAS **99**, 983 (1982).
129. L. S. Kaguni, R. A. DiFrancesco and I. R. Lehman, JBC **259**, 9314 (1984).
130. B. Beckman, F. Grosse, C. Urbanke, R. Frank, H. Blocker and G. Krauss, in press (1984).
131. H. Konig, H. D. Riedel and R. Knippers, EJB **135**, 435 (1985).
132. T. S-F. Wang and D. Korn, Bchem **19**, 1782 (1980).
133. M. L. DePamphhilis, S. Anderson, M. Cusick, R. May, T. Herman, M. Krokan, E. Shelton, L. Tack, D. Tappel, D. Weaver and P. M. Wasserman, in "Mechanistic Studies of DNA Replication and Genetic Recombination" (B. Alberts, ed.). Academic Press, New York, 1980.
134. K. Bose, P. Karran and B. Strauss, PNAS **75**, 794 (1978).
135. A. Matsukage, M. Nishizawa, T. Takahashi and T. Nozumi, J. Biochem. (Tokyo) **88**, 1869 (1980).
136. G. R. Banks and G. T. Yarranton, EJB **62**, 143 (1976).
137. A. G. McLennan and H. M. Meir, BJ **151**, 239 (1975).
138. M. Creran and R. E. Pearlman, JBC **249**, 3123 (1974).
139. E. Wintersberger, EJB **84**, 167 (1978).
140. D. Korn, P. A. Fisher and T. S.-F. Wang, in "New Approaches in Eukaryotic DNA Replication" (A.-M. de Recondo, ed.). Plenum, New York, 1983.
141. M. L. DePamphilis and P. M. Wasserman, ARB **49**, 627 (1980).
142. L. Rowen and A. Kornberg, JBC **253**, 770 (1978).
143. E. W. Benz, Jr., D. Reinberg, R. Vicuna and J. Hurwitz, JBC **255**, 1096 (1980).
144. L. J. Romano and C. C. Richardson, JBC **254**, 10483 (1979).
145. C.-C. Liu and B. M. Alberts, JBC **256**, 2821 (1981).
146. A. Kaftory and M. Fry, NARes **5**, 2679 (1978).
147. T. Kozu, T. Yagura and T. Seno, Cell Struct. Funct. **7**, 9 (1982).
148. T. Yagura, T. Kozu and T. Seno, JBC **257**, 11121 (1982).
149. T. Yagura, T. Kozu, T. Seno, M. Saneyoshi, S. Hiraga and H. Nagano, JBC **258**, 13070 (1983).
150. M. A. Wagar and J. A. Huberman, Cell **6**, 551 (1975).
151. H.-D. Riedel, H. Konig, H. Stahl and R. Knippers, NARes **10**, 5621 (1982).
152. S. Yoshida, R. Suzuki, S. Masaki and O. Koiwai, BBA **741**, 348 (1983).
153. R. M. Gronostatjski, J. Field and J. Hurwitz, JBC **259**, 9479 (1984).
154. T. Yagura, S. Tanaka, T. Kozu, T. Seno and D. Korn, JBC **258**, 6698 (1983).
155. M. Philippe, R. Sheinin and A.-M. de Recondo, in "Proteins Involved in DNA Replication" (U. Hubscher and S. Spadari, eds.). Plenum, New York, (1985).
156. U. Hubscher, EMBO J. **2**, 133 (1983).
157. B. Y. Tseng and C. N. Ahlem, JBC **257**, 9479 (1984).
158. B. Y. Tseng and C. M. Ahlem, JBC **258**, 9845 (1983).
159. P. Plevani, M. Foiani, P. Valsanini, G. Badaracco, E. Cheriathundam and L. M. S. Chang, JBC **260**, 7102 (1985).
160. S. M. Jazwinski and G. M. Edelman, JBC **260**, 4995 (1985).
161. F. E. Wilson and A. Sugino, JBC **260**, 8173 (1985).
162. C. G. Pritchard, D. T. Weaver, E. F. Baril and M. D. DePamphilis, JBC **258**, 9810 (1983).
163. B. Novak and E. F. Baril, NARes **5**, 221 (1978).
164. M. Brown, F. J. Bollum and L. M. S. Chang, PNAS **78**, 3049 (1981).
165. J. F. Burke, J. Plummer, J. A. Huberman and M. J. Evans, BBA **609**, 205 (1980).
166. P. C. Zamecnik, M. L. Stephenson, C. M. Janeway and K. Randerath, BBRC **24**, 91 (1966).

167. K. Randerath, C. M. Janeway, M. L. Stephenson and P. C. Zamecnik, *BBRC* **24**, 98 (1966).
168. E. Rapaport and P. C. Zamecnik, *PNAS* **73**, 3984 (1976).
169. C. Weinman-Dorsch, A. Medl, I. Grummt, W. Albert, F.-J. Ferdinand, R. R. Friis, G. Pierron, W. Moll and F. Grummt, *EJB* **138**, 179 (1984).
170. H. Probst, K. Hamprecht and V. Gekler, *BBRC* **110**, 688 (1983).
171. F. Grummt, *PNAS* **75**, 371 (1978).
172. F. Grummt, *CSHSQB* **43**, 649 (1978).
173. E. Rapaport, P. C. Zamecnik and E. F. Baril, *PNAS* **78**, 838 (1981).
174. E. Baril, P. Ronin, D. Burstein, K. Mara and P. Zamecnik, *PNAS* **80**, 4931 (1983).
175. E. Rapaport and L. Feldman, *EJB* **138**, 111 (1984).
176. E. Rapaport, P. C. Zamecnik and E. F. Baril, *JBC* **256**, 12148 (1981).
177. P. C. Zamecnik, E. Rapaport and E. F. Baril, *PNAS* **79**, 1791 (1982).
178. K. Ono, Y. Iwata, H. Nakamura and A. Matsukage, *JBC* **95**, 34 (1980).
179. K. G. Lawton, J. V. Wierowski, S. Schechter, R. Hilf and R. A. Bambara, *Bchem* **23**, 4294 (1984).
180. G. Herrick and B. Alberts, *JBC* **251**, 2124 (1976).
181. G. Herrick and B. Alberts, *JBC* **251**, 2133 (1976).
182. F. Cobinchi, S. Riva, G. Mastromei, S. Spadari, G. Pedrali-Noy and A. Falaschi, *CSHSQB* **43**, 639 (1978).
183. M. Duguet, C. Bonne and A.-M. de Recondo, *Bchem* **20**, 3598 (1981).
184. C. Bonne, P. Sautierre, M. Duguet and A.-M. de Recondo, *JBC* **257**, 2722 (1982).
185. S. R. Planck and S. H. Wilson, *JBC* **255**, 11547 (1980).
186. K. R. Williams, K. L. Stone, M. B. LoPresti, B. L. Merrill and S. R. Planck, *PNAS* **82**, 5666 (1985).
187. M. Fry, P. Lapidot and P. Weisman-Shomer, *Bchem* **24**, 7549 (1985).
188. M. Fry, P. Weisman-Shomer, P. Lapidot and R. Sharf, submitted (1986).
189. M. Sapp, H. Konig, H. D. Riedel, A. Richter and R. Knippers, *JBC* **260**, 1550 (1985).
190. B. Otto, M. Baynes and R. Knippers, *EJB* **73**, 17 (1977).
191. L. A. Loeb, B. Fansler, R. Williams and D. Mazia, *Exp. Cell Res.* **57**, 298 (1969).
192. B. Fansler and L. A. Loeb, *Exp. Cell Res.* **57**, 305 (1969).
193. W. E. Lynch, S. Surrey and I. Lieberman, *JBC* **250**, 8179 (1975).
194. A. Matsukage, N. Nishioka, M. Nishizawa and T. Takahashi, *Cell Struct. Funct.* **4**, 295 (1979).
195. K.-W. Knopf and A. Weissbach, *Bchem* **16**, 3190 (1977).
196. L. A. Loeb, *Nature* **226**, 448 (1970).
197. E. J. Schlaeger, H.-J. van Telgen, K.-H. Klempnauer and R. Knippers, *EJB* **84**, 95 (1978).
198. T. Yagura and T. Seno, *BBA* **608**, 277 (1980).
199. M. Fry, J. Silber, G. M. Martin and L. A. Loeb, *J. Cell. Physiol.* **118**, 225 (1984).
200. G. Martini, B. Tato, D. Attardi-Gandini and G. P. Tocchini-Valentini, *BBRC* **72**, 875 (1976).
201. G. Herrick, B. B. Spear and G. Veomett, *PNAS* **73**, 1136 (1976).
202. A. Weissbach, A. Schlabach, B. Firdlender and A. Bolden, *Nature NB* **231**, 167 (1971).
203. L. M. S. Chang and F. J. Bollum, *Bchem* **11**, 1264 (1972).
204. N. B. Hecht, *BBA* **312**, 471 (1973).
205. D. W. Sedwick, T. S.-F. Wang and D. Korn, *JBC* **250**, 7045 (1975).
206. K. G. Bench, S. Tanaka, S.-Z. Hu, T. S.-F. Wang and D. Korn, *JBC* **257**, 8391 (1982).

207. A. Matsukage, S. Yamamoto, M. Yamaguchi, M. Kuskabe and T. Takahashi, *J. Cell. Physiol.* **117**, 266 (1983).
208. H. Nakamura, T. Morita, S. Masaki and S. Yoshida, *Exp. Cell Res.* **151**, 123 (1984).
209. H. C. Smith and R. Berezney, *Bchem* **21**, 6751 (1982).
210. S. Yamamoto, T. Takahashi and A. Matsukage, *Cell Struct. Funct.* **9**, 83 (1984).
211. H. S. Smith and R. Berezney, *Bchem* **22**, 3042 (1983).
212. U. Hubscher, C. C. Kuenzle, W. Limacher, P. Schrerer and S. Spadari, *CSHSQB* **43**, 635 (1978).
213. K. S. Rao, G. M. Martin and L. A. Loeb, *J. Neurochem.* **45**, 1273 (1985).
214. M. Fry, P. Weisman-Shomer, Z. Renart and J. Lapidot, submitted (1986).
215. W. E. Lynch, S. Surrey and I. Lieberman, *JBC* **250**, 8179 (1975).
216. W. E. Lynch and I. Lieberman, *BBRC* **52**, 843 (1973).
217. M. Phillipe and P. H. Chevaillier, *BBA* **447**, 188 (1976).
218. K. Ono, F. Barre-Sinoussi, F. Rey, C. Jasmin and J.-C. Chermann, *Cell. Mol. Biol.* **30**, 303 (1984).
219. L. M. S. Chang and F. J. Bollum, *JBC* **247**, 7948 (1972).
220. E. F. Baril, M. D. Jenkins, O. E. Brown, J. Laszlo and H. P. Morris, *Cancer Res.* **33**, 1187 (1973).
221. P. B. Davis, J. Laszlo and E. F. Baril, *Cancer Res.* **36**, 432 (1976).
222. S. S. Agarwal and L. A. Loeb, *Cancer Res.* **32**, 107 (1972).
223. M. S. Coleman, I. J. Hutton and F. J. Bollum, *Nature* **248**, 407 (1974).
224. R. J. Mayer, R. G. Smith and R. C. Gallo, *Blood* **46**, 509 (1975).
225. R. K. Craig, P. A. Costello and H. M. Keir, *BJ* **145**, 233 (1975).
226. F. Hanaoka, K. Nagata, Y. Watanabe, T. Enomoto and M.-A. Yamada, *Cell Struct. Funct.* **6**, 357 (1981).
227. L. M. S. Chang and F. J. Bollum, *JMB* **74**, 1 (1973).
228. S. Spadari and A. Weissbach, *JMB* **86**, 11 (1974).
229. R. W. Chiu and E. F. Baril, *JBC* **250**, 7951 (1975).
230. C. Delfini, E. Alfani, V. DeVenezia, G. Oberholtzer, C. Tomasello, T. Eremenko and P. Volpe, *PNAS* **82**, 2220 (1985).
231. C.-C. Chang, J. A. Boezi, S. T. Warren, C. L. K. Sabourin, P. K. Liu, L. Glatzer and J. E. Trosko, *Somatic Cell Genet.* **7**, 235 (1981).
232. Y.-J. Tsai, F. Hanaoka, M. M. Nakano and M. A. Yamada, *BBRC* **91**, 1190 (1979).
233. Y. Murakami, H. Yasuda, H. Miyazawa, F. Hanaoka and M.-A. Yamada, *PNAS* **82**, 1761 (1985).
234. S. Spadari, F. Sala and G. Pedrali-Noy, *Adv. Exp. Med. Biol.* **179**, 169 (1984).
235. S. Ikegami, T. Taguchi, M. Ohashi, M. Oguro, H. Nagano and Y. Mano, *Nature* **275**, 458 (1978).
236. G. Pedrali-Noy and S. Spadari, *BBRC* **88**, 1194 (1979).
237. G. Pedrali-Noy, M. Belvedere, T. Crepaldi, F. Focher and S. Spadari, *Cancer Res.* **42**, 3810 (1982).
238. G. E. Wright, E. F. Baril and N. C. Brown, *NARes* **8**, 99 (1980).
239. G. E. Wright, E. F. Baril, V. M. Brown and N. C. Brown, *NARes* **10**, 4431 (1982).
239a. W. T. Ashton, J. D. Karkas, A. K. Field and R. L. Tolman, *BBA* **108**, 1716 (1982).
240. M. R. Miller, R. G. Ulrich, T. S.-F. Wang and D. Korn, *JBC* **260**, 134 (1985).
241. F. J. Bollum, *JBC* **234**, 2733 (1959).
242. M. Fry, L. A. Loeb and G. M. Martin, *J. Cell. Physiol.* **106**, 435 (1981).
243. R. T. Butt, W. M. Wood, E. L. McKay and R. L. P. Adams, *BJ* **173**, 309 (1978).

244. U. Bertazzoni, M. Stefanini, G. P. Noy, E. Giulotto, F. Nuzzo, A. Falaschi and S. Spadari, *PNAS* **73**, 785 (1976).
245. K. Suzuki, M. Miyaki, N. Akamatsu and T. Ono, *FEBS Lett.* **119**, 150 (1980).
246. V. M. Craddock and C. M. Ansley, *BBA* **564**, 15 (1979).
247. V. M. Craddock, *Carcinogenesis* **2**, 61 (1981).
248. V. P. Parker and M. W. Lieberman, *NARes* **4**, 2029 (1978).
249. U. Bertazzoni, A. I. Scovassi, M. Stefanini, E. Giulotto, S. Spadari and A. Pedrini, *NARes* **5**, 2189 (1978).
250. S. Seki, T. Oda and M. Ohashi, *BBA* **610**, 413 (1980).
251. G. Pedrali-Noy and S. Spadari, *Mutat. Res.* **70**, 389 (1980).
252. E. Giulotto and C. Mondello, *BBRC* **99**, 1287 (1981).
253. N. Hardt, G. Pedrali-Noy, F. Focher and S. Spadari, *BJ* **199**, 453 (1981).
254. F. Hanoka, H. Kato, S. Ikegami, M. Ohashi and M. Yamada, *BBRC* **87**, 575 (1979).
255. N. A. Berger, K. K. Kurohara, S. J. Petzold and G. W. Sikorsky, *BBRC* **89**, 218 (1979).
256. G. Ciarrocchi, J. G. Jose and S. Linn, *NARes* **7**, 1205 (1979).
257. R. D. Snyder and J. D. Regan, *BBRC* **99**, 1088 (1981).
258. S. Seki, M. Ohashi, H. Ogura and T. Oda, *BBRC* **104**, 1502 (1982).
259. R. Waters, K. Crocombe and R. Mirzayans, *Mutat. Res.* **94**, 229 (1981).
260. K. Yamada, K. Kato, F. Hanaoka and M.-A. Yamada, *Gann* **73**, 63 (1981).
261. S. L. Dresler, J. D. Roberts and M. W. Lieberman, *Bchem* **21**, 2557 (1982).
262. S. L. Dresler and M. W. Lieberman, *JBC* **258**, 9990 (1983).
263. S. L. Dresler, *JBC* **259**, 13947 (1984).
264. M. R. Miller and N. D. Chinault, *JBC* **257**, 46 (1982).
265. M. R. Miller and D. N. Chinault, *JBC* **257**, 10204 (1982).
266. A. R. S. Collins, S. Squires and R. T. Johnson, *NARes* **10**, 1203 (1982).
267. A. Collins, *BBA* **741**, 341 (1983).
268. A. A. Van Zeeland, C. J. M. Bussman, F. Degrassi, A. R. Filon, A. C. van Kesteren-van Leeuwen, F. Palitti and A. T. Matarajan, *Mutat. Res.* **92**, 379 (1982).
269. P. K. Liu, J. E. Trosko and C.-C. Chang, *Mutat. Res.* **106**, 333 (1982).
270. P. K. Liu, C.-C. Chang and J. E. Trosko, *Mutat. Res.* **106**, 317 (1982).
271. P. K. Liu, C.-C. Chang and J. E. Trosko, *Somatic Cell. Genet.* **10**, 235 (1984).
272. T. S.-F. Wang, B. E. Pearson, H. A. Suomalainen, T. Mohandas, L. J. Shapiro, J. Shroeder and D. Korn, *PNAS* **82**, 5270 (1985).
273. F. Hanaoka, M. Tandai, H. Miyazawa, T.-A. Hori and M.-A. Yamada, *Gann* **76**, 441 (1985).
274. L. A. Loeb and T. A. Kunkel, *ARB* **52**, 429 (1982).
275. O. Gotoh and Y. Tagashira, *Biopolymers* **20**, 1033 (1981).
276. M. Eigen, W. Gardiner, P. Schuster and R. Winkler-Oswatitsch, *Sci. Am.* **244**, 88 (1981).
277. J. W. Drake, *in* "The Molecular Basis of Mutation." Holden-Day, San Francisco, California, 1970.
278. T. A. Kunkel, R. Meyer and L. A. Loeb, *PNAS* **76**, 6331 (1979).
279. T. A. Kunkel, L. A. Loeb and M. F. Goodman, *JBC* **259**, 1539 (1984).
280. L. A. Loeb, C. F. Springgate and N. Battula, *Cancer Res.* **34**, 2311 (1974).
281. R. Lohrmann and L. E. Orgel, *JMB* **142**, 555 (1980).
282. R. Lohrmann, P. K. Bridson and L. E. Orgel, *J. Mol. Evol.* **17**, 303 (1981).
283. T. A. Kunkel and L. A. Loeb, *Science* **213**, 765 (1981).
284. A.-L. Lu, S. Clark and P. Modrich, *PNAS* **80**, 4639 (1983).
285. A.-L. Lu, K. Welsh, S. Clark, S.-S. Su and P. Modrich, *CSHSQB* in press (1986).
286. L. M. S. Chang and F. J. Bollum, *JBC* **248**, 3398 (1973).

287. S. S. Agarwal, D. K. Dube and L. A. Loeb, *JBC* **254**, 101 (1979).
288. L. M. S. Chang, *JBC* **248**, 6983 (1973).
289. J. Y. H. Chan and F. F. Becker, *PNAS* **76**, 814 (1979).
290. G. Seal, C. W. Shearman and L. A. Loeb, *JBC* **254**, 5229 (1979).
291. D. K. Dube, T. A. Kunkel, G. Seal and L. A. Loeb, *BBA* **561**, 369 (1979).
292. V. Murray and R. Holliday, *JMB* **146**, 55 (1981).
293. N. Battula and L. A. Loeb, *JBC* **249**, 4086 (1974).
294. M. Fry, C. W. Shearman, G. W. Martin and L. A. Loeb, *Bchem* **19**, 5939 (1980).
295. J. G. Salisbury, P. J. O'Connor and R. Saffhill, *BBA* **517**, 181 (1978).
296. S. W. Krauss and S. Linn, *Bchem* **19**, 220 (1980).
297. A. W. Holmes, I. P. Hessielwood and I. R. Johnston, *EJB* **43**, 487 (1974).
298. L. A. Loeb, J. Abbotts, P. K. Liu and J. R. Silber, in "Cellular Responses to DNA Damage" UCLA Symp. Mol. and Cell Biol. (E. C. Friedberg and B. A. Bridges, eds.). Liss, New York, 1983.
299. S. Brosius, F. Grosse and G. Krauss, *NARes* **11**, 193 (1983).
300. S. K. Krauss and S. Linn, *Bchem* **21**, 1002 (1982).
301. L. A. Weymouth and L. A. Loeb, *PNAS* **75**, 1924 (1978).
302. T. A. Kunkel, E. James and L. A. Loeb, in "DNA Repair: A Laboratory Manual of Research Procedures." Dekker, New York, 1983.
303. J. Abbotts and L. A. Loeb, *BBA* **259**, 6712 (1984).
304. J. Abbotts and L. A. Loeb, *JBC* **259**, 6712 (1984).
305. T. A. Kunkel and L. A. Loeb, *JBC* **254**, 5718 (1979).
306. A. R. Fersht and J. W. Knill-Jones, *JMB* **165**, 655 (1983).
307. F. Grosse, G. Krauss, J. W. Knill-Jones and A. R. Fersht, *EMBO J.* **2**, 1515 (1983).
308. T. A. Kunkel, *JBC* **260**, 5787 (1985).
309. T. A. Kunkel, *JBC*, in press (1985).
310. L. S. Kaguni, R. A. DiFrancesco and I. R. Lehman, *JBC* **259**, 9314 (1984).
311. J. Abbotts and L. A. Loeb, *NARes* **13**, 261 (1985).
312. T. A. Kunkel, R. M. Schaaper, R. A. Beckman and L. A. Loeb, *JBC* **256**, 9883 (1981).
313. D. Brutlag and A. Kornberg, *JBC* **247**, 241 (1972).
314. J. J. Hopfield, *PNAS* **71**, 4135 (1974).
315. J. Ninio, *Biochimie* **57**, 587 (1975).
316. L. K. Clayton, M. F. Goodman, E. W. Branscomb and D. J. Galas, *JBC* **254**, 1902 (1979).
317. S. M. Watanabe and M. F. Goodman, *PNAS* **79**, 6429 (1982).
318. J. J. Hopfield, *PNAS* **77**, 5248 (1980).
319. O. P. Doubleday, P. J. Lecomte and M. Radman, in "Cellular Responses to DNA Damage" (E. C. Friedberg and B. A. Bridges, eds.). Liss, New York, 1983.
320. P. T. Englund, J. A. Huberman, T. M. Jovin and A. Kornberg, *JBC* **244**, 3038 (1969).
321. D. L. Sloan, L. A. Loeb, A. S. Mildvan and R. J. Feldmann, *JBC* **250**, 8913 (1975).
322. L. J. Ferrin, R. A. Beckman, L. A. Loeb and A. S. Mildvan, in "Manganese in Metabolism and Enzyme Function" (F. C. Wedler and V. L. Schramm, eds.). Academic Press, New York, in press.
323. R. H. Scheuermann and H. Echols, *PNAS* **81**, 7747 (1984).
324. M. Radman, S. Spadari and G. Villani, *Natl. Cancer Inst. Monogr.* **50**, 121 (1978).
325. A. R. Fersht, J. W. Knill-Jones and W.-C. Tsui, *JMB* **156**, 37 (1982).
326. J. T. Hare and J. H. Taylor, *PNAS* **82**, 7350 (1985).
327. S. Aizawa, M. Ohashi, L. A. Loeb and G. M. Martin, *Somatic Cell. Genet.* **11**, 211 (1985).

328. D. Ayusawa, K. Iwata and T. Seno, *Somatic Cell Genet.* **7**, 27 (1980).
329. A. M. Holmes, *NARes* **9**, 161 (1981).
330. Y. Nishiama, S. Suzuki, M. Yamaguchi, K. Maeno and S. Yoshida, *Virology* **135**, 87 (1984).
331. L. H. Thompson, *in* "Molecular Cell Genetics" (M. M. Gottesman, ed.). Wiley, New York, 1985.
332. L. M. Johnson, M. Snyder, L. M. S. Chang, R. W. Davis and J. L. Campell, *Cell* **43**, 369 (1985).
333. T. Maniatis, *in* "Cell Biology: A Comprehensive Treatise" (D. M. Prescott and L. Goldstein, eds.). Academic Press, New York, 1980.
334. Y.-F. Lau and Y.-W. Kan, *PNAS* **80**, 5225 (1983).

Replication of Superhelical DNAs in Vitro

Kenneth J. Marians,
Jonathan S. Minden,
and Camilo Parada

Graduate Program in Molecular Biology
Memorial Sloan-Kettering Cancer Center
New York, New York 10021

Studies over the last 15 years have revealed that, enzymatically, DNA replication is a more complicated process than one would have thought when gazing at the structure of B-DNA in 1953. Whereas DNA replication per se requires the template-directed polymerization of nucleotides—something that can be accomplished by any number of different DNA polymerases—the duplication of any genome *in vitro* generally requires many more enzymatic activities. Most of these additional proteins have been purified and their activities characterized as a result of studies on the replication of small bacteriophage genomes that are either circular single-stranded or linear double-stranded DNAs. These additional proteins can be grouped into a number of different categories: (1) DNA-binding proteins that present the DNA template to the polymerase in an acceptable form, (2) DNA polymerase "accessory" proteins that convert relatively nonprocessive DNA polymerases (that polymerize only 10–100 nucleotides per binding event) to highly processive ones that can replicate genome-length tracts of DNA in one binding event (e.g., the processivity of the *E. coli* DNA polymerase III holoenzyme is essentially unmeasurable by current techniques) and also decrease the error frequency of the polymerase, (3) DNA "unwinding" proteins (helicases) that help maintain the separation of the parental DNA strands during replication, and (4) priming proteins that manufacture small ribo- or deoxyribooligonucleotides that serve as primers for the DNA polymerases.

The enzymes mentioned above and their mechanisms have been subjected to extensive review recently (*1, 2*) and thus are not reviewed in this article, which focuses on several new purified systems that can replicate closed-circular, double-stranded superhelical (form I) DNAs *in vitro*. Problems unique to the replication of DNAs under topological constraint are also discussed. Studies on the reconstituted pBR322 DNA (Table I) replication system (3) are used as a guide.

Recent studies on the replication of form I DNAs indicate that new classes of proteins are required to replicate these DNAs. The activities of these new enzymes help to solve some of the problems peculiar to the replication of superhelical DNAs. Replication of the two parental DNA strands without the introduction of a discontinuity in either strand presents a unique problem for the initiation of DNA replication of form I DNAs. The comparatively simple pathways that employ specific proteins, like the ϕX174 gene-A or M13 gene-2 proteins, to "nick" at a specific sequence in the DNA, thus generating a free 3'-OH end, cannot be employed (4–6). Instead, two pathways of initiation have been revealed. In the case of form I DNAs containing origins of DNA replication from either the chromosome of $E.$ $coli$ ($oriC$) or bacteriophage λ (λdv), special proteins interact with specific DNA sequences to form protein–DNA complexes subsequently recognized by other priming proteins that gain access to the DNA template at the site of the initial specific protein–DNA complex (7, 8).

In the initiation of pBR322 DNA replication, a different pathway operates. The formation of a RNA·DNA hybrid by a transcript manufactured by RNA polymerase serves to (1) prime leading-strand DNA synthesis after being processed by RNase H (EC 3.1.26.4) (9), and (2) allow a "primosome" assembly site (pas) sequence (2) on the lagging-strand template to assume an active configuration and catalyze the assembly of a primosome that subsequently primes lagging-strand DNA synthesis (10). This type of initiation of DNA replication has been termed "transcriptional activation" (10a). A similar mechanism, involving T7 RNA polymerase and the T7 gene-4 protein, accounts for initiation at the primary origin of DNA replication of bacteriophage T7 (11, 12).

Initiation of DNA replication of superhelical DNAs has been complicated by the observations that the proteins involved in initiation can be somewhat promiscuous (3, 13). Additional enzymes, not directly involved in the initiation process, are required to retain the specificity of the reactions (3, 14, 15). These enzymes have been termed "specificity" or "discriminatory" proteins (13), and play a major role in the initiation of replication at $oriC$ and the pBR322 origin of DNA replication. The mechanism by which they maintain specificity is still unknown, although a proposal to this end is set forth in Section II,C.

Cairns (16) was the first to note that replication of a closed-circular DNA molecule posed topological problems. Unwinding of the paren-

TABLE I
Specialized Terms

Term	Definition
Cruciform	The shape assumed by an inverted repeat sequence present in double-stranded DNA when base-paired in interstrand fashion
dnaA ... dnaK	Genetic loci that, when mutated, show defects in DNA synthesis
DNA-A protein ... DNA-K protein	The proteins encoded by the *dna* genes
D-loop	A displacement created in a double-stranded DNA caused by the introduction of a third (single) strand of DNA complementary to one of the original two strands
R-loop	The same, with the third strand being RNA
Form I DNA	A closed-circular, double-stranded, superhelical DNA molecule
Form II DNA	A circular, double-stranded DNA molecule containing at least one discontinuity in one of the strands
DNA gyrase	Type II DNA topoisomerase (EC 5.99.1.3) from *E. coli*, capable of introducing negative superhelical turns into a closed-circular, double-stranded DNA molecule, with hydrolysis of ATP
HU protein	A small (7 kDa) double-stranded-DNA binding protein from *E. coli*
Helicase	An enzyme capable of separating the strands of a duplex DNA in an ATP-dependent fashion
LRI	The *l*ast *r*eplicative *i*ntermediate containing two almost completed nascent strands and a short region of nonreplicated parental DNA
λdv	A circular DNA derived from phage λ that can be maintained as an extra chromosomal DNA in *E. coli*
λO, λP	λ genes required for DNA replication
oriC	The origin of chromosomal DNA replication in *E. coli*
pas	Primosome assembly site (or sequence)
pBR322 DNA	Plasmid pBR322 DNA, with a heavy (H) and a light (L) strand
Primase	The *dnaG* gene product from *E. coli* that can synthesize the short ribo- or deoxyribooligonucleotides that are used as primers during DNA replication
Primosome	A multienzyme conglomerate composed of the products of the *dnaB, dnaC, dnaG,* and *dnaT* genes and three genetically undefined enzymes, proteins n, n' (factor Y), and n" that assembles at a specific site on the DNA and can migrate unidirectionally (5' → 3') and processively, and can occasionally synthesize a primer
Proteins i, n, n', and n"	Components of the primosome (q.v.)

(continued)

TABLE I (*Continued*)

Term	Definition
RNA I and RNA II	RNA transcripts made off ColE1 (and pBR322) DNA. RNA II serves as the primer precursor of leading-strand DNA synthesis. RNA I interferes with the formation of a stable RNA–DNA hybrid between RNA II and the template
RNA-Pol	RNA polymerase holoenzyme
rom	A ColE1-encoded protein named *R*NA and *o*ne *m*odulator that accelerates the initial interaction between RNA I and RNA II during DNA synthesis
SS(c)	Single-stranded (circular)
Swivelase	The term used to describe enzymes that function to relieve positive superhelical pressure in a replicating form I DNA. Topoisomerase is a more accurate term (EC 5.99.1.2: DNA topoisomerase)
6sL fragment	The first nascent DNA detected during ColE1 DNA synthesis, containing only L-strand sequences; so-named because of its sedimentation pattern in alkaline sucrose gradients

tal strands during DNA replication causes first a loss of any existing negative superhelical turns and then an accumulation of positive superhelical turns. This situation is energetically unfavorable for the molecule, and if the strain induced by the positive superhelicity is not relieved, progress of the replication fork is arrested. Cairns deduced that the problem could be solved if a transient nick was introduced into one strand of the molecule, allowing that strand to rotate about the other and thus relieve the strain. Cairns called the putative enzyme that could accomplish this reaction "swivelase" (16).[1] Subsequently, topoisomerases [recently reviewed by Wang (17)] were discovered. These are enzymes that change the DNA linking number (the number of times the two strands of a closed-circle DNA molecule wind about each other) and thus can solve the problem that positive superhelical pressure creates during replication, although not by the mechanism proposed by Cairns. Both of the two major topoisomerases of *E. coli*, DNA gyrase[2] and topoisomerase I (Topo I), have major roles in the replication of form I DNAs *in vitro*.

[1] Recommended name (16a): DNA topoisomerase (EC 5.99.1.2). Other names in use (16a) are topoisomerase I, Type I DNA topoisomerase, untwisting enzyme, relaxing enzyme, nicking-closing enzyme, swivelase, ω-protein (Eds.).

[2] Recommended name (16a): DNA topoisomerase (ATP-hydrolyzing) (EC 5.99.1.3). Also called Type II DNA topoisomerase and DNA gyrase (Eds.).

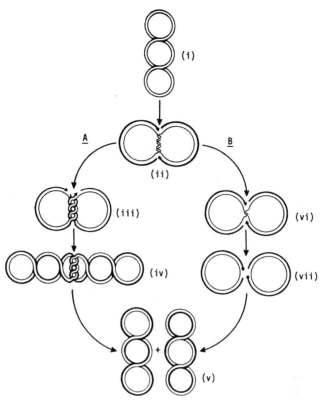

Fig. 1. Pathways of segregation of replicating form I DNA molecules. (i) Form I DNA template. (ii) Almost completely replicated DNA molecule. Pathway (A): replication of the final stretch of parental DNA without unlinking of the parental duplex strands. (iii) Multiply interlinked form II·form II DNA dimer. (iv) Multiply interlinked form I·form I DNA dimer. (v) Two newly synthesized form I DNA molecules. Pathway (B): replication of the final stretch of parental DNA coupled to the progressive unlinking of the parental duplex strands. (vi) Replicating DNA molecule with reduced linkages between the parental DNA strands. (vii) Two completely unlinked form II DNA molecules. Parental DNA, thin lines; nascent DNA, thick lines.

Another problem encountered during replication of form I DNAs is how segregation of the daughter DNA duplexes is accomplished (Fig. 1). If unwinding of the final nonreplicated region of parental DNA precedes their topological disengagement, daughter DNA molecules that are multiply intertwined will result. Once these molecules are sealed, they can only be unlinked by a type II topoisomerase [one capable of breaking and recombining two strands of DNA at a time, e.g., DNA gyrase (17)].[2] Unlinking of the daughter molecules before they are complete can be accomplished by either a type II topoisomerase or a type I [one capable of breaking and recombining one strand of DNA at a time, e.g., Topo I (*17*)]. This problem is identical to that encountered during the segregation of chromosome-sized DNA molecules after DNA replication and its solution is likely to be very similar to some of the enzymatic mechanisms required for the sorting of chromosomes during mitosis. In addition, identical problems are likely to exist during the replication of very long linear DNA duplexes that can be confined into topological domains by bound proteins. It is also possible that the strands of very long linear DNAs may simply get entangled with each other during replication and require topological resolution.

While the arguments advanced above for the involvement of topoisomerases during the terminal stages of DNA replication could be made based solely on their known activities and the topology of the replicating DNA molecules, little evidence exists that speaks to the enzymatic mechanism operating at this point during the replication process. The situation is more complex in a DNA molecule that is replicating bidirectionally. The events that occur when the forks meet are a mystery. The DNA replication system of pBR322 *in vitro* has given a surprising answer to some of these questions (J. Minden and K. Marians, unpublished; see Section III).

In the remainder of this article our knowledge of the enzymatic solutions to some of the problems outlined above is reviewed.

I. Initiation

The initiation of DNA replication is a process designed to provide a DNA polymerase with a 3'-OH terminus for polymerization. This can be accomplished on a single strand of DNA by either the action of RNA polymerase, which can start chains *de novo*, or one of the specialized priming enzymes (e.g., "primase"; *18a*). Either of two mechanisms can act in order to generate the initial 3' terminus during the replication of a form I DNA: an endonuclease can nick one strand of the DNA, or the duplex DNA can be unwound in some manner that allows an enzyme like RNA polymerase or primase to use the exposed single strand as a template and manufacture a primer (i.e., transcriptional activation). The simplest way to lend specificity to the initiation of form I DNA replication is through the use of a DNA sequence-specific endonuclease; however, none of the three DNA replication systems discussed here uses this mechanism.

A. oriC and λdv

Initiation of DNA replication at both *oriC* and *ori*λ is directed by nucleosome-type structures that form at specific repeated DNA sequences within the regions defined as the origins of DNA replication (8, 9). In the case of *oriC*, the *dnaA*-encoded protein (DNA-A protein), which is required for the initiation of chromosomal DNA replication, recognizes a 9 base-pair sequence, TTAT C(A)CACA that is repeated four times within the consensus *oriC* DNA sequence (*18*) in regions that are conserved from species to species (*19–21*). Binding of the DNA-A protein to *oriC* DNA appears to be cooperative, resulting in the formation of a structure estimated to contain 20 to 30 DNA-A protein monomers (52 kDa) and include 200 to 250 base-pairs of DNA (*7*). The DNA-A protein–*oriC* complex appears spherical in the electron microscope and condenses the associated DNA. DNase "footprinting" of this complex indicates that the DNA is generally accessible to the nuclease and that the DNA-A protein is probably bound to one side of the helix, suggesting that the minimal *oriC* DNA sequence (245 base-pairs) is wrapped around a core of DNA-A protein monomers (*7*).

A similar structure is formed at *ori*λ (*8*). The λO and λP proteins are the only phage-encoded proteins required for its DNA replication (*22*). The λO protein binds to the minimal *ori*λ sequence that lies within the λO gene. The target is four direct repeats of 19 base-pairs, AT CCCT CAAAA CAGGGGA, each of which is also an inverted repeat (*23*). Examination of the complex formed between the λO protein and *ori*λ in the electron microscope revealed a nucleosome-type structure (the O-some) estimated to contain eight λO protein monomers (32 kDa) and 125 base-pairs of DNA, including the direct repeats (*8*). Evidence that these protein–DNA complexes are actually involved in the initiation of DNA replication has yet to be obtained; however, additional data concerning the enzymological mechanisms operative during initiation of DNA replication on these templates are strongly supportive of this being the case.

The DNA-A protein acts early in the DNA replication reaction. The kinetics of DNA synthesis in a partially reconstituted system capable of replicating plasmid DNAs carrying *oriC* show a distinct lag that can be eliminated by incubation of the template with all the protein components in the absence of dNTPs (*13, 21*). Omission of either DNA gyrase or RNA polymerase from this incubation restores the full lag period, indicating that these proteins act very early during *oriC* DNA replication. Omission of the DNA-A protein from the incubation resulted in a reduction in the lag period, indicating that this protein acts early in the reaction, but perhaps after the action of DNA gyrase and RNA polymerase (EC 2.7.7.6) (13, 21). The influence of these latter two proteins on the formation of the DNA-A protein–oriC nucleosome-type structure has not yet been reported.

The reconstituted system for the replication of oriC DNA has been refined over the last 3 years. A mixture of the E. coli single-stranded DNA-binding protein (SSB), the DNA-A, DNA-B, DNA-C, and DNA-G proteins, DNA gyrase, ribonuclease H (EC 3.1.26.4), Topo I, protein HU, RNA polymerase holoenzyme (RNA-Pol), and the E. coli DNA polymerase III holoenzyme (Pol-III) will catalyze dnaA- and oriC-dependent DNA synthesis (24, 25).

The enzymes in the system have been divided into three classes: (1) initiation proteins (e.g., RNA-Pol, DNA gyrase, and the DNA-A protein) that allow priming of the leading-strand at oriC, (2) elongation proteins (e.g., priming proteins and Pol-III), that advance the fork, and (3) auxiliary (specificity) proteins (e.g., Topo I and RNase H), that suppress replication from origins other than oriC (24, 25). A number of different combinations of these proteins can catalyze extensive incorporation of deoxynucleoside triphosphate into acid-insoluble material using oriC-containing plasmid DNAs as templates (24). Early findings indicated, however, that low levels of RNase H (14) and Topo I (15) are required to maintain specificity for oriC-containing template DNAs [a phenomenon called "template discrimination" (3)]. Low levels (approximately 1/10 that required to saturate binding to the DNA) of the HU protein, a small double-stranded DNA binding protein (26), stimulate the oriC replication reaction 3- to 5-fold (27). A prepriming stage has been identified during oriC DNA replication that requires the template DNA, SSB, DNA gyrase, the DNA-A, DNA-B, and DNA-C proteins, ATP, and, depending on the priming system used (see below), RNA-Pol and the HU protein (24, 25). A prepriming intermediate could be isolated by filtration through BioGel A5-M columns; it can support DNA replication upon the addition of primase, Pol-III, SSB, ATP, and dNTPs (25). This prepriming complex is presumably reflected in the (DNA-A protein)·(oriC DNA) nucleosome structure.

Three distinct priming modes can operate during oriC DNA replication. Each mechanism gives DNA-A and oriC-dependent DNA replication. The levels of protein HU and Topo I govern which priming system operates (24). At very low levels of HU and Topo I, RNA-Pol alone can act; however, DNA synthesis is significantly reduced (24). At these levels of HU and Topo I, primase alone (the "solo primase" system) can also act, giving roughly 9-fold higher levels of DNA synthesis. The solo primase system is inhibited by increasing the levels of Topo I and HU. At high levels of HU (1:1 by weight with the DNA), both RNA-Pol and primase are required for extensive DNA synthesis (2-fold higher levels of DNA synthesis than the solo primase system). Previous requirements (13) for the other primosomal proteins (proteins i, n, n', and n") or for crude fractions are no longer apparent (24, 25).

In view of this, it is likely that the prepriming complex directed by the DNA-A protein acts both to unwind the DNA at the origin of replication and to provide access to the single-stranded template DNA for a mobile priming complex that may consist only of the DNA-B protein and primase. In the complete system (primase plus RNA-Pol), RNA-Pol may prime the leading-strand; alternatively, the first primer manufactured by primase in the mobile priming complex could serve as a primer for leading-strand DNA synthesis in the opposite direction. In this case, RNA-Pol may act only to aid in the exposure of the single-stranded template DNA through transcriptional activation.

Similar mechanisms have been proposed to account for initiation of DNA replication at $ori\lambda$. The most purified, partially reconstituted *in vitro* λdv replication system requires the DNA-B, DNA-G, DNA-J, and DNA-K proteins, SSB, DNA gyrase, RNA-Pol, and Pol-III as well as small amounts of crude fractions (28). A subsystem of the λdv DNA replication system, called the λSS (single-strand) system, has been discovered that will prime any single-stranded DNA (29). The combination of the λO and λP proteins and the DNA-J and DNA-K proteins act to transfer the DNA-B protein onto SSB-coated single-stranded DNA. This prepriming protein–DNA complex can be isolated and is recognized by primase, which subsequently catalyzes the synthesis of RNA primers that can then be elongated by Pol-III (30). Thus, this priming complex is very similar to the primosome (10), except that assembly of the λSS machinery is not dictated by a specific DNA region.

Addition of the λP protein and the DNA-B protein to the λO protein–*oriλ* DNA complex resulted in the production of a nucleoprotein structure significantly larger and somewhat more asymmetric than the O-some (8). This reflects previous evidence of interactions between the λP protein and the λO protein and with the DNA-B protein (22). It has been estimated that this larger structure is an equimolar complex of the λO, λP, and DNA-B proteins with 160 base-pairs of *oriλ* DNA (8). The authors proposed that the pathway for localized initiation of DNA replication at *oriλ* could be (1) recognition of *oriλ* by the λO protein and formation of the O-some, (2) formation of the large λO·λP·(DNA-B)·oriλ complex, (3) formation of a prepriming complex requiring the DNA-J and DNA-K proteins and RNA-Pol, and (4) recognition of the prepriming complex by primase and synthesis of the first primer (8).

B. pBR322 DNA

Initiation of DNA replication on pBR322 DNA proceeds by a different mechanism than the one that operates at oriC and oriλ. The model (Fig. 2) used for initiation at ori-pBR served as the basis for reconstitution of the synthesis of the DNA with purified proteins and was based on the considerable body of work that existed prior to 1984 on the replication of ColE1 DNA.

Several studies (9, 31, 32, 33) led to the expectation that RNA-Pol, Pol-I, and RNase H would be required for initiation of DNA synthesis of the leading-strand. [In pBR322 DNA replication, the plasmid H (for "heavy") strand is the leading-strand template, thus the nascent leading-strand is a plasmid L ("light") strand.] The early events on the template H-strand act to first form an R-loop (9) and then a stable D-loop (34) that should displace the "pas" sequence, located roughly 150 nucleotides downstream from the origin [the position of the first nascent deoxynucleoside (35)] of pBR322 DNA replication (36), from an inactive (with respect to primosome assembly), double-stranded form to an active, single-stranded form. In the presence of SSB, the primosomal proteins would be able to utilize the exposed pas to initiate lagging-strand DNA synthesis (in pBR322 DNA replication, the plasmid L-strand is the lagging-strand template, thus the nascent lagging-strand is a plasmid H-strand). Extensive DNA synthesis of either strand should require Pol-III. Unwinding of the parental duplex DNA would produce a topological strain in the molecule resulting in the generation of positive supercoils in the nonreplicated region of the DNA. A number of observations indicated that DNA gyrase would be the topoisomerase required to relieve this induced topological strain. (1) Only DNA gyrase can actively remove positive superhelical twists from a closed-circular, double-stranded DNA molecule (37). (2) The addition of DNA gyrase to reactions containing RNA-Pol, Pol-I, and RNase H resulted in extensive elongation of the nascent 6sL fragment (31). (3) ColE1 DNA synthesis in soluble extracts of E. coli cells was sensitive to the DNA gyrase inhibitors coumermycin and naladixic acid (38, 39). Based on this model, four criteria were developed for a bona fide purified pBR322 DNA replication system. (1) Replication should initiate in vitro at the region previously mapped in vivo (35).

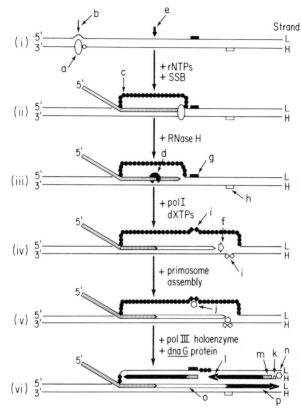

FIG. 2. Model for replication of pBR322 DNA. (i) (a) core RNA-Pol, plus σ subunit, ○, binds to the RNA II promoter (b). (ii) RNA-Pol catalyzes elongation of RNA II and DNA–RNA II heteroduplex formation and the binding of SSB (c) to exposed single-stranded DNA. (iii) RNase H (d) catalyzes cleavage of the DNA–RNA II heteroduplex at the origin of DNA synthesis (e). (iv) Pol I (f) extension of cleaved RNA II primer and the conversion of pas-BL and pas-BH (g, h) to their active forms (i). (v) The assembly of the primosome (j) at pas-BL and pas-BH. (vi) *dnaG* (k) catalyzes formation of RNA primers on the pBR322 L strand and Pol III (l) extension of primers formed by the *dnaG* protein and Pol I. RNA (m), parental DNA (n), nascent Pol-I-synthesized DNA (o), and nascent Pol-III-synthesized DNA (p).

(2) Replication should yield form I DNA product. (3) Based on the model presented in Fig. 2, only L-strand DNA should be made in the absence of the primosomal proteins. Thus, the third criterion was that the system should, when appropriately manipulated, be capable of synthesizing leading- and lagging-strand DNA independently. (4) The replication system should utilize pBR322 DNA as the template selectively.

One potential overlap between the control loops operating during pBR322 and *oriC* DNA replication is the role of the DNA-A protein. Control of the initiation of ColE1 DNA replication is exerted through regulation of the amount of RNA II that is available for use as the primer of leading-strand DNA synthesis (9, 40). This control feature, because it limits initiation frequency, also acts to control the copy number of the plasmid in the cell. Interestingly, the same control feature is the basis of the plasmid-derived incompatibility function (41). (Two plasmids that cannot coexist in the same cell are incompatible.) This regulation event involves both a protein and RNA species.

RNA I, a transcript 108-nucleotides long made off the L-strand of the plasmid and complementary to the 5' portion of RNA II, interferes with the RNase H-catalyzed processing of RNA II (40, 41). Unprocessed RNA II molecules can not be used as a primer for *E. coli* DNA polymerase I. Interference of RNA II processing is through a direct RNA–RNA interaction between the two species. Studies in a number of laboratories indicate that this interaction is the primary determinant of plasmid incompatibility and plasmid copy number (summarized in ref. 42).

RNA I can be folded into a tRNA-type of structure with three stems and loops. Evidence for this structure has been obtained (43, 44). The 5' end of RNA II can also, in theory, form a similar structure. The formation of the third (nearest to the 3' end) stem-and-loop structure in RNA II precludes the formation of another, very large stem-and-loop structure that is thought to be essential for primer processing. RNA I and RNA II are believed to first interact through the complementary loops of the three stem-and-loop structures in a phenomenon that has been termed "kissing" (45). In this mode there is no evidence for base-pairing between the two RNA species and the interaction is completely reversible (46). The RNA I–RNA II interaction then proceeds from the "kissing" stage to formation of a complete hybrid. Initiation of the annealing process occurs at the 5' end of RNA I and propagates toward its 3' end (45, 47). Formation of this RNA I–RNA II hybrid precludes the ability of RNA II to form a stable RNA–DNA hybrid with the template DNA. Thus, no RNase H-processed primer is available for leading-strand DNA synthesis.

The interaction between RNA I and RNA II is influenced by a small (63 amino acid) protein encoded by a region near the plasmid *nic* site (48, 49). Deletion of this region results in a 4- to 6-fold elevation of the copy number (50). This protein was originally thought to act as a classical repressor of RNA II transcription (48) and was thus named the "rop" protein (*r*epressor *o*f *p*rimer). However, recent evidence clearly indicates that the protein does not act in that fashion (49, 51, 52). Tomizawa's studies have shown that this protein, which he named "rom" protein (*R*NA *o*ne-inhibition *m*odulator), apparently acts by increasing the rate of stability of the initial transient contact between RNA I and RNA II (i.e., "kissing"). Thus, a decrease in the copy number of ColE1 DNA can be expected in the presence of rom because the increased efficiency of the RNA I–RNA II interaction will result in less frequent initiation of plasmid DNA replication. It is important to note that the interaction between RNA I and RNA II does not *require* the rom protein, it is merely made more efficient in its presence.

Fuller and Kornberg (21) have noted that there is a consensus sequence for DNA-A protein binding on the pBR322 L-strand between the origin of DNA replication and the pas sequence on that strand. They suggested that the DNA-A protein can exert negative regulation of pBR322 DNA synthesis by binding to this sequence. Abe (53) reported that the synthesis of pBR322 DNA ceased upon shift of *dnaA* temperature-sensitive strains to the nonpermissive temperature, but then resumed (while still at the nonpermissive temperature) 1 hour later. However, DNA replication had now shifted from a theta-type to a rolling-circle mechanism.

It remains to be determined how all these control features interact. However, it is interesting to note that the pBR322 DNA replication system can be manipulated to produce multigenome length, linear duplex synthetic DNA molecules as the major DNA product (see Section III).

Appropriate combinations of RNA-Pol, RNase H, DNA gyrase, DNA ligase, SSB, Pol-III, and the primosomal proteins [the DNA-B, DNA-C, and DNA-G proteins, proteins i, n, and n", and factor Y (protein n')] catalyze extensive incorporation of dNTPs into acid-insoluble product when form I pBR322 DNA is used as a template (3). Replication by this combination of proteins, however, does not meet any of the criteria outlined for a bona fide pBR322 DNA replication system. The products of the system are not form I DNA molecules, but nonsegregated replication intermediates. In addition, ϕX174 form I DNA is replicated as well as pBR322 DNA. This is particularly surprising considering the need for the ϕX174-encoded gene-A protein to catalyze normal ϕX174 form I DNA synthesis (4, 5). Thus, the system lacked the ability to discriminate between DNA templates. Finally, this system cannot synthesize leading- and lagging-strand DNA independently. The ability of a DNA replication system to synthesize either strand independently has been termed "strand discrimination" (3).

Remarkably, under the conditions of the replication reaction, the same enzyme, Topo I, was able to impart both strand and template discrimination at low levels (5–10 ng) and, at 30- to 50-fold higher levels, allow the completion and catalyze the segregation of the daughter DNA molecules in the complete system. The accumulation of incomplete replicative intermediate at low levels of Topo I was exploited to map the origin of DNA replication in the reconstituted system. Since the nascent DNA was incomplete, treatment with restriction endonucleases that cleave pBR322 DNA once should, if replication started at a unique point and proceeded unidirectionally, generate a series of DNA fragments with sizes corresponding to the distance between the restriction endonuclease cleavage site and the origin. An analysis of this type (Fig. 3) indicated that in the complete system, DNA replication was starting within 50 nucleotides of the *in vivo* origin of replication.

Thus, using the complete reconstituted system (i.e., with Topo I) it was possible to demonstrate replication specific for ColE1-type origins that initiated at the site previously mapped *in vivo*, showed the correct mechanism of initiation of the leading- and lagging-strands, and gave genuine form I DNA product.

C. Template Discrimination

An interesting new observation to emerge from the study of the replication of superhelical DNAs was the requirement for specificity factors. In the *oriC* DNA replication system, either high levels of protein HU or low levels of Topo I (roughly the same as the amount required to generate template and strand discrimination in the pBR322 DNA replication system) generate DNA template specificity (*13, 15, 27*). Surprisingly, in both the λdv (R. McMacken, personal communication) or pBR322 DNA replication systems (K. Marians, unpublished data) protein HU does not act as a discriminatory factor. In addition, Topo I is not required for template specificity in the λdv replication system (R. McMacken, personal communication) as it is for *oriC* and pBR322 DNA replication.

In pBR322 DNA replication, it is clear that all nondiscriminatory DNA synthesis requires RNA-Pol, thus if the mechanism of template discrimination is to inhibit nonspecific initiation of DNA replication, then it is likely that the action of RNA-Pol is being modulated in some fashion. This modulation could take a number of different forms. (1) The product of the RNA-Pol reaction could be participating directly as a primer for indiscriminate DNA synthesis. In this case, Topo I could act to (a) destabilize nonspecific primers that are annealed to the DNA

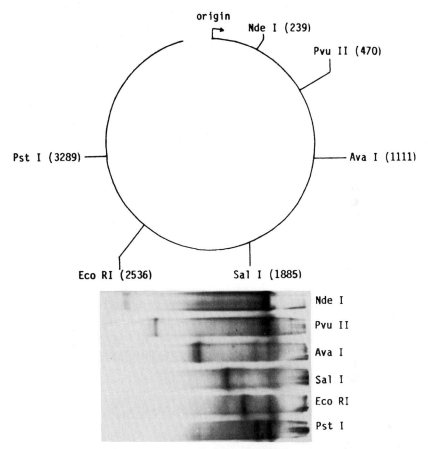

FIG. 3. Mapping of the origin of replication of pBR322 DNA *in vitro*. ^{32}P-Labeled reaction products from a standard pBR322 DNA replication reaction with the complete system were purified by extraction with phenol followed by ethanol precipitation. Aliquots were then digested with the indicated restriction endonucleases. The digested products were extracted with phenol, ethanol-precipitated, redissolved in 30 mM NaOH, and electrophoresed through a 1% alkaline agarose gel with 30 mM NaOH, 1 mM EDTA as the electrophoresis buffer. The gel was neutralized, dried, and autoradiographed. The figures in parentheses on the map are the expected sizes of the DNA fragments if replication started at the *in vivo* origin and proceeded unidirectionally.

by reducing the superhelical density of the DNA [because the formation of an RNA–DNA hybrid in form I DNA unwinds the molecule, superhelicity favors this reaction. Thus, the high superhelical density generated in the presence of DNA gyrase also stabilizes RNA–DNA hybrids (54)]. (b) Alternatively, in highly supercoiled DNA, regions that are normally double-stranded can become single-stranded. RNA-

Pol can bind to these regions and synthesize an RNA (not normally made from the DNA) that leads to the initiation of DNA synthesis. Modulation of the superhelical density of the DNA by Topo I may act to keep these "nonspecific" promoter sites cryptic. (2) On the other hand, it could be the presence of RNA-Pol itself on a region of the DNA where it normally is not (through the action of a nonspecific promoter), or because of an extension of the time that the enzyme remains on the DNA (binding of RNA-Pol is favored in highly supercoiled DNA) that leads to the transcriptional activation of an origin of DNA replication that is normally silent.

We initially tried to detect features of the DNA that would encourage any of the aberrant (in the context of the DNA replication system) actions of RNA polymerase described above. Thus, the topological states of pBR322 (as the specific template) and ϕX174 (as the nonspecific template) form I DNAs under discriminatory and nondiscriminatory conditions have been studied. These studies have utilized agarose gels containing different concentrations of chloroquine, an intercalating dye, to help resolve the different topoisomers (55). Using these gels (Fig. 4), and the band counting method of Keller (56) to determine superhelical density [and assuming a helical repeat of B-DNA in solution of 10.6 (57, 58)], superhelical densities of -0.0679 for ϕX174 form I DNA and -0.0425 for pBR322 form I DNA have been calculated. The value for pBR322 DNA is very close to that reported for ColE1 DNA [-0.043 (59)]. [The term "higher (or increased) superhelical density" is defined here as having a larger number of negative superhelical turns, whereas "lower (or decreased) superhelical density" is the opposite.]

Because of this significant difference in superhelical density, it was possible that Topo I would act to reduce the superhelical density of ϕX174 DNA to the value for pBR322 DNA, thus preventing nonspecific initiation by RNA-Pol by some of the mechanisms described above. Interestingly, however, whereas the superhelical density of ϕX174 is reduced under discriminatory conditions, that of pBR322 DNA is initially increased (Fig. 4). This is because DNA gyrase acts on form I pBR322 DNA to increase its superhelical density from -0.0425 to -0.0607); this does not occur with ϕX174 form I DNA (unpublished data), perhaps because this DNA is already too supertwisted to allow binding of DNA gyrase [the stability of DNA gyrase bound to the DNA is inversely related to the absolute superhelical density (60)]. From the data in Fig. 4 it can be determined that in the presence of DNA gyrase and Topo I the superhelical densities of the two different DNAs approach values that are remarkably similar

pBR322 DNA

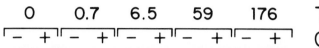

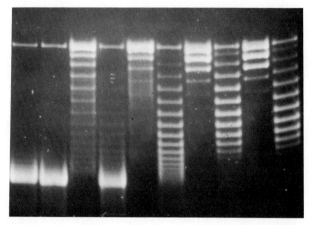

2 μg/ml Chloroquine

φX174 DNA

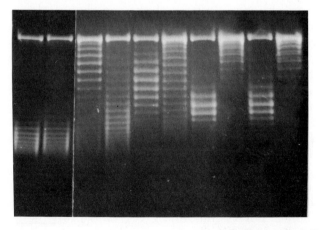

8.5 μg/ml Chloroquine

FIG. 4. Analysis of the effect of Topo I and DNA gyrase on form I DNAs. Form I DNAs (0.2 μg) were incubated with 50 ng DNA gyrase and the indicated amounts of Topo I for 20 minutes at 30°C. EDTA, SDS, and tracking dyes were added to 10 mM, 0.1%, and 0.1%, respectively, and the DNA was electrophoresed through 1% agarose gels containing the indicated amounts of chloroquine for 36 hours at 2 V/cm.

TABLE II
Superhelical Densities of pBR322 and φX174 DNAs in the Presence of Topo I and DNA Gyrase[a]

	pBR322	φX174
Form I	−0.0425	−0.0679
50 ng DNA gyrase +2.1 ng Topo I	−0.0607	−0.0462
50 ng DNA gyrase +19.6 ng Topo I	−0.0364	−0.0315
50 ng DNA gyrase +180 ng Topo I	−0.0267	−0.0266
50 ng DNA gyrase +545 ng Topo I	−0.0219	−0.0226

[a] Superhelical densities were calculated from the gels shown in Fig. 4 and similar gels containing different amounts of chloroquine.

(Table II). Maximum template discrimination occurs between 5 and 10 ng of Topo I, even though at very high levels of Topo I, DNA synthesis is maintained with pBR322 but not φX174 DNA as the template. These data suggested that template discrimination is not simply the result of a complete relaxation of the DNA induced by Topo I, but is the result of a slight modulation of the linking number of the DNA. This, perhaps, caused the loss of structures in the DNA required for the initiation of nonspecific DNA replication.

One structural feature of supercoiled DNA molecules very sensitive to superhelical density is the extrusion of an inverted repeat sequence as a cruciform structure. In general, the superhelical density must be greater than −0.05 for naturally occurring inverted repeat sequences to convert to cruciform structures (61–63). Since the formation of cruciform structures requires the energy stored in the superhelix, one can expect that if the DNA is relaxed, either by treatment with a restriction endonuclease (62) or by nicking with DNase I (64), the cruciform will be rapidly reabsorbed into the smooth, double-stranded structure. This has, in fact, been observed with a cruciform that has a 13–14 base-pair stem. After the DNA had been nicked with DNase, this cruciform relaxed with a half-life of 28 seconds at 37°C (64). Since it appears that naturally occurring cruciforms flip in and out of the extruded structure (65) if the DNA has a superhelical density greater than that required for extrusion, it is reasonable to expect that when a superhelical DNA is relaxed by Topo I to a superhelical density less than the critical point required for extrusion, any cruciforms present in the molecule will be rapidly readsorbed.

It is likely that RNA-Pol can be influenced by the existence of cruciforms in two different ways that could lead to nondiscriminatory DNA synthesis. (1) RNA-Pol may gain entry to the DNA by binding at the loop of the cruciform. This could lead to the synthesis of RNA at regions of the DNA that are not normally actively transcribed. This would be very similar to the synthesis by RNA-Pol of the RNA primer for fd SS(c) → RF DNA replication. This event occurs on a hairpin structure in the single-stranded DNA that does not normally act as a promoter (66). (2) During normal transcription, one of the reasons that the nascent RNA chain is forced out of the duplex is because of the rewinding of the DNA behind it. However, the same situation will not exist when an RNA-Pol molecule encounters a cruciform; the nontranscribed strand will not be available for rehybridization to the transcribed strand. Thus, the formation of RNA–DNA hybrids will be favored. These hybrids could readily be processed by the RNase H present in the reaction to give 3'-OH primer termini that can be extended by *E. coli* Pol-I. This would also encourage transcriptional activation of an origin of DNA replication.

Thus, one tentative hypothesis, currently under examination by us, regarding the action of Topo I as a discriminatory factor is as follows. Indiscriminate DNA replication—i.e., as described here, this would correspond to ϕX174 from I DNA-directed DNA synthesis or the lack of strand discrimination (presumably because of nonspecific initiation of the nascent H strand) in the absence of Topo I—is a result of an action of RNA-Pol at cruciform structures in the DNA that leads to the initiation of DNA synthesis. Topo I acts to eliminate this type of DNA synthesis by lowering the superhelical density of the DNA to a level where cruciform extrusion is not favored. However, at this level of superhelical density, strong genuine replication promotes (such as the one for RNA II) still operate. In the context of this theory, the HU protein would act as a discriminatory factor by removing negative superhelical turns by binding to the DNA, thereby having the same effect as Topo I.

II. Elongation

Relatively little is known about the mechanisms of elongation during the replication of superhelical DNAs. *E. coli* Pol-III is required for extensive elongation of any of the primers made by any of the mechanisms described in the previous section. However, the architecture of the polymerases at the replicating fork is unknown. Equally unclear are the mechanisms of unwinding of the parental duplex during replication. The *oriC* and pBR322 DNA replication systems exhibit an almost absolute requirement for DNA gyrase (3, 67). This probably is a reflection of the need for DNA gyrase to relieve the accumulated positive superhelical pressure. However, even replication of a linear *oriC*-containing DNA requires DNA gyrase (68), suggesting that DNA gyrase may contribute to the integrity of the protein complexes at the replication fork. In fact, DNA gyrase could actually contribute to unwinding by ensuring that the DNA always has a residual number of negative superhelical turns. Under these circumstances, there may be no need to have an actual unwinding enzyme at the fork. The parental strands could be separated by the forward momentum of the moving replication fork using the energy stored in the DNA as negative superhelical turns.

None of the form I DNA replication systems exhibits a requirement for any of the known DNA helicases. However, preliminary data indicate that the primosome is capable of unwinding duplex DNA. A significant percentage of replicating molecules (5 to 10%) in the reconstituted ϕX174 SS(c) → RF DNA replication reaction (requiring the primosomal proteins and *E. coli* Pol-III) generates DNA products that are extremely large (M. Mok and K. Marians, unpublished data). Treatment of these products with the *Pst*I restriction endonuclease (EC 3.1.21.3-.5) to yield a partial digest, followed by analysis by electrophoresis through 0.3% agarose gels, allowed us to estimate the size of these large products as at least 10^5 base-pairs. The susceptibility of these products to the restriction endonuclease and other nucleases (e.g., exonuclease III, EC 3.1.11.2) indicated that they are linear double-stranded DNAs of up to 20 unit lengths long.

These long DNA products undoubtedly are the result of a rolling-circle mechanism of DNA synthesis. Although the events leading to the establishment of this rolling-circle and the mechanism of priming of (+) strand synthesis remain, for the moment, obscure, it is clear that an active primosome is required. These products were not observed when phage G4 SS(c) DNA, which requires only primase and SSB for initiation of DNA replication (69), was used as a template, even in the presence of all the other primosomal proteins, or when the components of the general priming system (primase and the DNA-B protein) were used to prime ϕX174 complementary-strand DNA synthesis. This putative helicase activity of the primosome is presumably related in some way to the recently demonstrated helicase activity of the DNA-B protein (68a). The primosome unwinding activity may play a significant role in the unwinding of the parental duplex during pBR322 DNA replication and, as detailed in Section III, is proposed to play a primary role in the unwinding of the last stretch of nonreplicated parental pBR322 DNA duplex during the termination and segregation stage of pBR322 DNA replication.

III. Termination and Segregation of Daughter Molecules

As mentioned in the opening paragraphs, the last stages of the replication of a form I DNA present a final topological problem: ultimately, the linkages between the two parental strands of DNA must be completely eliminated. Failure to accomplish this will result in the production of daughter molecules that are intertwined (Fig. 1). Overall, two possible routes are available for the termination of replication of a form I DNA. (1) The replication fork(s) completely traverse the remaining parental duplex producing catenated gapped or nicked dimers (form II·form II "dimers"). These molecules are then sealed and supertwisted, yielding form I·form I dimers that subsequently can be decatenated to yield the two form I daughter duplexes. (2) In the second pathway, the almost completed molecules are segregated just as the final sealing and supercoiling reactions take place. Both pathways require topoisomerases; however, the former pathway can only be catalyzed by a type II topoisomerase (unless the substrate for separation is a form II·form I dimer), whereas the latter pathway can presumably be catalyzed by either a type I or type II topoisomerase. Note that the pathways for maturation of the replicative intermediate (RI, the structure containing two parental strands that are still linked and two almost completed nascent strands) differ. In pulse-chase experiments, one would expect to observe the sequence RI → form II·form II dimer → form I·form I dimer → form I, if the first pathway was operative, whereas the predicted observation if the second pathway were operative would be RI → form II → form I. For the sake of simplicity, the actual product of sealing a molecule of type II—i.e., a form I' molecule—has been omitted from this discussion.

Almost all enzymological studies of DNA replication have focused on the mechanisms of initiation and elongation of the nascent DNA chains. Thus, there are few data of this type concerning the events that occur during the terminal stages of replication. The classic studies, performed *in vivo*, of Sundin and Varshavesky (*70*, *71*), indicate that the first pathway described is probably the major pathway of termination and segregation of replicating SV40 form I DNA. They demonstrated that multiply intertwined form II–form II, form II–form I, and form I–form I SV 40 DNA dimers accumulate in SV40-infected cells that have been hypertonically shocked. They attributed the accumulation of these DNA molecules to the inactivation from the hypertonic shock of the type II topoisomerase required for separating these species.

On the other hand, both Oka and Inselburg (*72*) and Sakakibara *et* drawn in Fig. 1 and form II DNA. Sakakibara *et al.*, noted the formation of multiply intertwined dimers as a result of the replication process. However, these species seemed to be generated by a minor replicative pathway since they progressed from form II·form II to form I·form I dimers and were not further resolved.

None of the three reconstituted form I DNA replication systems discussed here yields form I DNA as the predominate final product. However, we indicated (3) that an activity could be detected in soluble extracts of *E. coli* that was capable of catalyzing the termination and segregation of pBR322 daughter DNA molecules. This activity has been identified as Topo I. Significantly higher levels of Topo I than those required for template and strand discrimination are necessary for segregation (Fig. 5). Interestingly, as the level of Topo I in the replication reaction is increased, the major products shift from RI and a species (band A) that has been identified as long, multigenome length, linear duplex DNA molecules, to form I·form I dimers, and finally to form I and form II DNA at the highest levels of Topo I tested. Pulse-chase experiments showed that at high levels of Topo I, the second pathway discussed at the beginning of this section was operative, i.e., the pathway of maturation was RI → form II → form I. At these levels of Topo I, no form I·form I dimers were evident, suggesting that the first maturation pathway described did not significantly contribute to the production of form I DNA product.

Figure 6 diagrams our interpretation of the events during segregation. (1) The DNA molecule prior to the initiation of DNA replication. (2) Replication initiates on the H strand, RNA II forms a persistent RNA–DNA hybrid. (3) Some time shortly after initiation of DNA synthesis of the lagging-strand. Extension of the first Okazaki fragment goes backward (compared to the overall direction of DNA replication). Since this fragment is simply undergoing extension by Pol-III (no other replication machinery should be associated with it), the length of the fragment should be dependent upon the extent of the parental L-strand that is single-stranded [Pol-III can not strand displace (74)]. Since the parental L-strand would be expected to rehybridize with the parental H strand after the RNA II·parental H strand DNA hybrid was digested by the combined action of Pol-I and RNase H, the length of the nascent H strand Okazaki fragment will be determined by the relative rates of the Pol-III elongation reaction compared to digestion of the RNA–DNA hybrid. It is likely that the Pol-III reaction is faster; thus the situation depicted in (3) could occur, i.e., the generation of a

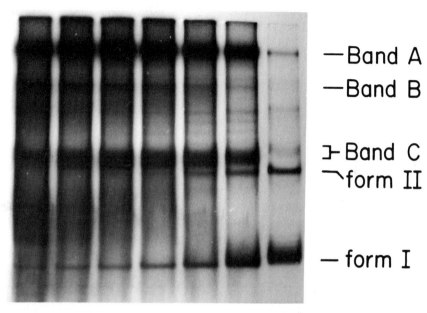

Fig. 5. Effect of Topo I on the products of pBR322 DNA replication. The DNA products of replication reactions containing various concentrations of Topo I were analyzed by electrophoresis through 1% agarose gels under native conditions. Lane 1, no Topo I; lane 2, 2 ng Topo I; lane 3, 7 ng Topo I; lane 4, 20 ng Topo I; lane 5, 59 ng Topo I; lane 6, 176 ng Topo I; lane 7, 520 ng Topo I.

gap on the parental H-strand upstream of the 5′ end of the nascent L-strand and opposite a complete daughter duplex. (4) Replication proceeds around the circle to the point illustrated. Replication now slows down considerably, or even stops completely because DNA gyrase is no longer capable of binding to the nonreplicated parental duplex and is thus unable to relieve the accumulated topological strain. This correlates well with the existing maps of DNA gyrase binding sites [detected by oxolinic acid-induced cleavage of the DNA by the enzyme (75, 76)] on pBR322 DNA (77, 78) that show a dearth of binding sites in the region from the origin of DNA replication to 500 nucleotides upstream. At this point segregation is accomplished by unwinding of al. (73) have determined that the major intermediates during ColE1 plasmid DNA replication are a species corresponding to the RI as

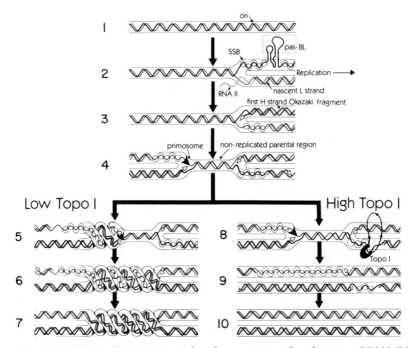

FIG. 6. Model for the Topo I-catalyzed segregation of replicating pBR322 DNA molecules. Represented in this model as (1) is a segment of pBR322 DNA including approximately 250 nucleotides downstream and 750 nucleotides upstream of the origin (ori) of DNA replication. The remainder of the molecule is not included in the drawing. The thin and thick lines represent the parental H and L DNA strands, respectively. (2) The initiation of DNA replication; small open circles represent SSB bound to single-stranded DNA; dotted open lines represent the RNA II primer; solid open lines represent nascent DNA with the arrowhead indicating the growing 3' end. Replication moves to the right as indicated, generating two daughter duplexes. (3) A single-stranded gap is generated on the parental H strand (described in detail in the text), creating a site of action for Topo I-catalyzed segregation of the daughter DNA molecules. (4) Elongation has continued around the circular DNA and slowed, giving the LRI DNA of which a section, including the regions of the daughter DNA duplexes immediately upstream and downstream of the residual nonreplicated parental region, is shown here. The large, solid triangular arrowhead represents the primosome. At this point one of two pathways of completing the replication cycle can be utilized. The concentration of Topo I determines the pathway taken. At low concentrations of Topo I, the primosome moves through the nonreplicated parental duplex, unwinding the DNA strands (5); however there is no concomitant unlinking of the daughter duplexes. Therefore, a form II·II dimer (6) is the product generated, which is eventually converted to a form I·I dimer (7). At high concentrations of Topo I, the enzyme (solid half-moon) can bind to the gap on the parental H-strand generated in step 3 and unlink (dotted arrow) the daughter duplexes as the primosome unwinds the residual parental duplex (8). Thus, the daughter DNA molecules are segregated to form II DNA (9) that is then converted to the final form I product (10).

the remaining stretch of nonreplicated parental DNA by the primosome, which should be idling up against the duplex region as illustrated in (4), coupled to strand passage by Topo I that is bound to the gap on the parental H-strand. This feature of Topo I is identical to its ability to catenate double-stranded DNA rings as long as there is a small percentage of form II DNA available (79, 80). The requirement for a high concentration of Topo I may reflect the fact that it must compete with SSB for binding to this region of the DNA. Topo I will have to pass strands once for each duplex turn the primosome removes.

At this stage (4), the relative rates of primosome-dependent DNA unwinding and Topo-I catalyzed strand passage will determine the segregation pathway taken. If unwinding is completed before separation, the DNA products will be multiply intertwined daughter molecules (5) that will progress through form II·form II (6) to form I·form I dimers (7). Under the conditions of the replication reaction, this appears to be a dead-end pathway. This is surprising, since one would expect the strand-passing activity of DNA gyrase to catalyze separation (81, 82) of the linked rings (all of the nodes formed by the linked double helices have a positive sign, so that action by DNA gyrase at these nodes should lead to separation). In the absence of any constraints, these linkages should equilibrate throughout the daughter DNA molecules and thus be available for binding by DNA gyrase. Therefore, one possible explanation for the failure of DNA gyrase to act is that the linkages are constrained to a particular region and are inaccessible to the enzyme. These constraints could be generated by the replication machines that were bound to and moving on the DNA. In a concerted process like DNA replication, where many different operations catalyzed by different enzymes are interdependent, the formal rules of DNA topology may, to a certain extent, be subverted.

Another possibility is that at the low levels of DNA gyrase used in the reactions (approximately one molecule per molecule of input plasmid DNA), when DNA gyrase binds to one of the rings of the form I'·form I' dimer, the favored reaction will be supertwisting and not separation. Binding of DNA gyrase to highly supertwisted DNA is relatively weak (60), so that at low concentrations the enzyme may not spend sufficient time on the DNA to find the two double helices and separate them. Therefore, as in Fig. 6, the supercoiled catenated dimers persist and are not further resolved.

If separation of the intermediate by Topo I is completed as the primosome finishes moving through the nonreplicated parental duplex (8), the expected form I product will result (10), evolving from

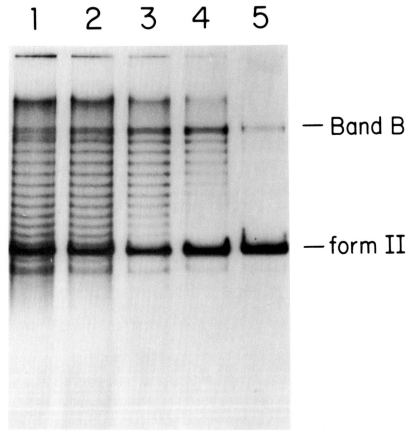

Fig. 7. Effect of Topo I on synthetic pBR322 DNA products in the absence of DNA ligase. The products of replication reactions that contained increasing amounts of Topo I in the absence of DNA ligase were separated by agarose gel electrophoresis under native conditions. Lane 1, 7 ng Topo I; lane 2, 20 ng Topo I; lane 3, 59 ng Topo I; lane 4, 176 ng Topo I; and lane 5, 520 ng Topo I.

form II (9). Additional support for this model can be found when the DNA products formed in the absence of DNA ligase are analyzed by electrophoresis through agarose gels. In the presence of DNA ligase, at low levels of Topo I, little form I DNA is evident, 30- to 50-fold higher levels of Topo I yield predominately form I DNA (Fig. 5). When DNA ligase is omitted (Fig. 7), at low levels of Topo I the predominate species are multiply intertwined form II–form II DNA molecules. At least 20 linkages can easily be counted (suggesting that the nonreplicated region is at least 200 nucleotides long and that even under these conditions, the DNA gyrase in the reaction is not capable of separating these molecules). Note that in the presence of high levels of Topo I, only a few species of linked form II–form II molecules are observed, while the majority of the product is form II. This conforms exactly to the prediction from the model in Fig. 6.

Analysis of the 5' and 3' ends of the nascent leading- and lagging-strands, respectively, at the region of the origin of DNA replication has confirmed the existence of a gap of 20 nucleotides (i.e., the nascent lagging-strand extends 20 nucleotides further upstream than the nascent leading-strand). Thus, it is likely that the model described in Fig. 6 is accurate under the conditions of the reconstituted pBR322 DNA replication system.

It is difficult, at the present time, to equate this model with events *in vivo*. Mutant yeast cells carrying a temperature-sensitive type II topoisomerase are clearly defective in chromosome segregation (83, 84), whereas mutants in the type I enzyme are perfectly viable (85, 86). Similarly, nucleoids (folded chromosomes) isolated at the nonpermissive temperature from *E. coli* mutants carrying a temperature-sensitive *gyrB* allele are roughly double the normal size and appear dumb-bell in shape, suggesting that these cells have been arrested in the process of chromosome segregation (87). *TopA* mutants of *E. coli* are also nonviable and rapidly accumulate compensatory mutations in the *gyrA* and *gyrB* genes that act to reduce the activity of DNA gyrase in the cell (88, 89). The pathway of segregation of replicating pBR322 DNA molecules has not been examined in these cells. It is likely that either overall pathway of segregation can occur. However, the critical step at which either pathway is committed is not clear.

IV. Conclusions

The lengths of the preceding three sections reflect our relative understanding of the processes of initiation, elongation, and termination and segregation of daughter molecules during the replication of a superhelical DNA. A considerable body of information now exists that deals with the different mechanisms of *de novo* initiation of DNA replication on form I DNAs. Even so, the last 2 years have seen the discovery of two novel replication operations: (1) the formation of specialized nucleoprotein complexes that direct other initiation proteins to origins of DNA replication, and (2) the maintenance of DNA template specificity through an apparent modulation of the superhelical density of the DNA. Presumably, an elucidation of the detailed sequence of steps required for initiation of DNA replication at *oriC* and *ori*λ and a determination of the mechanism of Topo-I catalyzed DNA template discrimination are imminent.

The existence of several form I DNA replication systems reconstituted with purified proteins ensures that our current lack of information about the elongation phase during form I DNA replication is only temporary. The ability to manipulate these purified systems will enable an investigation of the driving force for nascent chain elongation (i.e., how are the parental strands unwound) and the determination of whether coupling of the replication machinery that synthesize the leading- and lagging-strands occurs.

The first detailed investigation of the enzymatic activities involved in the segregation of daughter molecules in these purified systems has also dealt us a surprise: the activity responsible for segregation is a type I, and not a type II topoisomerase. The major question raised by this observation is why does not the type II enzyme catalyze segregation? It is possible that the answer may be derived from the likelihood that DNA replication must be considered a process. It is the result of the concerted action of many different enzymatic activities acting as one. The assemblages of these proteins on the DNA should, indeed, be considered machines. These replication machines could serve to constrain the DNA into conformations that are not immediately obvious from strictly topological considerations, yet that could channel the replicating molecules down one particular pathway of segregation. The coming years should reveal new and exciting information on the shape of the DNA during replication.

Acknowledgments

Studies from the author's laboratory were supported by NIH Grant GM34558. KJM is the recipient of an American Cancer Society Faculty Research Award and an Irma T. Hirschl, Monique Weill-Caulier Career Scientist Award.

References

1. N. Nossal, *ARB* **52**, 581 (1983).
2. K. J. Marians, *CRC Crit. Rev. Biochem.* **17**, 153 (1984).
3. J. Minden and K. J. Marians, *JBC* **260**, 9316 (1985).
4. S. Eisenberg, J. Scott and A. Kornberg, *PNAS* **73**, 1594 (1976).
5. C. Sumida-Yasumoto, A. Yudelevich and J. Hurwitz, *PNAS* **73**, 1887 (1976).
6. T. Meyer and K. Geider, *in* "The Single-Stranded DNA Phages" (D. Denhardt, D. Dressler and D. Ray, eds.), p. 389. Cold Spring Harbor Laboratory, Cold Spring Harbor, New York, 1978.
7. R. Fuller, B. Funnell and A. Kornberg, *Cell* **38**, 889 (1984).
8. M. Dodson, J. Roberts, R. McMacken and H. Echols, *PNAS* **82**, 4678 (1985).
9. T. Itoh and J.-I. Tomizawa, *PNAS* **77**, 2450 (1980).

10. K.-I. Arai and A. Kornberg, *PNAS* **78**, 69 (1981).
10a. W. F. Dove, H. Inokuchi and W. F. Stevens, in "The Bacteriophage Lambda" (A. D. Hershey, ed.), p. 747. Cold Spring Harbor Laboratory, Cold Spring Harbor, New York, 1971.
11. C. Fuller and C. Richardson, *JBC* **260**, 3185 (1985).
12. C. Fuller and C. Richardson, *JBC* **260**, 3197 (1985).
13. R. Fuller, L. Bertsch, N. Dixon, J. Flynn, J. Kaguni, R. Low, T. Ogawa and A. Kornberg, in "UCLA Symposia on Molecular and Cellular Biology, New Series" (N. Cozzarelli, ed.), Vol. 10, p. 275. Liss, New York, 1983.
14. T. Ogawa, G. Pickett, T. Kogoma and A. Kornberg, *PNAS* **81**, 1040 (1984).
15. J. Kaguni and A. Kornberg, *JBC* **259**, 8578 (1984).
16. J. Cairns, *JMB* **6**, 208 (1963).
16a. Nomenclature Committee of the International Union of Biochemistry, "Enzyme Nomenclature 1984." Academic Press, New York, 1984.
17. J. Wang, *ARB* **54**, 665 (1985).
18. J. Zyskind, J. Cleary, W. Brusilow, N. Harding and D. Smith, *PNAS* **80**, 1164 (1983).
18a. R. McMacken, K. Ueda and A. Kornberg, *PNAS* **74**, 4190 (1977).
19. T. Chakraborty, K. Yoshinaga, H. Lother and W. Messer, *EMBO J.* **1**, 1545 (1982).
20. F. Hansen, E. Hansen and T. Atlung, *EMBO J.* **1**, 1043 (1982).
21. R. Fuller and A. Kornberg, *PNAS* **80**, 5817 (1983).
22. M. Furth and S. Wickner, in "Lambda II" (R. Hendrix, J. Roberts, F. Stahl and R. Weinberg eds.), p. 145. Cold Spring Harbor Laboratory, Cold Spring Harbor, New York, 1983.
23. T. Tsurimoto and K. Matsubara, *NARes* **9**, 1979 (1981).
24. T. Ogawa, T. Baker, A. van der Ende and A. Kornberg, *PNAS* **82**, 3562 (1985).
25. A. van der Ende, T. Baker, T. Ogawa and A. Kornberg, *PNAS* **82**, 3954 (1985).
26. J. Rouviere-Yaniv and F. Gross, *PNAS* **72**, 3428 (1975).
27. N. Dixon and A. Kornberg, *PNAS* **81**, 424 (1984).
28. M. Zylicz, T. Yamamoto, N. McKittrick, S. Sell and C. Georgopoulos, *JBC* **260**, 7591 (1985).
29. J. LeBowitz and R. McMacken, *NARes* **12**, 3069 (1984).
30. J. LeBowitz, Z. M. Zylicz, C. Georgopoulos and R. McMacken, *PNAS* **82**, 3988 (1985).
31. T. Itoh and J.-I. Tomizawa, *CSHSQB* **43**, 409 (1979).
32. T. Itoh and J.-I. Tomizawa, *NARes* **10**, 5979 (1982).
33. G. Hillenbrand and W. Staudenbauer, *NARes* **10**, 833 (1982).
34. Y. Sakakibara and J.-I. Tomizawa, *PNAS* **71**, 1403 (1974).
35. J.-I. Tomizawa, H. Ohmori and R. Bird, *PNAS* **74**, 1865 (1977).
36. K. J. Marians, W. Soeller and L. Zipursky, *JBC* **257**, 5656 (1982).
37. P. Brown, C. Peebles and N. Cozzarelli, *PNAS* **76**, 6110 (1979).
38. W. Staudenbauer, *MGG* **145**, 273 (1976).
39. M. Gellert, M. O'Dea, T. Itoh and J.-I. Tomizawa, *PNAS* **73**, 4474 (1976).
40. J.-I. Tomizawa, T. Itoh, G. Selzer and T. Som, *PNAS* **78**, 1421 (1981).
41. J.-I. Tomizawa and T. Itoh, *PNAS* **78**, 6096 (1981).
42. J. Scott, *Microbiol Rev.* **48**, 1 (1984).
43. M. Morita and A Oka, *EJB* **97**, 435 (1979).
44. J. Tamm and B. Polisky, *NARes* **11**, 6381 (1983).
45. J.-I. Tomizawa, *Cell* **38**, 861 (1984).
46. J.-I. Tomizawa, *Cell* **40**, 527 (1985).
47. J. Tamm and B. Polisky, *PNAS* **82**, 2257 (1985).

48. G. Cesarini, M. Moesing and B. Polisky, *PNAS* **79**, 6313 (1982).
49. T. Som and J.-I. Tomizawa, *PNAS* **80**, 3232 (1983).
50. A. Twigg and D. Sherratt, *Nature* **283**, 216 (1980).
51. D. Moser, D. Ma, C. Moser and J. Campell, *PNAS* **81**, 4465 (1984).
52. J.-I. Tomizawa and T. Som, *Cell* **38**, 871 (1984).
53. M. Abe, *J. Bact.* **141**, 1024 (1980).
54. L. Liu and J. Wang, *ICN-UCLA Symp. Mol. Cell. Biol.* **3**, 38 (1975).
55. M. Shure, D. Pulleybank and J. Vinograd, *NARes* **4**, 1183 (1977).
56. W. Keller, *PNAS* **72**, 6458 (1975).
57. L. Peck and J. Wang, *Nature* **292**, 375 (1981).
58. D. Rhodes and A. Klug, *Nature* **292**, 378 (1981).
59. K. Shishido, *FEBS Lett.* **111**, 333 (1980).
60. N. Higgins and N. Cozzarelli, *NARes* **10**, 6833 (1982).
61. C. Singleton and R. Wells, *JBC* **257**, 6292 (1982).
62. A. Courey and J. Wang, *Cell* **33**, 817 (1983).
63. M. Gellert, M. O'Dea and K. Mizuuchi, *PNAS* **80**, 5545 (1983).
64. R. Sinden and D. Pettijohn, *JBC* **259**, 6593 (1984).
65. D. Lilley, *CSHSQB* **42**, 101 (1982).
66. C. Gray, R. Sommer, C. Polke, E. Beck and H. Schaller, *PNAS* **75**, 50 (1977).
67. J. Kaguni and A. Kornberg, *Cell* **38**, 183 (1984).
68. J. Kaguni, L. Bertsch, D. Bramhill, J. Flynn, R. Fuller, B. Funnell, S. Maki, T. Ogawa, K. Ogawa, A. van der Ende and A. Kornberg, *in* "Plasmids in Bacteria" (D. Helenski, S. Cohen, D. Clewell, D. Jackson and A. Hollaender, eds.), p. 141. Plenum, New York, 1985.
68a. J. LeBowitz and R. McMacken, *JBC* **261**, 4738 (1986).
69. J. Zechel, J.-P. Bouche and A. Kornberg, *JBC* **250**, 4684 (1975).
70. O. Sundin and A. Varshavsky, *Cell* **21**, 103 (1980).
71. O. Sundin and A. Varshavsky, *Cell* **25**, 659 (1981).
72. A. Oka and J. Inselburg, *PNAS* **72**, 829 (1975).
73. Y. Sakakibara, K. Suzuki and J.-I. Tomizawa, *JMB* **108**, 569 (1976).
74. M. O'Donnell and A. Kornberg, *JBC* **260**, 12884 (1985).
75. A. Sugino, C. Peebles, K. Kreuzer and N. Cozzarelli, *PNAS* **74**, 4767 (1977).
76. M. Gellert, K. Mizuuchi, M. O'Dea, T. Itoh and J.-I. Tomizawa, *PNAS* **74**, 4772 (1977).
77. D. Lockshon and D. Morris, *JMB* **181**, 63 (1985).
78. M. O'Connor and M. Mallamy, *JMB* **181**, 545 (1985).
79. Y.-C. Tse and J. Wang, *Cell* **22**, 269 (1980).
80. P. Brown and N. Cozzarelli, *PNAS* **78**, 843 (1981).
81. K. Mizzuuchi, L. Fisher, M. O'Dea and M. Gellert, *PNAS* **77**, 1847 (1980).
82. K. Kreuzer and N. Cozzarelli, *Cell* **20**, 245 (1980).
83. S. DiNardo, K. Voelkel and R. Sternglanz, *PNAS* **81**, 2616 (1984).
84. C. Holm, T. Goto, J. Wang and D. Botstein, *Cell* **41**, 553 (1985).
85. C. Thrash, K. Voelkel, S. DiNardo and R. Sternglanz, *JBC* **259**, 1357 (1984).
86. T. Goto and J. Wang, *PNAS* **82**, 7178 (1985).
87. T. Steck and K. Drlica, *Cell* **36**, 1081 (1984).
88. G. Pruss, S. Manes and K. Drlica, *Cell* **31**, 35 (1982).
89. S. DiNardo, K. Voelkel, R. Sternglanz, A. Reynolds and A. Wright, *Cell* **31**, 43 (1982).

Aspects of the Growth and Regulation of the Filamentous Phages

> WILDER FULFORD,
> MARJORIE RUSSEL, AND
> PETER MODEL
>
> *The Rockfeller University*
> *New York, New York 10021*

Intensive study of how bacteriophage or eukaryotic viruses interact with their hosts can lead directly to some of the most basic of cell processes, and the growth and assembly of viruses provide lucid examples of the characteristics of large molecular systems. Prokaryotic DNA replication was first understood in a significant way with the admirable work on the replication of ϕX174 and the filamentous phages, and detailed analyses of the steps of DNA replication still often rely on the use of T4 or λ as substrates (1–5). Many current ideas about the packaging of DNA by proteins and the assembly of supramolecular structures emerged from work on the assembly of P22, T4, λ, and other phages (4, 6, 7), and concepts such as "antitermination" and "retroregulation" are almost entirely derived from phage work. Although phages are not living cells, nor infected cells the same as healthy ones, experience suggests that even the most apparently recondite and specific of phage processes can provide a clear window on parallel events in the whole cell, with all the genetic and physical versatility that phage provide. In this review, we describe two series of studies of the physiology and genetics of bacteriophage f1, with implications beyond the phage itself.

We consider first the regulation of filamentous phage DNA synthesis. Phages in a lytic cycle do not need to limit their overall expression—indeed, their main interest is precisely to gain maximum expression at as great an expense to the host as possible. Any control they exhibit is meant to manage their own course of development and streamline the exploitation of the hapless host. Filamentous phages (f1, M13, fd), on the other hand, are nonlytic; instead they grow continuously as long-term parasites (8–10), and so must consider the appropriate balance of host and phage metabolism. These phages are particularly adept at this kind of control. After an initial, rapid build-

Unfamiliar or Nonstandard Terms[a]

Term	Definition
Gyrase	DNA topoisomerase (ATP-hydrolyzing), EC 5.99.1.3. This enzyme, required in f1 DNA synthesis, introduces negative supercoils in closed circular DNA (1). Its subunits are the products of the gyrA and gyrB genes of E. coli (at 48 and 83 minutes on the linkage map)
Helicase	Enzymes that use the energy of nucleoside triphosphate hydrolysis to "melt" duplex DNA ahead of a replication fork, and move relative to the separated strands (1). Examples include the E. coli rep protein (required for f1 and ϕX174 DNA synthesis), and the T7 gene-4 protein
Primase	An enzyme or assembly of enzymes that produce short RNA primers for DNA replication (1). Primases are not, properly speaking, classical RNA polymerases. They include for the filamentous phage, E. coli RNA polymerase and single-strand-binding protein (1, 2, 5); for E. coli G4, and ϕX174, the dnaG protein and various other factors (1); for T7, the gene-4 protein (1, 59, 60); and for Col-I or RP4 and related plasmids, the proteins of the sog (61, 61a) or pri (61b) loci
Transposase	Enzymes that bind the ends of insertion elements and transposons and catalyze their insertion at random new sites in DNA (161). The transposase gene of the kanamycin-resistance transposon Tn5 encodes both the transposase and the smaller transposase inhibitor, identical to the large C-terminal section of the transposase (62, 63)
ilv	The isoleucine–valine biosynthesis operon, at 85 minutes on the E. coli genetic map
rep	The gene for the rep helicase, at 85 minutes on the E. coli map
rho	The gene for transcription termination factor Rho, also at 85 minutes on the map
trxA (= fipA)	The gene for E. coli thioredoxin, at 85 minutes on the map
trxB	The gene for E. coli thioredoxin reductase (EC 1.6.4.5), at about 20 minutes on the genetic map (162)

[a] This table defines any such term that appears more than once in the text, or that is not defined in the text. E. coli genetic loci are according to the standard E. coli linkage map (160). All terms having to do with DNA replication are discussed by Kornberg (1).

up of phage DNA and protein, the infected cells settle into a phage-producing steady state that persists indefinitely (10, 10a). Only a small fraction of the cells (perhaps 10^{-3} per generation; 10) lose the phage if reinfection is blocked. In some sense, this is a failure of the control system, but in another it is a testimonial to its effectiveness. In its broadest context, the steady state is the manifestation of the phage's solution to a universal problem in biology—the problem of growth control or limitation.

The second topic we discuss is the unexpected participation of thioredoxin in phage morphogenesis. Filamentous phage assembly takes place principally at the infected cell's inner membrane. Assuming that host mutations that prevent or affect the insertion or processing of membrane proteins (phenomena of interest to us) would lead to the inability to support phage growth, we sought *E. coli* mutants unable to produce phage. Surprisingly, the clearest cut mutant, and the most genetically tractable, turned out to be defective in the gene for thioredoxin, a small, abundant, ubiquitously distributed protein whose function in most cells is still unclear (indeed has become more unclear as it has been more intensively studied). In this instance, using the phage to approach one aspect of the host metabolism led us, instead and unexpectedly, to quite another, of interest nonetheless.

I. The Elements of the f1 Self-Regulatory Circuit

A. Introduction to the Filamentous Phages

1. f1 DNA Replication

Filamentous phage DNA enters its host as a single (plus)-stranded circle (ss), of the same polarity as all the viral mRNAs (*11–14*). Figure 1 shows the 10 genes and single large intergenic region of f1. The genome is a model of economical design. The genes are arranged in three groups: the genes of DNA replication (II, X, and V); the coat protein genes (VII, IX, VIII, III, and VI), arrayed in the order in which they appear in the mature phage particle; and the genes involved in phage assembly (I and IV). There are virtually no gaps in the coding region—indeed, many of the genes overlap slightly at their ends. There is a transcription terminator between genes VIII and III (*15–18*), and another at the end of gene IV (*19*), and they define the 3' ends of two sets of nested transcripts, one initiating at various points from gene II to gene VIII, and another at points from gene III to gene IV (*18–20, 22*). The intergenic region (*2*), of about 500 bases between genes IV and II, contains, starting at the left (5') end: a transcription terminator, which overlaps almost entirely with a sequence, the morphogenetic signal, required for phage packaging; the origin of minus, or complementary, strand synthesis; the origin of plus, or viral, strand synthesis; and the gene II promoter.

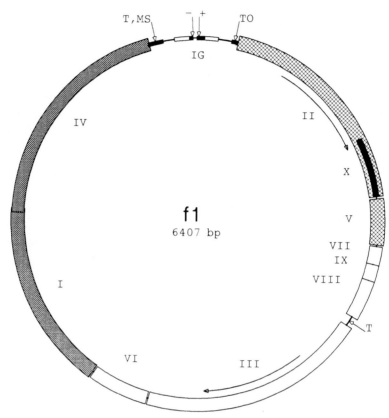

FIG. 1. Genetic map of f1. The 6407 bases of phage f1 are shown with the plus strand arranged 5' to 3' in the clockwise direction. The genes are labeled with roman numerals, and divided into three classes. The genes of DNA replication (II, X, and V) are shown cross-hatched (with the in-frame overlapping gene, X, in black); the virion protein genes (VII, IX, VIII, III, and IV) are in white; and the genes for the two proteins involved in particle assembly (I and IV, which overlap slightly) are in grey. In black, and indicated by small arrows, are terminators (T), and, in the intergenic region (IG), the morphogenetic signal (MS), minus-strand origin (−), plus strand origin (+), and gene II translational operator (TO). For the minus-strand origin, the arrow indicates the beginning of the complementary strand RNA primer, the black region the extent of the primer, and the white region additional required sequences. For the plus-strand origin, the arrow indicates the gene-II protein "nicking" site, the black region the extent of the binding site on double-stranded DNA, and the white region other sequences that contribute to efficient origin activity. The large arrows inside the map show the direction of transcription, and denote the two major transcription units. The map is discussed further in the text (Section I, especially I,A,1).

In the course of an infection, DNA synthesis appears to proceed in three stages (23–26; recently reviewed by Baas, 5). (1) Immediately on infection, host enzymes synthesize a second (minus) strand on the infecting DNA (ss to RF synthesis). (2) For the next few minutes, the resultant negatively supercoiled molecule of replicative form DNA (RF-I) replicates to produce a pool of RF-I (RF to RF synthesis). (3) Later, replication switches almost entirely to the concerted, asymmetric synthesis of new viral strands for export (RF to ss synthesis). As shown in Fig. 2, two mechanisms can account for all three stages (24–26): (1) minus, or complementary, strand synthesis, initiated by host RNA polymerase on single plus strands at the minus strand origin, and carried to completion by host DNA polymerases, ligase, and gyrase. (2) Plus, or viral, strand synthesis from RF-I, initiated at a specific "nick" introduced in the plus strand by the gene-II protein (27). Elongation proceeds from the 3′ OH of the nick [requiring the *rep* helicase, *E. coli* single-stranded DNA binding protein (SSB), and *E. coli* DNA polymerase III; 23], displacing the old strand, until the gene-II protein terminates the synthesis after one round. The products of this reaction are one molecule of relaxed covalently closed RF-DNA (RF-IV), which is rewound by gyrase to reenter the RF-I pool, and one of single-stranded viral DNA (26). Early in infection, the new plus strands are doubled up to enter the pool of RF-DNA; later on, as phage proteins accumulate, they are most often sequestered by the gene-V single-stranded DNA binding protein for transport to the membrane and assembly into virion particles (28–31). At this point, any gene-V protein not attached to single-stranded DNA can act to repress the translation of the gene-II and -X proteins, thereby linking the rate of DNA synthesis to the level of accumulated DNA (32–34).

2. The Need for Regulation

The filamentous phages are the only known phages capable of coexisting indefinitely with their *E. coli* host in a state of persistent productive infection (3, 8–10, 10a, 35), not unlike many eukaryotic viruses. To reach this state, a phage faces two tasks: it must stage a rapid coup, and then install a stable new regime. That is, it has to infect its host efficiently and quickly, and then tread lightly, as a long-term parasite must, taking from the cell only what it requires for a steady flow of new phage (about 100/cell/generation) through the cell wall (31). It needs to maintain a pool of template DNA adequate to supply the proteins and viral single-stranded DNA to make more phage, but not so large as to mortally impair the host. The phage develop fast enough to produce progeny as soon as 10 minutes after

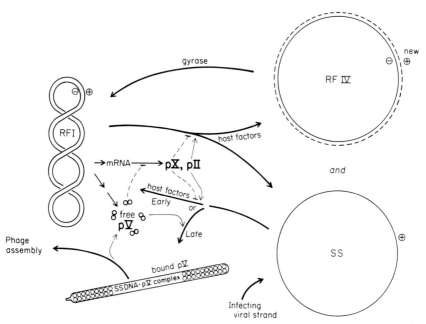

FIG. 2. f1 replication. A summary of events discussed in Section I of the text. The thickest arrows represent the transformations of the DNA itself. RF-I is shown producing a molecule each of relaxed RF-IV (which is then rewound by gyrase) and viral ssDNA (+). The ssDNA can undergo either complementary (−) strand synthesis (early in infection, or if gene-II protein is overrepresented), or sequestration by gene-V protein (pV, shown as small double rings, representing dimers), for eventual assembly and export. The picture of the ssDNA · pV complex is purely schematic. The arrows of medium thickness represent transcription and translation, and the thinnest arrows (solid or broken) are the proteins in action. The thin solid arrows indicate where the proteins act positively, and the broken lines ending in a bar are where they have inhibitory activities. Some of the arrows do not necessarily imply particular mechanisms, but simply represent observed phenomena. Gene-II protein (pII), is shown catalyzing or facilitating both viral and complementary strand synthesis, and gene-X protein (pX) is shown inhibiting these activities. pV is shown entering and leaving the pV · ssDNA complex. In its free form, it can also repress the translation of pII and pX.

infection, and in rich media they reach their peak of productivity after some 30 minutes. This is accompanied by the rapid synthesis of phage DNA, and the concomitantly rapid synthesis of phage proteins. The rate of increase levels off fairly sharply at about 30 minutes, after which the amount of intracellular DNA remains fairly constant, declining slowly until a steady state is reached (10). Then, if conditions are favorable, the cells continue to grow and divide and produce phage at a constant rate for as long as one has the patience to measure. If mutations allow overproduction of template DNA, or prevent the export of phage, the accumulation of phage products is lethal to the host (35, 37).

It might be thought that the phage's steady state is the passive consequence of the intrinsic rates of the metabolic reactions involved in phage synthesis. Perhaps in some ancestral phage this was true, but recent work has shown that the modern filamentous phages regulate their steady state through a system of checks and balances operating in complicated ways on both protein and DNA synthesis (32, 33, 37–39). The known protagonists in this elegant interplay are three viral proteins (the products of genes II, X, and V), and several cis-acting elements (the origins of replication, and "translational operators" in gene II). These elements constitute a sensitive homeostatic circuit that may provide a simple analogue of the many complex steady states that every cell must maintain. We discuss some of what is known about the activities and interactions of the proteins of this system [the cis-acting elements have been recently reviewed (2)] and a few of the implications and applications of this knowledge.

B. Gene-II Protein

1. ACTIVITIES

The multifunctional gene-II protein is intimately involved in all RF synthesis other than the initial doubling up of the infecting strand (40–42). *In vivo* evidence from gene-II mutants suggested that the protein was an endonuclease able to nick the plus strand of RF-I (41, 42). The 46-kDa enzyme was isolated on the basis of its ability to nick phage RF-DNA, and its ability to stimulate fl DNA synthesis *in vitro* (43, 44). The unique nicking site defines the origin of plus strand synthesis, in an area of dyad symmetry in the phage intergenic region (27). The nicking activity, which requires Mg^{2+}, is completely specific for a single site on RF-I DNA, yielding, *in vitro*, some RF-II (nicking activity) and some RF-IV ("nicking-closing" activity) (43–46). Given Mn^{2+}, the gene-II protein can linearize RF-I to RF-III (43, 44, 46, 47). The ends of this linear molecule have not been characterized. This activity, which has also been used as the basis for a purification (46, 47), has been taken to imply that gene-II protein interacts with DNA as a dimer, but can usually cut only one side of the double strand (23, 44). However, the stoichiometry of interaction is not yet known.

fl DNA synthesis has been studied *in vitro* with several reconstituted systems (23, 45, 48, 49). After nicking the RF-I, gene-II protein forms a complex with the resultant RF-II that is the substrate for unwinding by *E. coli rep* helicase and single-stranded binding protein, and elongation by DNA polymerase-III. Gene-II protein is required for the elongation reaction on RF-II, perhaps by providing an entry site for the helicase, and is also responsible for the termination of synthesis after each round. It is suspected but not known that it remains associated with the moving replication fork.

Although the enzyme does not cut free ssDNA, it does cleave and circularize the plus strand displaced during each round of rolling circle replication (23, 47, 48). In vitro, a bacteriophage T4 DNA replication system can spin out long, concatenated, linear, single strands of f1 DNA. These are processed to unit-length linear and circular molecules only if gene-II protein is present from the start of the reaction. This implies that the gene-II protein tracks the replication fork and recognizes some structure that occurs as the fork reaches the plus strand origin. In practice, complete processing of nascent ssDNA in this system requires a considerable molar excess of purified gene-II protein, making it difficult to judge how many molecules act at any step, or what structures they recognize.

Despite the great complexity of the activities already described, we have recently found (in experiments in which excess protein was provided from plasmids carrying gene II) that gene-II protein may also encourage the doubling up of plus strands in the presence of gene-V protein (W. Fulford, K. Horiuchi, and P. Model, unpublished). Such an activity is evidently not required for the initial doubling of the infecting viral strand (which occurs in the absence of all phage proteins). It could be imagined that gene-II protein interacts with ssDNA in the region of the minus strand origin, thereby helping to keep it clear of the initiation-inhibiting gene-V protein.

2. SPECIFICITY

The DNA sequence recognized by the gene-II protein nicking activity has been studied by assessing the ability of purified enzyme to nick plasmids carrying variously deleted f1 plus-strand origins (46, 50). The region extends from not more than four nucleotides on the 5' side of the nicking site to between 11 and 29 nucleotides on its 3' side. The ability of these mutated origins to terminate synthesis has been determined in vivo by testing whether they can terminate synthesis begun at a second, complete, origin on the same plasmid (50–54). By this test, the termination signal extends from about 12 bases 5' to 11 to 29 bases 3' of the nicking site (50, 52). This is also about the extent of the region (on both DNA strands) that has been shown to interact with gene-II protein by filter-binding (54a) and DNase "footprinting" (D. Greenstein and K. Horiuchi, personal communication).

The complete signal for initiation, as determined by the ability of plasmids carrying f1 sequences to enter the f1 mode of replication in the presence of superinfecting phage, and by an initiation assay like the termination assay described above, extends for an additional 100 bases 3′ to the nicking site (46, 50, 55, 56). The requirement for most of these last 100 nucleotides can be relieved by excess or mutant gene-II protein (38, 39, 57).

In summary, gene-II protein binds to the plus-strand origin of f1 RF-I, introduces a specific nick in the plus strand, facilitates entry of host replication enzymes, and finally mediates cleavage and circularization of unit length ssDNA, and perhaps the initiation of new complementary strands. Current studies of this unusually complex enzyme are examining the specificity and mechanism of DNA binding and nicking, and the mechanism of the possible role in minus-strand synthesis.

C. Gene X

1. Overlapping Genes

Gene II actually encodes two proteins, one from the full length of the gene, and another that is separately initiated and translated in the same reading frame to produce a protein identical to a C-terminal portion of the larger protein. The gene-X protein is identical to the C-terminal 13 kDa of the 46-kDa gene-II protein (11, 58).

Various other in-frame overlapping genes have been described, such as the bacteriophage T7 DNA primase helicase, the product of gene 4 (59, 60), the two genes of the *sog* or *pri* (primase) loci of plasmids such as Col I or RP4 (61, 61a, 61b), the transposase and transposase inhibitor of Tn5 (62, 63), the overlapping genes at the *che*A locus of *E. coli*, one of the loci specifying the chemotactic response (64), the C and *Nu*3 overlapping morphogenetic genes of phage λ (65), the phage T4 gene 69, involved in DNA replication (possibly in priming; 65a), and, perhaps most interestingly, the A and A* proteins of phage φX174 (66, 67). The multifunctional φX174 gene-A protein initiates and terminates φX174 ssDNA synthesis in much the same way as the apparently unrelated gene-II protein, with the difference that the gene-A protein remains covalently bound to the 5′ end of the displaced strand until the termination event (68–74). The A* protein comprises the C-terminal 37 kDa of the 56-kDa gene-A protein. Both the A and A* proteins have been purified and extensively studied for nucleolytic and DNA-binding activities. These studies are summarized by Brown *et al.* (75). The A* protein, though it cannot replace gene-A protein for *in vitro* DNA synthesis (76, 77), echoes many of its activities, often with changed specificity or potency (76, 78–82). Functions suggested for the A* protein include the shutdown of host DNA synthesis (83) by DNA binding (79), or DNA cleavage (80), as well as active roles in DNA synthesis or DNA packaging (71), or in the regulation of the late switch to viral strand synthesis (78, 80, 83). However, its *in vivo* role has resisted biochemical identification.

It is intriguing that many of the known overlapping genes are involved, like gene II, in specific DNA recognition and synthesis initiation events. Such pairs of proteins, one a subdomain of the other, suggest interesting regulatory and functional interactions. The approach we used to dissect genes II and X may prove useful for the future study of such genes.

2. The Function of Gene X

It is not possible to construct a structural mutation in the overlapping region of a pair of in-phase genes that is certain (or even likely) to affect one but not the other (as can be done for out-of-frame overlapping genes). However, one can affect only the larger protein by mutating upstream from the second translational start, and produce a down or null mutation of the smaller protein by attacking either its ribosome binding site, or its initiator codon. By appropriate combinations of such mutants, the functions of the genes can be rescued, but split genetically (84).

To do this for genes II and X, we constructed a plasmid encoding gene-X, but not gene-II, protein, and phage in which the gene-X initiator AUG has been changed to UAG (amber) or UUG (leucine) (84). The UUG mutant was able to make gene-X protein, presumably initiated at the UUG, and grew normally, indicating that gene II with a substitution at the position of the methionine codon was functional. The amber mutant could not make gene-X protein, and could not make healthy plaques on any suppressor (including leucine), unless gene-X protein was supplied from the plasmid. In other words, f1 cannot grow unless it has the products of both gene II and gene X; neither is sufficient on its own.

Amber-suppressing cells infected with the amber mutant in the absence of complementing gene-X protein accumulate normal amounts of RF at a normal rate; however, they accumulate no ssDNA and export no phage. In labeling experiments, it was found that the RF DNA made by the amber mutant late in infection incorporated label evenly into plus and minus strands, in contrast to wild-type phage, which primarily make viral strands. Thus, without gene-X protein, viral-strand synthesis and accumulation are deficient, and RF synthesis is correspondingly enhanced, just as they are if gene-II protein is overproduced. That is, gene-II protein, if by itself or in excess, promotes RF synthesis, that ssDNA is an intermediate in this process notwithstanding. Gene-II protein must be involved in both the plus- and minus-strand stages of DNA replication, and gene-X protein must have a role, along with gene-V protein, in the late switch to concerted single-strand synthesis.

Recent experiments (involving the provision of gene-X protein from a plasmid-borne gene) suggest that gene-X protein can also block initiation of plus-strand synthesis by gene-II protein (W. Fulford, and P. Model, unpublished). If, as we have suggested, gene-II protein is involved in facilitating complementary strand synthesis on single strands, as well as initiating plus-strand synthesis on RF-I, then the effects of the absence or overabundance of gene-X protein can be neatly explained by positing that it can prevent or modulate both functions of gene-II protein. In this model, gene-II protein exists to initiate and expedite every stage of RF synthesis, and gene-X protein to keep it in check and frustrate its ambitions, by helping to sequester completed single strands, and slow the rate of plus-strand initiation.

D. Gene-V Protein

1. Biochemistry

The final member of the f1 DNA replication equation is the abundant (10^5 copies/cell) 9.7-kDa gene-V-encoded ssDNA-binding protein (85, 86). Since it was first isolated (85, 86), this protein has been an important model for protein–DNA interactions. A great deal is known about its physical characteristics (87, 88). The sequence of its 87 amino acids has been determined (89, 90), and confirmed from the DNA sequence (12–14). In solution, the protein exists mainly as a dimer (91, 92); it appears to bind DNA entirely in this form. It binds strongly and cooperatively to nucleic acids in the order of affinity ssDNA > ssRNA > dsDNA and dsRNA, and spans 4 or 5 nucleotides per monomer (87, 88, 91, 93, 94). The binding coefficient for ssDNA is about 10^4 to 10^5 M^{-1} for noncooperative binding, and perhaps 200- to 500-fold higher for cooperative binding (93, 94). Chemical modification and spectroscopic studies implicate three tyrosines, one phenylalanine, and several arginines and lysines in DNA binding (92, 95–99). Binding is thought to involve intercalation of the aromatic residues between the bases of the DNA, and ionic interactions of the charged residues with the sugar-phosphate backbone. Poly(dA) is the least well bound of the homopolymers, perhaps because base stacking interferes with intercalation (93). In fact, the protein displays a higher affinity for both poly(rU) and poly(rI) than for poly(dA) (94). X-Ray diffraction studies (to 2.3 Å resolution) of the crystallized protein show that the monomer consists of a three-stranded β-sheet, a two-stranded β-ribbon, and a broad connecting loop (100). In the crystal, monomers are dimerized tightly around a dyad axis to form a β-barrel with antiparallel DNA-binding grooves on one face. The arrangement of amino-acid residues in these grooves supports the mechanism of DNA binding suggested by other studies (101).

2. THE GENE-V PROTEIN–ssDNA COMPLEX

In vivo, gene-V protein is accumulated for the first several minutes of infection, until it reaches a concentration sufficient to displace *E. coli* ssDNA-binding protein from nascent single strands and coat them to help prevent complementary strand synthesis (102–104). Phage mutants lacking active gene-V protein do not accumulate ssDNA or export phage, and display an elevated rate of RF synthesis. *In vitro*, gene-V protein prevents complementary-strand synthesis on single strands when it is present at the concentration needed to completely coat available ssDNA (105). We have recently shown that gene-V protein produced in a cell with a gene-V carrying plasmid renders the cell immune to f1 infection by preventing the doubling of the infecting DNA (unpublished observations). The rod-like complex of ssDNA and gene-V protein consists of one ssDNA viral strand, and perhaps 1500 monomers of gene-V protein arranged as a long stack of dimers with 6-fold symmetry around the central axis, wrapped in two antiparallel spiralling loops of DNA (28–30, 88). One may think of it as a rubber band stretched and then wound around the length of a pencil. In Fig. 2, the DNA is drawn with a "hairpin" at one end; although it is not certain, it is thought that the hairpin of the morphogenetic signal is not melted by gene-V protein (31, 94), and is found at one end of the complex, perhaps so that it may be efficiently presented to the packaging functions. The gene-V protein is recycled after it is displaced from the DNA by coat proteins during virion assembly at the membrane (85, 86). Because of the very high affinity of the interaction, most of the gene-V protein in the cell is found in these ssDNA complexes. Only the fraction that remains unbound is available to repress the translation of gene-II and -X proteins (32, 33).

3. Specific Repression of Gene-II and Gene-X Translation

The first indication of this regulation was the observation that gene-II protein is overproduced in cells infected with phage unable to make active gene-V protein (25, 106). The regulation can take place in the absence of RNA synthesis (32), and the synthesis of gene-II protein from gene-II mRNA *in vitro* can be repressed by added gene-V protein (33), indicating that the repression is at the level of translation. *In vitro*, repression has a steep concentration threshold at about 17 μM gene-V protein, not far from the estimated intracellular concentration of free gene-V protein late in infection (33–34).

A number of other cases of translational repression of specific mRNAs are known (reviewed by Campbell *et al.*, ref. 107). Of these, the most analogous to the gene-V protein is the T4 gene-32 protein, a cooperative ssDNA-binding protein required for phage DNA replication, repair, and recombination (108), that controls its own translation (109–113). Gene-V protein and gene-32 protein are rather similar in their nucleic acid binding characteristics (34, 87). It has been suggested that gene-32 protein acts by binding highly cooperatively to a particularly "unstructured" region in the 5' untranslated leader of its own mRNA (111–115). However, the fact that gene-V protein and gene-32 protein do not mimic each other's regulatory effects *in vitro* suggests that their mechanisms of repression have at least some element of sequence specificity (34). Indeed, any single-strand-binding protein whose interaction is not limited to the sugar phosphate side of the nucleic acid must have some intrinsic sequence preference. Any such preference could be taken advantage of by a repressor site, and magnified by binding cooperativity (34, 115). The untranslated leader regions of both gene-32 and gene-II mRNAs contain unusual sequences that make excellent candidates for target sites.

Recently, some f1 mutants have helped to define the site of action of gene-V protein. It was found that the poor-growth (turbid plaque) phenotype of certain f1 replication origin mutants could be relieved (to give clear plaques) by second-site mutations elsewhere in the genome (38, 39, 57). Two of the mutants isolated were found to overproduce gene-II and gene-X proteins. One mapped in the untranslated 5' leader of the gene-II RNA (G to U at the third G of the unusual sequence $U_5G_4CU_4C$; 39). This mutation is presumably at a site of interaction of the gene-V protein with the gene-II mRNA. Recent studies have also shown that deletion of this region of the DNA renders the resultant transcript insensitive to repression by gene-V protein (unpublished observations). The other mutation was a change of Arg[16] to Cys in the putative gene-V protein DNA-binding groove (38, 100, 101, 88). Since the phage with the protein mutated at this position is perfectly viable, it seems probable that the change affects either the sequence specificity or the intrinsic RNA affinity required to effect repression. Little is yet known about gene-V protein repression of gene X. Although it seems that genes II and X are independently regulated (deletion of the gene-II 5' leader does not affect gene-X repression; unpublished observation), there are no obvious homologies between sequences in or near gene X and the putative gene-II translational operator, and as yet no mutants to shed light on the question. Genetic and biochemical studies are in progress to better define the sites binding gene-V protein.

E. The Regulatory Circuit

After the first burst of activity following infection, the f1-infected cell settles into an essentially permanent steady state (10), in which a fixed pool of about 20 RF molecules divides only as fast as the host, and synthesizes about 100 phage per generation. This is an ideal parasitic state: enough template DNA to produce progeny and prevent curing by segregation, but not so much that the host's health—its continued ability to reproduce itself and make more phage—is seriously compromised. Phage mutants that stray from this balance (either up or down) are usually at a growth disadvantage; revertants and second-site suppressors arise at a startling frequency (38, 39, 57; K. Horiuchi and W. Fulford, unpublished observations). That such mutants occur at all itself implies that the phage is a self-regulating entity, rather than a hitchhiker in the back seat of the cell's metabolism. This then begs two questions. First, how does the phage gauge its rate of DNA synthesis and particle assembly; and, second, how does it translate this into regulatory action? That is, what agents tug DNA and protein synthesis in one direction or another?

Figure 2 is an outline of f1 DNA synthesis and the key players that run it. It is clear that at the center of what must be controlled are the absolute levels of both RF and ssDNA, and their ratio. Because it is made from RF DNA but binds to ssDNA, gene-V protein is the one phage element that can be used as a measure of all phage DNA in the cell. The total amount of the protein reflects template (RF-I) concentration; the amount of gene-V protein *not* bound in high affinity ssDNA complexes is therefore a measure of the ratio of ssDNA to RF DNA. If the ratio is high, there will be very little free gene-V protein (it will be made slowly, and quickly bound by ssDNA), and if the ratio is low, free gene-V protein will abound (it will be made quickly, and with nowhere to go). Free gene-V protein is a subtle gauge of phage replication, but to harness it the phage requires a means of distinguishing the unbound fraction of the protein, and allowing it to nudge DNA synthesis away from RF and toward single strands, and then only if the absolute level of RF DNA is high enough. That is, any regulatory site meant for free gene-V protein to act upon must be less attractive than ssDNA, but more so than the bulk of nucleic acids in the cell. The phage handles this requirement very elegantly by placing translational operators on the mRNA of genes II and X. Only after gene-V protein has fully titrated available ssDNA will it seek out these targets, and the targets will be abundant enough and have a low enough absolute affinity for the protein that they will not be repressible below a fairly high concentration of gene-V protein. The binding of gene-V protein to ssDNA is accelerated and strengthened by cooperative interactions, which are, of course, highly concentration-dependent. Although it is not understood exactly how gene-V protein binds to gene-II RNA, it is probably also a cooperative event, as is suggested by the high concentration required for repression *in vitro* and *in vivo* (32, 33, 115). In other words, repression is not possible below a certain level of template DNA, and even then only occurs when the ratio of ssDNA to RF DNA is too low.

The finding, described above, that gene-II protein can encourage all steps to RF synthesis and accumulation, at the expense of ssDNA, makes possible a model to explain how controlling genes II and X can affect the relative levels of RF and ssDNA. The synthesis of f1 DNA would be controlled by the tension between gene-II protein, which exists to make RF DNA, and gene X- and V-proteins, which exist to moderate its activity and sequester single strands. If the level of free gene-V protein rises—due to increased RF or decreased ss—gene-II and -X translation slows down, skewing DNA synthesis away from RF to ssDNA, which, as it is sequestered, will soak up the excess gene-V protein. Conversely, if the amount of unbound gene-V protein is too low (indicating insufficient RF or too much ssDNA), gene-II translation will rise to encourage the synthesis of compensating RF DNA. Gene-X protein exists to mute the activity of gene-II protein, and discourage the overproduction of RF DNA.

A detailed understanding of the tuning and sensitivity of this self-damping feedback circuit is currently being worked out by independently varying its different elements using phage mutants and various cloned phage genes and cis-acting elements. Elucidation of this relatively simple circuit may provide insights into the kinds of self-regulating feedback networks that operate in whole cells, such as the complicated multilevel controls of ribosome biosynthesis (*116*), or the macromolecular synthesis (*mms*) operon of *E. coli* (*117*). Ultimately, when the exact roles and quantitative relationships of the elements in the f1 regulatory circuit are understood, it may be an excellent system for developing the techniques of mathematical and computer modeling that will be necessary for the complete understanding of homeostases of all kinds.

II. The Role of Thioredoxin in Phage Assembly

A. Filamentous Phage Assembly

Presumably, the purpose of the circuit discussed in the preceding section is ultimately to assure efficient and steady assembly and export of new phage. In our discussion of the circuit, we largely ignored phage assembly as a potential rate-limiting step, and concentrated on the control elements directly involved in DNA synthesis. However, assembly obviously does count, and fluctuations in its rate can feed back on conditions in the rest of the cell by way of the complex of gene-V protein and ssDNA. If assembly slows (due, perhaps, to a paucity of coat constituents), the complex will back up in the cell, promoting, as we have seen, new synthesis of gene-II protein, and therefore of RF DNA, which can relieve the shortages. If assembly happens to run too far ahead (possibly as a result of excess RF, and, therefore, coat proteins), more gene-V protein is released, and it can feed back to reduce the level of RF DNA, and skew the balance of DNA synthesis back toward ssDNA. And when assembly is blocked completely (generally by a mutation in the coat or assembly genes), available gene-V protein is quickly bound up by ssDNA, allowing RF synthesis to run away, and eventually kill the cell. To understand the life cycle and self-regulation of f1, it is obviously essential to understand the rates and mechanisms of its assembly, still a mysterious process. Since the topic of filamentous phage morphogenesis has recently been well reviewed (*31*), we discuss only briefly what is known of the process, before describing our recent work on part of the process.

Filamentous phage particles contain five different virion structural proteins: 2700 copies of the major coat protein (*118*) [product of gene VIII (*119*)], about 5 copies each of the products of genes III and VI (*120, 121*), and about 10 copies of gene-VII protein and gene-IX protein (*123, 124*). The gene-III and -VIII proteins are classical integral membrane proteins before they are incorporated into the phage; they are synthesized with N-terminal extensions (signal peptides), which are cleaved by host enzymes as the proteins are inserted into the membrane (*125–127*), and they exist transiently as transmembrane proteins with their N-termini exposed toward the cell exterior (*106, 128, 129, 131*). If phage morphogenesis is blocked, or if they are expressed from a plasmid carrying the appropriate gene, these proteins remain indefinitely as trans-membrane proteins (*131*). They are not detected as free cytoplasmic proteins (*131, 132*). Much less is known of the products of genes VI, VII, or IX, but they also appear to be membrane-associated (*31*).

The cytoplasmic complexes of single-stranded DNA and gene-V protein are the immediate precursors of the mature virions. Careful examination of these complexes suggests that they contain no other phage-encoded proteins, although they may carry a few molecules of the host ssDNA-binding protein, a necessary participant in phage DNA synthesis (*30, 133*). No virions can be found within the infected cell (*9*), and the two coat proteins for which there are good assays are found exclusively in the cell's inner membrane (*106, 131, 132*). The inference from these observations is that the replacement of the gene-V ssDNA-binding protein takes place at or within the cell membrane. By a process that is not at all understood, all of the DNA-binding protein is removed and replaced by the major coat protein, the product of gene VIII. The available evidence suggests that the remaining four proteins are found at the ends of the virion, gene-VII and -IX proteins at one end (*133, 124*), and gene-III protein together with gene VI-protein at the other end (*122–124*), the last to emerge (*134*). In the absence of the gene-III and gene-VI proteins, viable phage are not formed, but many noninfectious polyphage, consisting of several unit length DNA molecules encapsidated in an appropriately longer protein sheath, are (*119, 134, 135*). Absence of the gene-VII and -IX proteins leads to the production of a very small number of infectious polyphage (*134*). In the absence of the major coat protein, no particles are formed (*119*).

B. A Host Mutant in Assembly

As an experimental approach to the processes of f1 assembly, we screened heavily mutagenized *E. coli* for those unable to produce progeny phage at high temperature. One such mutant, named *fip* (for filamentous phage production), and later *fip*A and *fip*B–E were discovered, had the following properties (*136*).

1. The cells grew normally when not infected, both at 37 and at 42°C.

2. Cells infected with phage f1 produced neither viable phage nor nonviable particles at 42°C.

3. The cells could be productively infected with the F-pilus-specific, RNA-containing phage f2 at 42°C. The cells also were competent to mate at 42°C.

4. Infection of a non-*amber* suppressing (wt) host with phage containing an *amber* mutation in any of several genes is lethal (37). Infection of these mutant (*fip*) cells at 42°C by wild-type or *amber* phage also is lethal, which shows that there is at least some expression of phage genes.

5. Synthesis of phage DNA and those phage proteins that are readily measured is normal.

6. Cells infected at low temperature and shifted to 42°C ceased making phage very rapidly (within 1 minute). In similar fashion, cells initially infected at high temperature started to make phage at 1 minute after removal to a lower temperature. This suggested that Fip protein participates directly in phage assembly, rather than via an indirect effect on the cells.

The *fip* gene was located on the *E. coli* chromosome by conventional mapping of a linked transposon. It lies near *ilv* (the isoleucine-valine biosynthetic operon), just upstream of *rho*, the gene encoding the transcription termination factor Rho (136), at about 85 minutes on the *E. coli* genetic map. The wild-type allele of the gene was isolated directly, by screening a library of random *E. coli* DNA inserted in a derivative of f1 useful as a cloning vector (137). Recombinant phage newly able to grow on *fip* mutant hosts were examined, on the presumption that they would carry sequences able to complement the host defect. All such phages carried a common fragment of DNA. The *fip*-encoding region of this fragment was identified by subcloning and transposon mutagenesis. A suitable fragment was subcloned into a high copy number plasmid, and the protein products were examined in the minicell system. From such experiments, and experiments with DNA-containing transposon insertions, it was possible to show that the *fip* gene encoded a protein of about 12.5 kDa. The protein (extracted from acrylamide gels) was coupled to a solid support, and used to prepare *fip* antiserum in rabbits. Later, a substantial amount of the protein was purified by conventional means (DEAE-Sephadex chromatography followed by gel filtration) from cells carrying the *fip* plasmid. The N-terminal amino-acid sequence of the protein thus isolated was determined.

The region of the E. coli chromosome extending from the ilv operon through the rho gene has been cloned and characterized by several groups (138–140). By aligning the restriction maps it was possible to precisely locate the fip gene. Shigesada and Imai, as a part of their studies of the control of the expression of the rho gene, sequenced a substantial length of DNA upstream of rho (140). Their DNA sequence, which they kindly provided, included an open reading-frame, located about 650 base-pairs upstream from rho, that predicted the 10-amino-acid amino-terminal sequence of the purified Fip protein. The protein sequences were used to query the Protein Data Bank of the National Biomedical Research Foundation; the fip gene is in fact the structural gene encoding E. coli thioredoxin (141).

C. Thioredoxin

Thioredoxin is a heat stable, ubiquitous protein containing active-site sulfhydryl groups that readily and reversibly form a disulfide on oxidation. It was originally isolated as a necessary, heat-stable cofactor for the reduction of nucleoside diphosphates to deoxynucleoside diphosphates and triphosphates by ribonucleotide reductase (EC 1.17.4.1 and .2) (142). Thioredoxin is oxidized as the deoxyribonucleotides are formed, and is then reduced by a highly specific enzyme, thioredoxin reductase (EC 1.6.4.5), which derives its reducing power from the oxidation of NADPH (142a).

In addition to extensive biochemical studies [recently reviewed by Holmgren (143)] thioredoxin has also been studied by X-ray crystallographic methods (144, 145). The structure of the oxidized form has been determined to 2.8 Å resolution, and shows that the redox-active S–S bridge is protuberant but accessible from only one side of the structure. Significantly enough (see below), Gly^{92} lies on this face and Gly^{74} close by. Optical methods show that there are differences in the conformation of the oxidized and reduced molecule (146, 147). Thus far the structure of reduced thioredoxin has not been obtained.

Thioredoxin is also a constituent of phage T7 DNA polymerase, an enzyme that, when isolated, contains two polypeptide chains, one the product of T7 gene 5 and the other thioredoxin (148, 148a). Very recent evidence suggests that thioredoxin acts to confer processivity on the polymerization reaction carried out by the T7 enzyme. The phage-encoded polypeptide can, by itself, carry out some DNA synthesis, but the products are short oligonucleotides and the extent of synthesis is rather limited (H. Huber and C. C. Richardson, unpublished).

Other than this role in T7 DNA synthesis (and, as we now recognize, filamentous phage assembly), the exact function of thioredoxin, which is usually quite abundant, remains something of a mystery. Prokaryotes mutant in the thioredoxin gene are able to grow even if exogenous deoxyribonucleotides are not supplied (*149, 150*). Indeed, this observation led Holmgren and co-workers to discover a second cofactor for ribonucleotide reductase, a protein called glutaredoxin, which can channel the reducing power from reduced glutathione to ribonucleotide reductase (*151*). Glutaredoxin is about an order of magnitude more effective than thioredoxin as a cofactor for ribonucleotide reduction, but its concentration is about two orders of magnitude less (*151*). It is not yet clear which protein is the usual cofactor *in vivo* (*143*).

D. Mutant Thioredoxins

The role of thioredoxin in the reduction of ribonucleotides led us to expect that it might act in similar fashion in filamentous phage morphogenesis. Two additional observations suggested that this might be the case. First, the original *fip* mutation changed a proline to a serine in the center of the 4-residue active site, Cys^{32}-Gly^{33}-Pro^{34}-Cys^{35} (*141*). Second, it was found that cells with mutations in the thioredoxin reductase gene (*trx*B) produce only about a tenth as many phage as do wild-type hosts (*141, 152*).

This hypothesis turned out to be incorrect. By site-directed oligonucleotide mutagenesis we have replaced either or both of the active site cysteine residues with alanine or serine. In all cases filamentous phage are still produced in hosts specifying only the mutant thioredoxin (M. Russel and P. Model, unpublished). If the host strain, by contrast, contains an inactivating insertion or deletion, or a chain-terminating (amber) mutation, in the thioredoxin gene, none of the filamentous phage can be propagated (M. Russel and P. Model, unpublished). Thus thioredoxin is absolutely required for filamentous phage production, but the cysteine residues needed for thioredoxin's redox activity are not.

We have also isolated additional mutations in the thioredoxin gene in an effort to define the portions of thioredoxin critical for its role in filamentous phage assembly. Several independent isolates consisted of mutations in which UGG, encoding tryptophan, was changed to UGA, which is a chain-terminating mutation in a wild-type host; additional mutations were of Gly^{74} to Asp, of Gly^{92} to Ser, and of Gly^{92} to

Asp. This last is a reisolate of a mutation originally termed 7004, isolated as unable to support T7 growth by Chamberlin (153) and characterized by Holmgren et al. (154).

A commonly used assay for thioredoxin is the reduction of 5,5′-dithiobis(2-nitrobenzoic acid) (Nbs_2) in the presence of thioredoxin and thioredoxin reductase, with NADPH as electron donor. Thioredoxin reductase itself does not reduce the Nbs_2 effectively; thus the assay measures the ability of thioredoxin to accept electrons from thioredoxin reductase. All of the mutants in which an active site cysteine has been replaced by another residue are inactive in this assay (M. Russel and P. Model, unpublished). The Asp^{74} mutant is inactive in this assay, the Asp^{92} mutant is partially active, and the Ser^{92} mutant is as active as the wild type. Furthermore, the original *fip* mutant ($Pro^{34} \rightarrow Ser$), which does not support f1 growth at 42°C, makes a protein that is not temperature sensitive in the Nbs_2 assay (unpublished results). Thus there is no correlation between the ability of the mutant proteins to support filamentous phage growth and their enzymatic activity.

There is some correlation, however, between assembly and the quantity of different forms of thioredoxin. Cells with a multicopy thioredoxin-encoding plasmid make 30–40 times as much thioredoxin as cells with a single chromosomal copy. Several mutant thioredoxins will only support f1 growth if overexpressed in this way. Certain other thioredoxin mutants will only support filamentous phage growth if thioredoxin reductase is present. Because filamentous phage assembly is inherently somewhat temperature sensitive (11, 42, 136), the inhibition of assembly by mutations in thioredoxin reductase is more pronounced at 42 than at 37°C. However, the effects of such reductase mutations on phage assembly are overcome if the thioredoxin in question is mutated at either (or both) of its active site cysteines. These thioredoxin active site mutants are, in effect, second-site suppressors of mutations in thioredoxin reductase.

From the observations we have just described, it seems clear that what the phage requires is a reasonable quantity of *reduced* thioredoxin, or, more precisely, of thioredoxin in its reduced conformation. We infer that the mutants in which the active site cysteines have been replaced fold into the reduced conformation; that is why they become independent of thioredoxin reductase. It is unlikely that they all mimic the conformation of reduced thioredoxin perfectly; that is why the products of different alleles are required in different quantities. The other mutant data suggest that increasing the total quantity of thioredoxin can compensate for inferior "quality," since some of the mutants, such as Pro^{34} → Ser will support efficient filamentous phage assembly at high temperature if supplied from a high copy number plasmid, but only at 37°C or below if supplied from a single chromosomal copy, and will not support phage growth at all if combined with a thioredoxin reductase mutation.

E. The Role of Thioredoxin

What can thioredoxin be doing in phage assembly? To start analyzing this, we selected mutant phage that can grow better on thioredoxin mutant hosts. Starting with wild-type f1, we obtained phage (at a frequency of about 10^{-6}) that can grow on the Pro^{34} → Ser mutant strain at 42°C. Several independent isolates had all incurred the same mutation, a change at amino-acid 142 in gene I from Asn to Tyr. M13, a closely related filamentous phage that differs from f1 at only four (*13*) or eight (*12*) sites that cause amino-acid changes over the entire genome, contains histidine at this position, as does fd, another closely related filamentous phage. M13 and fd grow, albeit poorly, on the Pro^{34} → Ser mutant (*136*). That gene I is involved in an interaction with thioredoxin is also suggested by the observation that several gene-I-missense or suppressed-*amber* mutants will not grow on thioredoxin mutant strains under conditions in which they will grow on wild-type cells, and in which wild-type phage will grow on the mutant cells (*136*). Missense and suppressed-*amber* mutants in other phage genes did not show this phenotype. Unfortunately, little is known of how gene I acts. Mutants in gene I are assembly-defective, and temperature-sensitive mutants cease phage production very quickly after shift-up (*37*). Attempts to study the gene have been hampered because it seems to be very lethal to cells: even when the gene is placed under inducible control, cells die so fast after induction that its physiological effects, both on the host cell and on the phage, remain elusive (*156*).

Direct examination of the role of thioredoxin will require a reasonable *in vitro* system for phage assembly. No satisfactory system yet exists, in spite of fairly determined efforts in a number of laboratories.

We can, however, use the phage T7 DNA replication system, which does work very well *in vitro*, to account for some of the possible effects of thioredoxin. As mentioned above, T7 DNA polymerase seems to use thioredoxin as a way of conferring processivity on the elongation reaction. At first glance, DNA synthesis and phage assembly have little in common, but in this instance they do share the property that proteins must move relative to DNA. T7 DNA-polymerase activity, whether assayed *in vivo* by the ability of thioredoxin mutant strains to support T7 growth, or assayed *in vitro* with purified mutant thioredoxin and T7 gene-5 protein, closely mimics the filamentous phage observations; like f1, T7 can utilize thioredoxins with active-site cysteine changes, is temperature sensitive on the Pro^{34} → Ser strain, grows poorly on the Gly^{92} → Ser strain, and not at all on the Gly^{92} → Asp or deletion strain (H. Huber, C. C. Richardson, M. Russel, and P. Model, unpublished). It might not be unreasonable to speculate that phage assembly, like DNA synthesis, may utilize thioredoxin to confer processivity on the assembly reaction.

What is thioredoxin really doing in the bacterial (or higher) cell? The initially perceived function for thioredoxin was as a cofactor in nucleotide reduction. That may still be one of its key roles, particularly when glutaredoxin, or glutathione metabolism, is insufficient. A second identified function is in sulfate metabolism (*143*). Here also, however, other systems must exist, since thioredoxin null mutants can still utilize sulfate for growth, and do not require either cysteine or methionine. The reduction of methionine sulfoxide, however, seems to be absolutely dependent on thioredoxin: no other protein substitutes *in vivo* (M. Russel and P. Model, unpublished). In higher cells, thioredoxin appears to participate in the light activation of chloroplasts (*157*), and in the activation of the glucocorticoid receptor (*158*). It has also been described as an essential cofactor for *in vitro* protein synthesis in reticulocyte extracts (*159*). All of these functions, however, are almost surely related to the use of thioredoxin as an oxidation–reduction cofactor. There are as yet no cellular analogs to the way in which filamentous phage or T7 use the protein. It will be interesting to see whether one can find cellular uses for the protein that are similar to the way in which these parasitic organisms have coopted it.

III. Concluding Remarks

The most fascinating aspects of the filamentous phage are the unusual requirements of their unique style of parasitism. Rather than usurping an entire cell, briefly and lethally, as the lytic phages do, the filamentous phage pursue the more subtle course of infiltrating the cell and working inside the system. In this they more closely resemble self-mobilizing elements like the F factor, which colonizes the cell and uses its resources to ensure its own replication and propagation. Such entities all face the same tasks: entering and establishing themselves in a cell, ensuring their own continued and stable replication, and arranging their propagation through the export of progeny in some form, all without killing the cell, and often under the constraint of a small genome. The intimacy of the phage–host interactions implicit in this lifestyle suggests that by their study we can shed light on homeostatic and replicative mechanisms, the design of membrane-associated assembly and polymerization reactions, and potential functions for a number of proteins (such as thioredoxin, as we have seen).

We have concentrated here on a few new details of the total story of how the filamentous phage manage their free ride: in particular, how they regulate their replication, and, concentrating on one member of the equation, how they manage to export the products of that replication. The story is by no means complete. The goals—still remote—are, first, a detailed picture of all the steps in viral propagation, from the beginning of DNA replication, through the intermediary sequestration of viral DNA, to the final mysterious collaboration of phage and host proteins, including thioredoxin, to create a leakproof assembly port for phage extrusion, and, second, an accurate and flexible quantitative description of the coupling and control of these steps, as functions of time and concentrations. We have described some of the coupling mechanisms of the system; how well they work is manifest in the stability of the open steady state of these phage, finely balanced between the vertical transmission of replicative DNA, and the horizontal export of phage. To understand this circuit will be to appreciate the requirements of the kind of global control of synthesis so basic to all cells. And to understand phage entry and export will be to know a great deal about the nature of membranes and the complicated processes for which they are the venue.

Acknowledgments

We thank N. D. Zinder, K. Horiuchi, K. Jakes, and J. Brissette for critical readings of the manuscript, and D. Greenstein and K. Horiuchi for access to unpublished results. The work in our laboratory is supported by grants from the National Science Foundation, and the National Institutes of Health.

References

1. A. Kornberg, "DNA Replication." Freeman, San Francisco, California, 1980.
2. N. D. Zinder and K. Horiuchi, *Microbiol. Rev.* **49**, 101 (1985).
3. D. Denhardt, D. Dressler and D. Ray (eds.), "The Single-Stranded DNA Phages." Cold Spring Harbor Laboratory, Cold Spring Harbor, New York, 1978.
4. C. K. Mathews, E. M. Kutter, G. Mosig and P. B. Berget (eds.), "Bacteriophage T4." Amer. Soc. Microbiol., Washington, D.C., 1983.
5. P. D. Baas, *BBA* **825**, 111 (1985).
6. R. W. Hendrix, J. W. Roberts, F. W. Stahl and R. A. Weisberg (eds.), "Lambda II." Cold Spring Harbor Laboratory, Cold Spring Harbor, New York, 1983.
7. S. Casjens, (ed.), "Virus Structure and Assembly." Jones and Bartlett, Boston, Massachusetts, 1985.

8. P. H. Hofschneider and A. Preuss, *JMB* **7**, 450 (1963).
9. H. Hoffman-Berling and R. Maze, *Virology* **22**, 305 (1964).
10. T. J. Lerner and P. Model, *Virology* **115**, 282 (1981).
10a. M. V. Merriam, *J. Virol.* **21**, 880 (1977).
11. K. Horiuchi, G. F. Vovis and P. Model, in "The Single-Stranded DNA Phages" (D. Denhardt, D. Dressler and D. Ray, eds.), p. 113. Cold Spring Harbor Laboratory, Cold Spring Harbor, New York, 1978.
12. E. Beck and B. Zink, *Gene* **16**, 35 (1981).
13. D. F. Hill and G. P. Petersen, *J. Virol.* **44**, 32 (1982).
14. P. M. G. F. van Wezenbeek, J. J. M. Hulsebos and J. G. G. Schoenmakers, *Gene* **11**, 129 (1980).
15. L. Edens, R. N. H. Konings and J. G. G. Schoenmakers, *NARes* **2**, 1811 (1975).
16. K. Sugimoto, H. Sugisaki, T. Okamoto and M. Takanami, *JMB* **111**, 487 (1977).
17. M. J. Rivera, M. A. Smits, W. Quint, J. G. G. Schoenmakers and R. N. H. Konings, *NARes* **5**, 2895 (1978).
18. J. S. Cashman, R. E. Webster and D. A. Steege, *JBC* **255**, 2554 (1980).
19. P. B. Moses and P. Model, *JMB* **172**, 1 (1984).
20. M. A. Smits, J. G. G. Schoenmakers and R. N. H. Konings, *EJB* **112**, 309 (1980).
21. M. Lafarina and P. Model, *JMB* **164**, 377 (1983).
22. K. J. Blumer and D. A. Steege, *NARes* **12**, 1847 (1984).
23. T. F. Meyer and K. Geider, *Nature* **296**, 828 (1982).
24. S. Eisenberg, J. F. Scott and A. Kornberg, *PNAS* **73**, 3151 (1976).
25. K. Horiuchi, J. V. Ravetch and N. D. Zinder, *CSHSQB* **43**, 389 (1979).
26. K. Horiuchi and N. D. Zinder, *PNAS* **73**, 2341 (1976).
27. T. F. Meyer, K. Geider, C. Kurz and H. Schaller, *Nature* **278**, 365 (1979).
28. R. E. Webster and J. S. Cashman, *Virology* **55**, 20 (1973).
29. D. Pratt, P. Laws and J. Griffith, *JMB* **82**, 425 (1974).
30. R. A. Grant and R. E. Webster, *Virology* **133**, 315 (1984).
31. R. E. Webster and J. Lopez, in "Virus Structure and Assembly" (S. Casjens, ed.), p. 235. Jones and Bartlett, Boston, Massachusetts, 1985.
32. T. S. B. Yen and R. E. Webster, *Cell* **29**, 337 (1982).
33. P. Model, C. McGill, B. Mazur and W. D. Fulford, *Cell* **29**, 329 (1982).
34. W. Fulford and P. Model, *JMB* **173**, 211 (1984).
35. D. A. Marvin and B. Hohn, *Bacteriol. Rev.* **33**, 172 (1969).
37. D. Pratt, H. Tzagoloff and W. S. Erdahl, *Virology* **30**, 397 (1966).
38. G. P. Dotto and N. D. Zinder, *PNAS* **81**, 1336 (1984).
39. G. P. Dotto and N. D. Zinder, *Nature* **311**, 279 (1984).
40. D. Pratt and W. S. Erdahl, *JMB* **37**, 181 (1968).
41. H. M. Fidanian and D. S. Ray, *JMB* **72**, 51 (1972).
42. N. S.-C. Lin and D. Pratt, *JMB* **72**, 37 (1972).
43. T. F. Meyer and K. Geider, *JBC* **254**, 12636 (1979).
44. T. F. Meyer and K. Geider, *JBC* **254**, 12642 (1979).
45. K. Geider and T. F. Meyer, *CSHSQB* **43**, 59 (1979).
46. G. P. Dotto, K. Horiuchi, K. S. Jakes and N. D. Zinder, *JMB* **162**, 335 (1982).
47. G. P. Dotto, V. Enea and N. D. Zinder, *PNAS* **78**, 5421 (1981).
48. T. F. Meyer, I. Baumel, K. Geider and P. Bedinger, *JBC* **256**, 5810 (1981).
49. G. Harth, I. Baumel, T. F. Meyer and K. Geider, *EJB* **119**, 663 (1981).
50. G. P. Dotto, K. Horiuchi and N. D. Zinder, *JMB* **172**, 507 (1984).
51. K. Horiuchi, *PNAS* **77**, 5226 (1980).
52. G. P. Dotto and K. Horiuchi, *JMB* **153**, 169 (1981).
53. G. P. Dotto, K. Horiuchi and N. D. Zinder, *PNAS* **79**, 79 (1982).

54. G. P. Dotto, K. Horiuchi, K. S. Jakes and N. D. Zinder, *CSHSQB* **47**, 717 (1983).
54a. K. Horiuchi, *JMB* **188**, 215 (1986).
55. J. M. Cleary and D. S. Ray, *PNAS* **77**, 4638 (1981).
56. S. Johnston and D. S. Ray, *JMB* **177**, 685 (1984).
57. M. H. Kim and D. S. Ray, *J. Virol.* **53**, 871 (1985).
58. T. S. B. Yen and R. E. Webster, *JBC* **256**, 11259 (1981).
59. C. C. Richardson, *Cell* **33**, 315 (1983).
60. J. J. Dunn and F. W. Studier, *JMB* **166**, 477 (1983).
61. B. M. Wilkins, G. J. Boulnois and E. Lanka, *Nature* **290**, 217 (1981).
61a. G. J. Boulnois, B. M. Wilkins and E. Lanka, *NARes* **10**, 855 (1982).
61b. E. Lanka, R. Lurz, M. Kroger and J. P. Furste, *MGG* **194**, 65 (1984).
62. R. R. Isberg, A. L. Lazaar and M. Syvanen, *Cell* **30**, 873 (1982).
63. R. C. Johnson, J. C. P. Yin and W. S. Reznikoff, *Cell* **30**, 873 (1982).
64. R. A. Smith and J. S. Parkinson, *PNAS* **77**, 5370 (1980).
65. J. E. Shaw and H. Murialdo, *Nature* **283**, 30 (1980).
65a. P. M. Macdonald and G. Mosig, *EMBO J.* **3**, 2863 (1984).
66. E. Linney and M. Hayashi, *Nature NB* **245**, 6 (1973).
67. F. Sanger, G. M. Air, B. G. Barrell, N. L. Brown, A. R. Coulson, J. C. Fiddes, C. A. Hutchison III, P. M. Slocombe and M. Smith, *Nature* **265**, 687 (1977).
68. B. Francke and D. S. Ray, *JMB* **61**, 565 (1971).
69. T. J. Henry and R. Knippers, *PNAS* **71**, 1549 (1974).
70. J.-E. Ikeda, A. Yudelevich and J. Hurwitz, *PNAS* **73**, 2669 (1976).
71. H. Fujisawa and M. Hayashi, *J. Virol.* **19**, 416 (1976).
72. S. Eisenberg, J. Griffith and A. Kornberg, *PNAS* **74**, 3198 (1977).
73. S. A. Langeveld, A. D. M. van Mansfeld, P. D. Baas, H. S. Jansz, G. A. van Arkel and P. J. Weisbeek, *Nature* **271**, 417 (1978).
74. E. S. Tessman, *JMB* **17**, 218 (1966).
75. D. R. Brown, J. Hurwitz, D. Reinberg and S. L. Zipursky, *in* "Nucleases" (S. M. Linn and R. J. Roberts, eds.), p. 187. Cold Spring Harbor Laboratory, Cold Spring Harbor, New York, 1982.
76. S. Eisenberg and M. Finer, *NARes* **8**, 5305 (1980).
77. J.-E. Ikeda, A. Yudelevich, N. Shimamoto and J. Hurwitz, *JBC* **254**, 9416 (1979).
78. L. Dubeau and D. T. Denhardt, *BBA* **653**, 52 (1981).
79. S. Eisenberg and R. Ascarelli, *NARes* **9**, 1991 (1981).
80. S. A. Langeveld, A. D. M. van Mansfeld, A. van der Ende, J. H. van de Pol, G. A. van Arkel and P. J. Weisbeek, *NARes* **9**, 545 (1981).
81. A. van der Ende, S. A. Langeveld, G. A. van Arkel and P. J. Weisbeek, *EJB* **124**, 245 (1982).
82. A. D. M. van Mansfeld, H. A. A. M. van Teeffelen, J. Zandberg, P. D. Baas, H. S. Jansz, G. H. Veeneman and J. H. van Boom, *FEBS Lett.* **150**, 103 (1982).
83. D. F. Martin and G. N. Godson, *BBRC* **65**, 323 (1975).
84. W. Fulford and P. Model, *JMB* **178**, 137 (1984).
85. B. Alberts, L. Frey and H. Delius, *JMB* **68**, 139 (1972).
86. J. L. Oey and R. Knippers, *JMB* **68**, 125 (1972).
87. S. C. Kowalczykowski, D. G. Bear and P. H. von Hippel, *in* "The Enzymes" (P. D. Boyer, ed.), Vol. 14, p. 373. Academic Press, New York, 1981.
88. A. McPherson and G. D. Brayer, *in* "Biological Macromolecules and Assemblies. Volume 2: Nucleic Acids and Interactive Proteins" (F. A. Jurnak and A. McPherson, eds.), p. 523. Wiley, New York, 1985.
89. T. Cuypers, F. J. van der Ouderaa and W. W. de Jong, *BBRC* **59**, 557 (1974).

90. Y. Nakashima, A. K. Dunker, D. A. Marvin and W. Konigsberg, *FEBS Lett.* **40**, 290 (1974).
91. S. J. Cavalieri, K. E. Neet and D. A. Goldthwait, *JMB* **102**, 697 (1976).
92. H. T. Pretorius, M. Klein and L. A. Day, *JBC* **250**, 9262 (1975).
93. N. C. M. Alma, B. J. M. Harmsen, E. A. M. de Jong, J. v. d. Ven and C. W. Hilbers, *JMB* **163**, 47 (1983).
94. H. Bulsink, B. J. M. Harmsen and C. W. Hilbers, *J. Biomol. Struct. Dyn.* **3**, 227 (1985).
95. R. A. Anderson, Y. Nakashima and J. E. Coleman, *Bchem* **14**, 907 (1975).
96. J. E. Coleman, R. A. Anderson, R. G. Ratcliffe and I. M. Armitage, *Bchem* **15**, 5419 (1976).
97. J. E. Coleman and I. M. Armitage, *Bchem* **17**, 5038 (1978).
98. G. J. Garssen, G. I. Tessor, J. G. G. Schoenmakers and C. W. Hilbers, *BBA* **607**, 361 (1980).
99. N. C. M. Alma, B. J. M. Harmsen, J. H. van Boom, G. van der Marel and C. W. Hilbers, *EJB* **122**, 319 (1982).
100. G. D. Brayer and A. McPherson, *JMB* **169**, 565 (1983).
101. G. D. Brayer and A. McPherson, *Bchem* **23**, 340 (1984).
102. B. J. Mazur and P. Model, *JMB* **78**, 285 (1973).
103. B. J. Mazur and N. D. Zinder, *Virology* **68**, 490 (1975).
104. J. S. Salstrom and D. Pratt, *JMB* **61**, 489 (1971).
105. K. Geider and A. Kornberg, *JBC* **249**, 3999 (1974).
106. R. E. Webster and M. Rementer, *JMB* **139**, 393 (1980).
107. K. Campbell, G. Stormo and L. Gold, in "Gene Function in Prokaryotes" (J. Davies, J. Gallant and J. Beckwith, eds.), p. 185. Cold Spring Harbor Laboratory, Cold Spring Harbor, New York, 1983.
108. D. Doherty, P. Gauss and L. Gold, in "Multi-Functional Proteins" (J. F. Kane, ed.), p. 45. CRC Press, Cleveland, Ohio, 1982.
109. H. M. Krisch, A. Bolle and R. H. Epstein, *JMB* **88**, 89 (1974).
110. L. Gold, P. Z. O'Farrell and M. Russel, *JBC* **251**, 7251 (1976).
111. M. Russel, L. Gold, H. Morrissett and P. Z. O'Farrell, *JBC* **251**, 7263 (1976).
112. L. Gold, G. Lemaire, C. Martin, H. Morrissett, P. O'Connor, P. Z. O'Farrel, M. Russel and R. Shapiro, in "Nucleic Acid-Protein Recognition" (H. J. Vogel, ed.), p. 91. Academic Press, New York, 1977.
113. G. Lemaire, L. Gold and M. Yarus, *JMB* **126**, 73 (1978).
114. H. M. Krisch and B. Allet, *PNAS* **79**, 4937 (1982).
115. P. H. von Hippel, S. C. Kowalczykowski, N. Lonberg, J. W. Newport, L. S. Paul, G. D. Stormo and L. Gold, *JMB* **162**, 795 (1982).
116. M. Nomura, S. Jinks-Robertson and A. Miura, in "Interaction of Translational and Transcriptional Controls in the Regulation of Gene Expression" (M. Grunberg-Manago and B. Safer, eds.), p. 91. Elsevier, Amsterdam, 1982.
117. J. R. Lupski and G. N. Godson, *Cell* **39**, 251 (1984).
118. L. A. Day and R. L. Wiseman, in "The Single-Stranded DNA Phages" (D. T. Denhardt, D. Dressler and D. S. Ray, eds.), p. 605. Cold Spring Harbor Laboratory, Cold Spring Harbor, New York, 1978.
119. D. Pratt, H. Tzagoloff and J. Beaudion, *Virology* **39**, 42 (1969).
120. J. L. Woolford, H. M. Steinman and R. E. Webster, *Bchem* **16**, 2694 (1977).
121. M. E. Goldsmith and W. H. Konigsberg, *Bchem* **16**, 2868 (1977).
122. T.-C. Lin, R. E. Webster and W. Konigsberg, *JBC* **255**, 10331 (1980).
123. R. A. Grant, T.-C. Lin, W. Konigsberg and R. E. Webster, *JBC* **256**, 539 (1981).

124. G. F. M. Simons, R. N. H. Konings and J. G. G. Schoenmakers, *PNAS* **78**, 4194 (1981).
125. C. N. Chang, G. Blobel and P. Model, *PNAS* **75**, 75 (1978).
126. H. Schaller, E. Beck and M. Takanami, in "The Single-Stranded DNA Phages" (D. T. Denhardt, D. Dressler, and D. S. Ray, eds.), p. 139. Cold Spring Harbor Laboratory, Cold Spring Harbor, New York, 1978.
127. K. H. Sugimoto, H. Sugisaki, T. Okamoto and N. Takanami, *JMB* **111**, 487 (1977).
128. W. Wickner, *PNAS* **72**, 4749 (1975).
129. T. Date, J. M. Goodman and W. T. Wickner, *PNAS* **77**, 4669 (1980).
131. J. D. Boeke and P. Model, *PNAS* **79**, 5200 (1982).
132. H. Smilowitz, J. Carson and P. W. Robbins, *J. Supramol. Struct.* **1**, 8 (1972).
133. R. R. Paradiso and W. Konigsberg, *JBC* **257**, 1462 (1982).
134. J. Lopez and R. E. Webster, *Virology* **127**, 177 (1983).
135. J. R. Scott and N. D. Zinder, in "The Molecular Biology of Virus" (J. S. Colter and W. Paranchych, eds.), p. 212. Academic Press, New York, 1967.
136. M. Russel and P. Model, *J. Bact.* **154**, 1064 (1983).
137. M. Russel and P. Model, *J. Bact.* **157**, 526 (1984).
138. S. Brown, B. Albrechtsen, S. Pedersen and P. Klemm, *JMB* **162**, 283 (1982).
139. J. L. Pinkham and T. Platt, *NARes* **11**, 3531 (1983).
140. Y. Matsumoto, K. Shigesada, M. Hirano and M. Imai, *J. Bact.* **166**, 945 (1986).
141. M. Russel and P. Model, *PNAS* **82**, 29 (1985).
142. T. C. Laurent, E. C. Moore and P. Reichard, *JBC* **239**, 3436 (1964).
142a. E. L. Moore, P. Reichard and L. Thelander, *JMB* **239**, 3445 (1964).
143. A. Holmgren, *ARB* **54**, 237 (1985).
144. A. Holmgren, B.-O. Soderberg, H. Eklund and C.-I. Branden, *PNAS* **72**, 2305 (1975).
145. H. Eklund, C. Cambillau, B.-M. Sjoberg, A. Holmgren, H. Jornvall, J.-O. Hoog and C.-I. Branden, *EMBO J.* **3**, 1443 (1984).
146. L. Stryer, A. Holmgren and P. Reichard, *Bchem.* **6**, 1016 (1967).
147. H. Reutimann, B. Straub, P.-L. Luisi and A. Holmgren, *JBC* **256**, 6796 (1981).
148. D. F. Mark and C. C. Richardson, *PNAS* **73**, 780 (1976).
148a. P. Modrich and C. C. Richardson, *JBC* **250**, 5508 (1975).
149. A. Holmgren, I. Ohlsson and M.-L. Grankvist, *JBC* **253**, 430 (1978).
150. M. Russel and P. Model, *J. Bact.* **159**, 1034 (1984).
151. A. Holmgren, *PNAS* **73**, 2275 (1976).
152. M. Russel and P. Model, *J. Bact.* **163**, 238 (1985).
153. M. Chamberlin, *J. Virol.* **14**, 509 (1974).
154. A. Holmgren, G.-B. Kallis and B. Nordstrom, *JBC* **256**, 3118 (1981).
156. J. I. Horabin and R. E. Webster, *JMB* **188**, 403 (1986).
157. B. B. Buchanan, R. A. Wolosiuk and P. Schurmann, *TIBS* **4**, 93 (1979).
158. J. F. Grippo, W. Tienrungroj, M. K. Dahmer, P. R. Housley and W. B. Pratt, *JBC* **258**, 13658 (1983).
159. T. Hunt, P. Herbert, E. A. Campbell, C. Delidakis and R. J. Jackson, *EJB* **131**, 302 (1983).
160. B. J. Bachmann, *Microbiol. Rev.* **47**, 180 (1983).
161. N. D. F. Grindley and R. R. Reed, *ARB* **54**, 863 (1985).
162. B. L. Haller and J. A. Fuchs, *J. Bact.* **159**, 1060 (1984).

Roles of Double-Strand Breaks in Generalized Genetic Recombination

FRANKLIN W. STAHL

Institute of Molecular Biology
University of Oregon
Eugene, Oregon 97403

I. The Role of Phage λ in Recombination Studies

In 1961, Meselson and Weigle demonstrated the excellence of phage λ for studies on genetic recombination (1). Among the progeny of a lytic-cycle phage-cross there are viable particles whose chromosomes did not replicate during that cycle (nonreplicated chromosomes, or nonreplicates).[1] When the infecting phages are density-labeled (^{13}C and ^{15}N), the nonreplicates can be detected in cesium density gradients as particles with new (light) protein coats but fully heavy (HH) chromosomes (Fig. 1). Among them are particles genetically recombinant between markers (in genes cI and R; Fig. 2) located near λ's right end (as the virion chromosome is conventionally displayed). The existence of these particles permits the conclusion that DNA as well as genetic specificity is donated from parents to recombinants; copy-choice (Fig. 3) can be eliminated as the (exclusive) mechanism for genetic recombination in λ. The strength of this conclusion lies in the ability to assess both the material (density label) and genotypic constitution of individual progeny particles and to do so for great numbers of particles.

Because of the terminal disposition of the markers used by Meselson and Weigle, their crosses failed to reveal whether the observed nonreplicated recombinants were the products of breaking and joining, or whether they might have arisen as a result of a "break-copy" act (Fig. 3) (see Section IV,B,1). Were the latter mechanism operating, one of the recombinants would have only a minor content of newly incorporated atoms and might be indistinguishable from an HH particle, while the complementary recombinant would be so light as to be indistinguishable from the replicates. In 1964, through the use of markers located more symmetrically (in cI and J; Fig. 2), Meselson

---

[1] Called "free-loaders" in ref. 11 [Ed.].

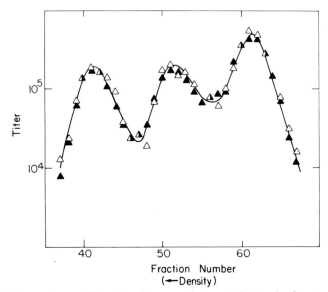

FIG. 1. Isopycnic centrifugation of progeny from an infection by density-labeled λ. The three peaks correspond to particles carrying a fully-heavy (HH) chromosome (on the left), a half-heavy (HL) chromosome and a fully-light (LL) chromosome (on the right). This progeny was from a cross in which DNA replication was partially blocked so that the nonreplicated (HH) peak is a large fraction of the total progeny. The two infecting genotypes, distinguishable by a marker in cI, were separately enumerated and were recorded by different symbols.

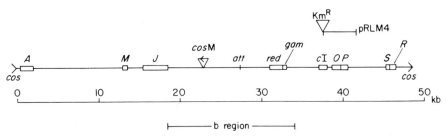

FIG. 2. The chromosome of phage λ showing only the features to which reference is made in this review. The chromosome is shown in the form present in mature phage particles, i.e., open at cos. The boxes show the locations and sizes of the indicated genes. cosM is a cos site cloned in the b region in inverse orientation. The polarity of cos and of cosM indicated by the arrowheads corresponds to the direction of injection and packaging dictated by that cos. Plasmid pRLM4 is a piece of λ (from cI through P) bearing a gene for kanamycin resistance (Km^R). att is the site of action of the Int recombination system (see Table I). The scale is in units of thousands of nucleotide pairs (kb).

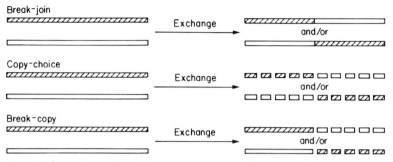

FIG. 3. Three (nonmolecular) concepts of generalized recombination. In "break-join," recombinant molecules are made primarily of DNA derived from the two interacting parents. The scheme allows modest amounts of "repair synthesis" at the joint depending on molecular details. In "copy-choice," recombinant molecules inherit only information from the parent molecules. "Break-copy" schemes call for a substantial contribution of DNA from one parent with the other parent providing only information. The duplex character of DNA and the semiconservative nature of its replication limits the usefulness of this historically valuable classification.

sought to determine which mechanism, "break-join" or "break-copy," is operating (2). Both parents were density labeled, and fully heavy recombinants were found in the nonreplicated peak. Thus it was concluded that recombination can occur by breaking and rejoining segments of duplex DNA. Subsequent investigation revealed that *att*, the site of action of the specialized Int recombination system of λ, is in the *J*–*c*I interval. Thus the applicability of Meselson's conclusion to generalized recombination was put into doubt.

Several developments in our understanding of λ growth and recombination permitted a deeper investigation of density transfer crosses. (1) Techniques were developed that reduce DNA replication so that nonreplicates become a major part (3–5) or even the only part (6) of a cross progeny. (2) Mutants defective in recombination functions were discovered (7–9). The former development, by reducing or even eliminating replicated phage, made it possible to ask whether recombination ever goes by routes that involve major amounts of DNA synthesis, while the latter development allowed the investigation of generalized recombination via its several pathways separately and free of contributions by Int. These studies show that double-chain breaks figure prominently in the processes of generalized recombination in λ, and that they do so differently for different pathways. Before we examine the experiments leading to these conclusions, we shall set their context with a brief description of the behavior of λ's chromosome replicating in the lytic cycle.

II. The Lytic Cycle of Phage λ (*10*)

The chromosome in a mature λ particle is a Watson–Crick duplex of 48.5 kilobase-pairs (kb). At its 5' end, each of the two chains projects beyond the duplex. These projecting 5' termini are complementary to each other ("sticky ends"). When the chromosome is injected into a bacterial cell, these ends anneal, and the resulting circular duplex is covalently closed by bacterial ligase. The region of union is *cos*. Following the expression of λ genes *O* and *P*, whose protein products adapt the *E. coli* replication complex to λ, the DNA replicates in either of two modes. In the "theta" mode, diverging replication forks are initiated at *ori*, in gene *O*.[2] When the forks meet, the sister duplexes are separated as individual, monomeric circles. In the "sigma" or "rolling circle" mode, one replication fork operates continuously, generating a linear "tail" of iterated λ chromosomes (a "concatemer")[3] as a result of repeated replication of the λ circle. Theta replication is short-lived, so that DNA replication later in the cycle is normally in the sigma mode. The expression of λ's late genes provides the structures and enzymes for packaging λ DNA into phage heads. Head-precursors free of DNA ("proheads") and an enzyme ("terminase") that cuts λ DNA at *cos* are the principal actors.[2]

The concatemeric products of sigma replication are the normal precursors of packaged DNA. Terminase binds at *cos* and nicks each chain so as to create a pair of sticky ends. The terminase remains bound to the right of the cut (i.e., near the left end of λ as conventionally drawn; see Fig. 2) where it provides an anchor for a prohead. As the DNA is moved past terminase into the prohead, terminase scans for a second *cos* located approximately the length of one λ chromosome from the initiating *cos*. When it finds such an intact *cos* site, it cuts there, and maturation of the now-filled head proceeds while the terminase guides packaging of the next chromosome in the concatemer.

Nonreplicates (by definition) are not packaged from the products of DNA replication. Consequently, they are dependent on genetic recombination (either generalized or Int-mediated) to achieve a dimeric state (*11*). One or the other monomer can then be packaged from the dimer (*12*).

---

[2] See Table I for definitions of terms peculiar to λ recombination.

[3] "Concatemer," which is not in standard dictionaries, has been illustrated by "multiple units of λ joined covalently head to tail" and "an immediate precursor of the DNA of functional phage particles" (*10*, p. 151). The term seems to have originated in 1967 (*10a*) and was discussed and further defined in 1968 (*10b*) [Eds.].

The rolling-circle replication of λ can be blocked by exonuclease V (Exo V), one of the activities of the product of *E. coli's recB* and *recC* genes (*13*). Wild-type λ produces an inhibitor of ExoV; *gam* mutants of λ fail to do so (*14*). As a consequence, λ *gam* is dependent upon recombination genes for viability (*15*); only chromosomes that have become components of concatemers by the activity of recombination gene products can be packaged. Consequently, Gam⁻ phages that are defective for λ recombination (Red⁻) and are growing in recombination-defective *E. coli* (Rec⁻ by virtue of a *recA* mutation) give a burst too small to support plaque formation. The burst is even less if the phages are additionally Int⁻.

Because of the dependence of packageability on recombination functions, recombinant frequencies among mature phage progeny do not directly measure recombination activity. This is true both for nonreplicate particles and for total particles from crosses in which rolling-circle replication is inhibited by unbridled ExoV activity.

The involvement of the Rec and the Red systems in the production of concatemeric DNA may differ. As we explain below, the Red system is dependent on DNA replication for its proper functioning, while the principal pathway (RecBC) of the Rec system fails to give similar evidence of being replication dependent. For Red, there may be a mutual interdependence between recombination and replication, which renders the significance of recombinant frequencies yet harder to assess (*16*).

III. Recombination of Nonreplicated λ Chromosomes

DNA replication can be severely blocked with a *dnaB-ts* mutation at high temperature (*3, 17*) or, at lower temperatures, with the *ts* mutation and an amber mutation in genes *O* or *P* of λ (*6*). When replication is so blocked, density-labeled particles produce a progeny that is unimodal in a cesium gradient. The single peak corresponds to an essentially fully heavy chromosome in a new, light capsid; all the progeny particles are nonreplicates. Since the chromosome density is well conserved (little contribution from the growth medium), crosses between heavy and light parents yield progeny particles, the density of each directly revealing its material heritage from the two parent classes. When the parents are genetically marked as well, each particle whose material heritage is given by its position in the cesium gradient can be scored for its informational heritage (genotype). When the two parents carry selectable markers near opposite ends of the chromosome, the density distribution of the selected recombinants is a direct display of the distribution of exchanges along the chromosome (18, 19).

Among the features of these distributions, two are especially relevant to this review. (1) When the *red* genes of λ are wild-type and the *recB/C* gene product is inactive (by mutation or by Gam inhibition), exchanges are strongly clustered near the ends. In Red⁺ RecBC⁻ crosses that were RecA⁻, exchanges were exclusively near the right end, while in *recA*⁺ crosses both ends enjoyed a high density of exchanges. Furthermore, the total exchange rate (yield of phage) in the RecA⁺ is about 100 times that in the RecA⁻ crosses (19). (2) On the other hand, when the RecBC pathway was operating (λ *red gam* in *rec*⁺ cells), there was no tendency for exchanges to be located near the termini (19).

We turn now to an analysis of Red-dependent recombination near λ's ends. In several respects Red-dependent recombination in *rec*⁺ cells is similar to that in RecA⁻ cells. However, analysis of the RecA⁻ crosses has proceeded farther and the data have yielded to a simple interpretation. Therefore, we confine our discussion to the RecA⁻ picture.

IV. Recombination by λ's Red System

A. Heteroduplexes

Among nonreplicating chromosomes, Red-dependent recombination in RecA⁻ bacteria is confined to λ's 3′ end. Most of the DNA and most of the genetic information is contributed from one parent, while the other parent makes a minor contribution, which is typically confined to one of the two chains of the duplex. This single-chain contribution results in chromosomes that are heteroduplex for markers located at λ's right end. The heteroduplexes are signaled by the production of genetically mixed progeny from individual recombinant particles (20, 21). The application of a method devised by White and Fox (21) revealed that most or all of these recombinant heteroduplex molecules have the structure diagrammed in Fig. 4. The chromosomes have a smaller, single-chain contribution from one parent, and that contribution is always on the polynucleotide chain that has a 5′ overhang at the right end (22).

B. Involvement of *cos*

The high concentration of Red-dependent recombination of nonreplicated chromosomes at λ's right end suggests that it is the end itself

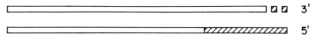

FIG. 4. Structure of a Red-mediated recombinant arising at the right end of a nonreplicating λ. Most of the DNA comes from one parent. The other parent makes a single-chain DNA contribution at the right end. This contribution is always on the chain that ends 5' at the right. As described in Section IV,D,3, there is sometimes newly synthesized DNA on the other chain near the right end.

that stimulates recombination. This view was supported by means of crosses with λ strains in which the standard *cos* sites were deleted and their function replaced by *cos* sites cloned in the *b* region (see Fig. 2). In these crosses, the high rate of nonreplicator recombination normally seen between genes *P* and *S* near the right end was abolished (*23*).

By what mechanism does *cos* increase the exchange rate in its vicinity? Two classes of models were considered.

1. THE "BREAK-COPY" MODEL

The Red-mediated distribution of exchanges on nonreplicated chromosomes is different from that on replicated chromosomes. The latter cannot be visualized with density labels, but can be inferred from relative frequencies of recombination in various genetically marked intervals. Whereas recombinants for central and left intervals are lacking among nonreplicates, they enjoy at least roughly proportional representation among progeny particles bearing replicated chromosomes (*24*). This statement can be rephrased to say that exchanges occurring far from the right end of the chromosome are replication-dependent, while those occurring close to the right end are not so, or are less so. The "break-copy" model (Fig. 3) was invoked to account simultaneously for the replication-dependence of Red-catalyzed recombination and the feature that this dependence decreased with proximity to the (right) end.

Initiation (cutting) is imagined to occur with uniform probability along the chromosome. Only recombinants initiated near *cos* are recovered as nonreplicates, since only those have negligible amounts of new DNA incorporated in the finished chromosome. Experiments involving total blocks on replication have used P⁻ λ and *dnaB*ts *E. coli*. Polymerases in these crosses have been wild-type. Thus, these crosses are blocked for replication-fork formation but not for chain extension. Those recombinants that initiate near *cos* can be completed (i.e., copied to *cos*) by the chain-extension activity of polymerase alone. Recombinants initiated further from *cos* need replication forks to get to *cos*. In this model, the recombination-stimulating activity of *cos* is not to initiate exchange but to signal the successful completion of one. When replication (or chain extension) has reached *cos*, then (and only then) can the recombinant be packaged and scored as a member of the progeny.

2. Initiation Model

In these models, recombination is initiated by a double chain cut at *cos*. Such a cut and the ensuing exchange are independent of replication fork activity and account for the high rate of exchange at the right end among nonreplicators. Exchange remote from the right end is replication-dependent for either of two reasons: the exchange mechanism may be break-copy, or replication may promote the initiation of recombination (for instance, the tip of the rolling-circle tail might be recombinogenic in the same fashion that *cos* is proposed to be).

3. Distinguishing the Models

The initiation model was supported by showing that a *cos* stimulates nearby exchange even when that *cos* is not used to package the resulting recombinant. The critical experiment, described below, revealed other features of *cos*-stimulated Red-mediated recombination as well (25).

The parents (Fig. 5) were marked near their right ends. Each contained a medially located, cloned *cos*; in one parent (Experiment 2) or the other (Experiment 3) the standard *cos* was deleted. The cross was in a *dnaB*ts host so that many of the progeny particles were nonreplicators. In both crosses many more such recombinants were seen than were seen when both standard *cos* sites were deleted. Essentially all of the recombinants produced in Experiments 2 and 3 were deleted for *cos* at the standard locus. The recombinants, evidently, were all packaged beginning at the cloned *cos* in one parent and ending at the cloned *cos* in the other. This set of experiments revealed not only that the recombination-stimulating role of *cos* is separable from its role in packaging, but also that Red-mediated recombination *can* proceed by a break-join mechanism. Thus, break-copy models for Red recombination lose on two counts, and one may incline to the view that the replication-dependence of non-*cos*-stimulated recombination reflects a replication-generated initiation structure.

ROLES OF DOUBLE-STRAND BREAKS IN RECOMBINATION 177

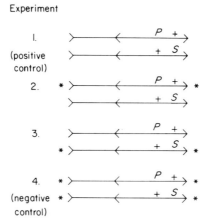

FIG. 5. Crosses between mutants in genes P and S that demonstrate a recombination-stimulating role for cos in Red-mediated recombination. In Experiments 2 and 3, only one parent has a functional standard cos. Nevertheless, nonreplicator P+S+ recombinants are stimulated, so that these two crosses give more nonreplicator P+S+ than does Experiment 4. The asterisk denotes nonfunctional cos.

C. A Detailed Model for Initiation by cos

The properties of cos-stimulated exchange described above allowed us to postulate a sequence of events that produces those exchanges (Fig. 6). The protein components of the Red system are an exonuclease (λ exo, EC 3.1.11.3) and a helix-destabilizing protein (beta). The exonuclease digests DNA in the 5′ to 3′ direction and is specific for double-chain DNA. It can operate slowly on the 12-bp overhangs at the ends of chromosomes from mature particles. Our model (25) presumes that intracellular λ, in the circular, nonreplicating state (Fig. 6a), is cut at cos by terminase. Terminase remains bound to the left end of the linearized chromosome. (A prohead may become bound to terminase, and packaging may proceed rightward.) The left end of the chromosome is, consequently, unavailable for recombination (Fig. 6). At the 3′ end, λ exo initiates digestion, which leads to a 3′ overhang (Fig. 6c). The overhang is taken up by a circular, supercoiled λ. As digestion continues, the resulting D-loop enlarges (leftward in Fig. 6d). Beta protein, complexed with exo and moving with it, facilitates the loop enlargement by promoting the transfer of a chain from the recipient duplex to the growing heteroduplex (26). The loop enlargement ends "accidentally" upon encountering a nick (n) in the chain being transferred (Fig. 6e). The λ exo is then dislodged by the transferred chain, since the enzyme is unable to bind

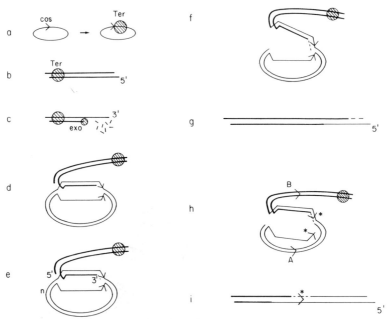

FIG. 6. A model for *cos*-stimulated, Red-mediated exchange (25, 26) (see Section IV,C). The sequence of events depicted appears to be the simplest one that is consonant with the λ life cycle, the properties of the relevant proteins, and the structures of *cos*-stimulated, Red-dependent recombinant chromosomes. The two interacting λ chromosomes are distinguished by heavy and thin lines; dashed lines represent newly synthesized DNA. *cos*, the packaging origin of λ, is symbolized by an arrowhead whose direction indicates the direction of packaging of the λ chromosome. *cos* with an asterisk is a mutant, nonfunctional *cos*. Proteins terminase (Ter) and λ exonuclease (exo) are shown as shaded circles.

at a nick (Fig. 6f). The accidental nature of termination results in the observed heteroduplex length distribution, which is more or less exponential from the right end leftward.

The recombinant becomes completed and packageable upon acquisition at the right end of the D-loop of a complete, cuttable *cos* site covalently continuous with it. The easiest way to do this appears to be chain extension primed by the invading 3' end (Fig. 6e, f, and g). As we describe below, breakage of the displaced chain followed by joining to the invading 3' end is an alternative.

When the invaded parent has a nonfunctional, deleted *cos*, packaging cannot be completed using that *cos*. Thus chain extension from the 3' invading end, though it may reach the mutant *cos*, does not suffice to give a completed packageable nonreplicated recombinant. Instead, breakage of the displaced chain with joining to the invading end must occur (Fig. 6h). A plausible enzyme for the breakage is the nick-translating activity of polymerase I (26). The resulting recombinant (Fig. 6i) can be packaged from secondary *cos* sites (A and B) cloned centrally in the two parents. This recombinant is a conserved Red-mediated recombinant with a medially placed point of exchange induced by the double-chain cut site, *cos*. The gene order of the recombinant is a permutation of the normal order. The enzymatic postulates of this model are yet to be tested. However, as described below, genetic predictions have been tested, and the model remains intact.

D. Tests of the Model

1. Exchange Stimulated by Restriction Enzyme Cuts

The central feature of the model for *cos*-stimulated Red-mediated recombination is that the initiating event is the double-chain cut at *cos*. Thus the model predicts that double chain cuts made by enzymes other than terminase will be recombination-stimulating in λ's Red pathway. This prediction has been verified by D. S. Thaler (personal communication) with two types of crosses. (1) Crosses between genetically marked λ strains were conducted in a bacterial host carrying one or another restriction/modification system (on a plasmid). One parent in the cross had previously been modified, so that it was uncuttable during the cross. The other parent, which had just one restriction site, was not modified.

In these crosses, recombination was elevated in the neighborhood of the restriction site. As with *cos*, the stimulation spread from the site for many thousands of nucleotides. (2) Crosses between unmodified λ and the λ-derived plasmid KC31 (similar to pRLM4; 27) were conducted in the restricting host. The λ carried a mutation (in gene *P*) in the part of its chromosome homologous with KC31. Among the progeny, particles that had picked up the wild-type allele were detected. Some particles carried chromosomes that had done so by splicing the entire λdv into themselves (creating a tandem duplication). Others had simply patched in the wild-type P allele without incorporating the KC31. When the λ carried an unmodified restriction site within the region homologous with KC31, both splices and patches were stimulated. When an unmodified site was outside the region of homology, no stimulation of recombination with the KC31 was seen. The recombinational behavior of this system was like that reported for double-chain-break repair in yeast (Section IV,E3). By showing that *Eco*RI, which makes double-chain cuts, can initiate Red-mediated recombination, these experiments support the assumption that terminase initiates recombination at *cos* via a double-chain cut.

2. Nonreciprocal Exchange at *cos*

According to the model, terminase initiates recombination by binding to *cos* and cutting, just to its left. Terminase may then remain bound to *cos* or it may, itself, bind a prohead and track the DNA as the DNA is packaged, from left to right, into the bound head. In either case, the left end of the chromosome is unable to participate in recombination. This feature of the model was suggested by the failure to see Red-mediated recombinants near the left end of λ (to the right of gene A) (see Fig. 2) among nonreplicating phage. That failure, however, could mean, instead, that left-end recombination among nonreplicators is restricted to the short interval between *cos* and A, where they would have been undetected.

Replication-blocked crosses involving genes M and cI (see Fig. 2) were performed to distinguish the two possibilities. In these crosses, one parent or the other was mutant for standard *cos*. Recombinants arising by exchanges stimulated by the wild-type *cos* were packaged (*in vivo*) by cloned *cos* sites (*cos*M, see Fig. 2). In each cross, complementary recombinants differed by a factor of 20. The predominant recombinant was the one in which the cI gene came from the parent with the functional standard *cos*, while the M gene came from the parent with the mutant *cos* (26).

3. DNA Synthesis at *cos*

As cited above, *cos* in one of the two parents can stimulate exchange to produce nonreplicating recombinants that are packaged by other, remote *cos* sites. Within the framework of the model, this result demonstrates the break-join (Fig. 3) feature of the model illustrated in Fig. 6h. Since the right end of the D-loop can apparently be resolved by break-join, there is no logical requirement for the synthesis that resolves the structure (i.e., reaches *cos*) in Fig. 6f. However, resolution via such synthesis seems simple; the experiments described below give substance to the idea.

In Fig. 6f it is apparent that the amount of 3'-primed synthesis required to acquire *cos* will depend on the extent to which the invading 3' end has been degraded, presumably prior to its successful invasion of the homologue. For a marker located near λ's right end, its appearance in the heteroduplex state will depend on whether degradation of the 3' chain progressed leftward beyond the marker. Thus the probability of homozygosity will correlate positively with the amount of DNA synthesis manifested by a progeny particle. Such a correlation was found when nonreplicator phage were examined at loci R and cI. cI heterozygotes were more frequent on the heavy than on the light side of a nonreplicator peak. Among cI heterozygotes, R-heterozygotes were likewise more frequent on the heavy than on the light side (28). It is unlikely that the correlation of DNA synthesis with homozygosity is a result of mismatch correction. Our model (Fig. 6g) predicts that cI heterozygotes that are homozygous at J (to the left of cI) and at R (near the right end) will be recombinant for J and R; the explanation of mismatch correction makes no such prediction. Experiments support the prediction of the model (28).

E. Double-Chain Breaks as Initiators Elsewhere

1. RED-LIKE PATHWAYS IN *E. coli*

In *E. coli*, mutations in the *recB* or *recC* genes reduce recombination to about 1% of the rec^+ level, as measured in bacterial conjugation or in phage-mediated transduction. Two kinds of extragenic suppressor can restore recombination ability. *sbcA* mutations restore recombination by allowing the production of an otherwise unexpressed recombination protein from a cryptic lambdoid prophage. *sbcB* mutations inactivate exonuclease I of *E. coli* (EC 3.1.11.1). [For a summary of the properties of these substitute recombination pathways, see the review by A. J. Clark (29).] Of relevance to this review are the observations that each of these pathways (called RecE and RecF, respectively) responds to double-chain cuts as does Red. Each shows a high rate of recombination near *cos* among nonreplicators (30; M. M. Stahl, personal communication), and each responds to restriction enzyme cuts by undergoing exchange in the vicinity of the cut (D. S. Thaler, personal communication). It is no surprise that the RecE pathway mimics Red (30)—these two pathways are homologous. However, the observation that RecF is similar to the other two pathways legitimately generalizes the conclusion that double-chain cuts initiate exchange.

The properties of λ cited earlier make λ an unusually good subject for establishing the properties and determining the mechanism(s) of recombination initiated by double-chain breaks. However, clear evidence that double-chain breaks can initiate recombination was obtained in several systems prior to its investigation in λ.

2. Phage T4

In phage T4, the recombination-stimulating nature of DNA duplex ends (i.e., double-chain cuts) has long been evident. The work of Doermann (31) showed that terminal redundancies in T4 recombine at a higher rate than do those in the nonterminal portion of the same chromosome. Other work (32, 32a) examined recombination in abnormally short chromosomes, which lack redundancies. From one particle to another, the ends of these defective chromosomes are in different parts of the map. For each particle, the rate of recombination near its ends was high, leaving no doubt that the ends themselves were responsible for the recombination.

3. Yeast

The recombination-stimulating nature of double-chain breaks in eukaryotes was argued by Resnick (33) on the basis of quantitative analysis of X-ray-induced recombination in yeast. A more compelling demonstration was achieved in investigations of recombinations between yeast chromosomal DNA and transforming plasmids, incapable of replicating autonomously in yeast, into which yeast fragments had been cloned (34, 35). In the most telling experiments (35) the plasmid carried two different yeast fragments. Transformation occurred by incorporation of the plasmid at the locus of one fragment or the other with comparable frequency. However, when a double-chain cut was introduced into one fragment in the plasmid, the overall rate of transformation was increased, and the plasmid incorporation was invariably at the locus of the cut fragment. In a variation of this experiment, the transforming plasmid was cut at two places in a cloned yeast gene. The piece of the gene released from the plasmid was discarded. Following transformation, the plasmid was found at the locus of the cut fragment, and it was flanked by a pair of fully intact copies of the chromosomal region corresponding to the cloned fragment. Thus, in association with the exchange that resulted in plasmid incorporation into the chromosome, the gap in the cloned gene was repaired. A reaction more-or-less like that shown in Fig. 7 is strongly implied. A feature of this reaction, whatever the mechanism, is that the gap is filled by DNA derived from the yeast chromosome. The loss of information from the plasmid (by restriction enzyme cutting) is compensated for by a donation from the chromosome in a reaction whose net result is the formal equivalent of gene conversion. This association of conversion with exchange is reminiscent of meiotic recombination

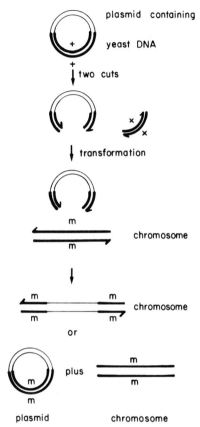

FIG. 7. Double-chain break repair of a plasmid in yeast (35). *m* is a mutant gene, and + is its wild-type allele.

and led to the proposal that double-chain breaks are normal meiotic initiation events (36).

In meiosis, conversion is not always accompanied by the recombination of DNA flanking the converted site. Half (or more) of the time, converted loci retain parental configurations of flanking markers. The parallelism between plasmid gap repair and meiotic conversion was extended by the demonstration that plasmid gap repair can occur without incorporation of the plasmid into the chromosome, an outcome equivalent to the retention of a parental configuration of flanking markers (37).

In yeast, a "normal" role for double-chain breaks as initiators of recombination is illustrated by mating-type switching (38). A double-chain cut at *MAT*, the mating type locus, is made by the sequence-specific nuclease product of the *HO* gene. The cut locus "repairs" itself from one of the silent copies of mating-type information stored at the *HMR* and *HML* loci, each at some distance from *MAT* on the same chromosome. In so doing, *MAT* is converted from the allele for one mating type to that for the other. This conversion of *MAT* by a silent-copy locus differs from meiotic conversion in that it is rarely, if ever, accompanied by exchange, which would lead to a gross chromosomal rearrangement. If the silent copies of mating-type information have been deleted, then *HO*-induced cutting of *MAT* in diploids stimulates mitotic conversion of the cut chromosome by the uncut homologue. This isolocal conversion is frequently accompanied by exchange (*38a*).

Meiotic conversion at *MAT* is likewise inducible by *HO*, and when induced conversion occurs it is frequently accompanied by exchange (A. Kolodkin, A. Klar, and F. W. Stahl, unpublished; A. Klar, unpublished). Thus, double-chain breaks can initiate meiotic conversion and crossing over that is reminiscent of the spontaneous events, and it may be that double-chain breaks are routinely the meiotic initiating events. Reasons for suspecting so have been presented elsewhere (*36*).

However, Hastings (*39*) has called attention to shortcomings of this hypothesis. He argues that the facts of yeast recombination (and those of other fungi) are better understood by supposing that exchanges are usually initiated by single-chain breaks, as in the Aviemore variation (*40*) of Holliday's (*41*) scheme (Fig. 8). The nonreciprocal transfer of a segment of one chain of a nicked duplex to the other duplex results, after appropriate DNA synthesis and degradation, in a Holliday structure in which, in yeast, one and only one of the two participating duplexes has a segment of biparental origin (Fig. 8). If the biparental segment happens to lie over a genetic marker, then it will be heteroduplex DNA. Hastings then proposes that most mismatches in yeast are recognized by a zealous correction system that degrades both chains of a heteroduplex. The result is a duplex with a double-chain gap. That duplex, Hastings proposes, then invades its homologue, and the gap repair reaction results in its conversion. The overall process results in exchange of flanking markers about half the time. This scheme differs from the original double-chain-break repair model (*36*) by postulating that double-chain breaks arise primarily in the course of mismatch repair. For any event in which the biparental segment of the Aviemore model does not fall upon a marker (or on a cryptic point of sequence difference between the participants), recombination proceeds without the involvement of double-chain breaks.

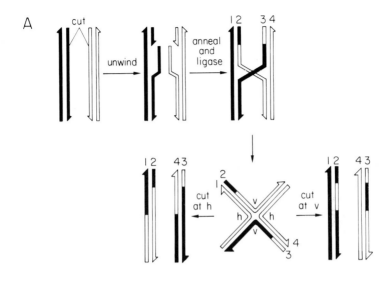

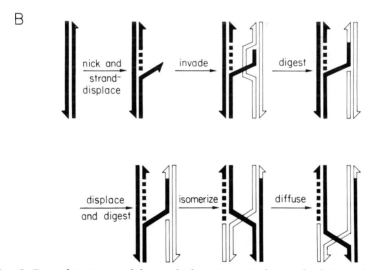

Fig. 8. Recombination models in which initiation is by single-chain nicks. (A) Holliday model. (B) Aviemore model. The Aviemore model (40) is an asymmetric version of Holliday's model (11). The final, cross-strand structure shown in B is resolved like the fully symmetric one in A.

4. Other Suspects

As described above (Sections IV,B, C, and F,3), when recombination is initiated by a double-chain break, the site of the break is lost—it undergoes conversion by repair off a homologue. One may speculate that the reverse is true—when a recombinator is characteristically converted during recombination, that recombinator is acting as an initiator of double-chain-break recombination. When so viewed, *cog* of *Neurospora crassa* (42) and *M26* of *Schizosaccharomyces pombe* (43) are indicted as double-chain cut sites. Chi, on the other hand (Section V), has no such self-sacrificing tendency (43a).

V. *E. coli's* RecBC Pathway

As described in Section II, λ *red gam* DNA replicates in $recBC^+$ cells in the theta mode and is consequently dependent for packaging upon dimerization by crossing over. Only a small fraction of intracellular DNA is packaged, and the burst size and plaque size are correspondingly small. In a cross, however, the frequency of recombinants within that small yield is understandably high. The conclusions of the following section are two: (1) RecBC acts poorly on λ, because λ lacks a sequence (Chi) that stimulates the RecBC pathway, and (2) the ability of RecBC to act on λ depends strongly on the occasional linearization of λ due to cutting of *cos* by terminase, i.e., the RecBC pathway requires double-chain breaks.

A. Discovery of Chi

Chi (see Table I), which has a conspicuous phenotype in λ only in *red gam* double mutants, was first encountered in the following two circumstances. (1) One class of λ-transducing phage arises *in vivo* when λ prophage is excised so as to lose its *red* and *gam* genes and to pick up in their place a segment of bacterial DNA. This well-known class of phage (λ*pbio*) makes large plaques on rec^+ bacteria in contrast to the small plaques made by other *red gam* mutants. (2) When λ *red gam* that forms small plaques is repeatedly subcultured on rec^+ bacteria, variants forming large plaques accumulate. These variants are the result of mutation (Chi) at any one of several spots on the λ chromosomes (44, 45). The common nature of the large-plaque phenotype for these two classes of λ *red gam* was shown as follows. λ*pbio* was observed to have an extraordinary rate of RecBC-mediated recombination in and near the *bio* substitution, while each of the selected large-plaque variants of λ *red gam* has a high RecBC-mediated recom-

TABLE I
Glossary of Terms Peculiar to the Study of Genetic Recombination in Phage λ

Term	Definition
Sites on λ DNA	
cos	Packaging origin for the λ chromosome; site of action of the enzyme "terminase"; sequence that separates one λ chromosome from another in the concatemer
att	Site of action of the specialized recombination enzyme "integrase," which acts to recombine one λ with another, or with the bacterial chromosome
ori	The origin of replication of λ DNA
Chi (χ)	Site of action of the generalized *E. coli* "recombination pathway" that depends on *recA* protein and on the *recB* and *recC* proteins (the RecBC pathway)
Enzymes and other proteins	
Terminase	Sequence-specific endonuclease (product of λ genes *A* and *Nu1*) that cuts λ DNA at *cos* sites to generate the "sticky" ends of the linear virion chromosome (see 58)
Integrase	Sequence-specific polynucleotide transferase (product of λ gene *int*) that recombines λ chromosomes only at site *att* (see 10, p. 211)
Exonuclease V (Exo V)	Exodeoxyribonuclease V (EC 3.1.11.5): one of the activities of the multifunctional *recBC* protein
Excisionase	Product (enzyme) of the λ gene *xis*. Part of the "Int system" (see below). With integrase, promotes the separation of the λ chromosome from the bacterial chromosome (see 10, p. 211)
recBC protein	The protein (product of the *recB* and *recC* genes of *E. coli*) whose activity defines the RecBC recombination pathway. The protein has nucleolytic activities and a DNA-unwinding (topoisomerase) activity
recA protein	The product of the *recA* gene of *E. coli*. This protein promotes Watson–Crick pairing between single-stranded DNA and double-stranded DNA. Elimination of this protein (by mutation) eliminates essentially all *E. coli* recombination activity as measured in bacterial conjugation or phage-mediated transduction
red exonuclease (λ exodeoxyribonuclease, EC 3.1.11.3)	Product of the *redα* gene of λ; a double-chain-specific, 5′-to-3′ processive exonuclease

(*continued*)

TABLE I (Continued)

Term	Definition
red beta protein	Product of the *redβ* gene of λ. It is a DNA-helix-destabilizing protein isolated from λ-infected cells as a complex with *red* exonuclease
gam protein	Product of the *gam* gene of λ; it inhibits all activities of the *recBC* protein so far examined
Systems and pathways	
Rec system	The ensemble of *E. coli* recombination pathways described below
RecBC pathway	The principal enzymatic pathway for genetic recombination in wild-type *E. coli*. The enzyme components identified are *recBC* and *recA* proteins
RecF pathway	In bacteria deficient for *recBC* protein, mutations (*sbcB*) that eliminate *E. coli*'s exonuclease I restore recombination proficiency by activating this pathway, whose enzymatic components include the *recA* protein
RecE pathway	Some strains of *E. coli* carry a silent fragment of a lambdoid phage. Activation of transcription of this cryptic, defective prophage (by *sbcA* mutations) turns on the RecE pathway, whose properties are similar to those of the Red system
Red system	The recombination system elaborated by phage λ. Its component proteins are *red* exonuclease and *red* beta protein. Its ability to act on unreplicated λ DNA is increased by *recA* protein
Int system	The specialized recombination system of λ that promotes recombination at site *att*. It includes the enzymes "integrase" and "excisionase," products of the λ genes *int* and *xis*, respectively. It is independent of all the Rec and Red system genes

bination rate in the neighborhood of the mutation that caused the large plaque phenotype (45–47). It was concluded that *bio* contains Chi, and that Chi is a nucleotide sequence that stimulates RecBC recombination in its neighborhood. None of the other recombination pathways that can operate on λ (Red, RecE, RecF) was influenced by Chi (30, 48).

B. Rules of Chi Activity in λ

An attractive possibility for Chi's role in recombination was that it could provide a point at which homologous pairing is initiated with high probability. Some versions of this simple view were eliminated by the demonstration that Chi promotes recombination even when present in only one of the two parents (47). Furthermore, the Chi⁻ parent can be deleted opposite the Chi in the other parent so that the possibility of homologous interaction within several thousand nucleotides of Chi is eliminated (49). In such a cross no recombination occurs in the immediate neighborhood of Chi but instead is concentrated near one edge of the deletion (or heterologous substitution). This edge was always the left edge, as represented on the conventional linkage map of λ (50). To test whether this directionality is a property of Chi or of λ, we inverted several Chi in λ and found that their activities (almost) entirely disappeared; thus, the phenotype of Chi depends on its orientation (51). This orientation-dependence of Chi activity has two implications: Chi is an asymmetric sequence; Chi interacts with an element of λ such that Chi is active only when oriented leftward.

The asymmetry of the Chi sequence was detailed by the observation that Chi is (5')-GCTGGTGG (52). Several sequences differing by one base from Chi have detectable recombination-stimulating activity (53).

The element in λ with which Chi interacts is cos (Chi–cos interaction). This observation depended upon the availability of a λ strain that had a second cos, cloned in inverse orientation into the λ chromosome near the mid-point. When introduced into this strain, Chi was active in each orientation. When the standard cos was deleted, Chi was active only when oriented rightward. When the cloned cos in this deletion phage was inverted so as to assume the standard (rightward) orientation, Chi was active only when oriented leftward (54). Thus, cos and Chi interact in such a way that Chi functions only when oriented antiparallel to cos. Since Chi is active in a phage carrying cos in both orientations, it is likely that cos activates an antiparallel Chi, rather than suppressing a parallel Chi.

The orientation of cos relative to the rest of the λ chromosome determines the direction in which that chromosome is to be packaged and the direction in which it is to be injected. In standard λ, the chromosome is packaged left to right and is injected right to left ("last gene in is the first gene out"). Thus, Chi–cos interaction might be mediated by (1) packaging or (2) injection. The following experiments eliminate any major role for injection, and show also that packaging, sensu strictu, is not responsible for the interaction.

(1) The antiparallel cos that activates Chi need not be employed in the packaging of the resulting recombinant. The recombinant can, instead, be packaged by parallel cos carried in the two parents (55).

(2) Chi is active on a chromosome injected via a parallel *cos* as long as the chromosome carries an antiparallel one (56). In fact, the Chi–*cos* interaction occurs in the absence of injection (57). The elimination of these two routes for the Chi–*cos* interaction forced consideration of a possibility that previous studies (for review, see 58) on λ packaging had implied was unlikely: perhaps *cos* opens reversibly *in vivo* through the cutting activity of terminase. Such a reversible and heretofore undetected opening might allow *cos* to activate Chi. In order to account for orientation-dependence of the Chi–*cos* interaction, we proposed that opened *cos* is the entry site for a recombination enzyme, and that terminase, remaining bound to its substrate, blocks entry of the enzyme at one of the two ends (54, 56). The demonstration that terminase remains bound *in vitro* to λ's left end (58) implied that the recombination enzyme enters λ's right end. Several kinds of *in vivo* observations support this model.

1. In order for an antiparallel *cos* to activate Chi, the infected cell must produce a terminase capable of cutting that *cos* (56).

2. A Chi in λ that is inactive by virtue of being parallel to *cos* becomes active if the chromosome is cut by a restriction endonuclease *in vivo* (59).

3. Chi in transposon Tn5 (located in the *gam* gene of λ) stimulates exchange less than does χ^+B, located near it, to its left. The double Chi$^+$ phage has the low activity characteristic of the Chi of Tn5. This suggests that a recombination enzyme entering λ at the right and traveling leftward sees only the Chi that it first encounters (60).

4. When λ carries two *cos* oriented in the same direction, Chi-stimulated recombinants as well as "ordinary" RecBC-mediated recombinants tend to be packaged from the nearer *cos* to the right of the exchange (61). The relationship of this observation to the direction of enzyme travel needs an explanation.

When Chi is present in only one of two parents in a cross, complementary Chi-induced recombinants are recovered in the progeny in unequal numbers (50). The favored recombinant, arising to the left of Chi, is the one that fails to inherit Chi. This was taken to mean that Chi acts nonreciprocally, i.e., one of the two possible crossover products is, at least temporarily, defective. However, an alternative explanation proved correct—the *cos* that is opened to activate the Chi (admits the recombination enzyme) tends to be the *cos* that is used to initiate left-to-right packaging from the resulting dimer (62). Since only one moiety of a dimer can be packaged (12), one of the two complementary recombinants appears in the progeny in excess of the other. This correlation between Chi activation and packaging allows us to deduce which of two *cos*, each antiparallel to a Chi, more often activated that Chi by determining which was more often used to package the resulting crossover chromosome. The demonstration that the *cos* nearer the right of the exchange was used preferentially was understandable only on the assumption that the recombination enzyme travels leftward (61).

C. In Vitro Studies

The conclusion that a RecBC recombination enzyme travels leftward in λ, derived from *in vivo* genetic analysis, is supported by *in vitro* studies on interactions between purified *recBC* gene product and DNA that contains Chi (63). *recBC* protein, which will invade DNA at a double-chain cut, will not invade a duplex end with a long overhang (longer than λ's "sticky ends"). This feature of the enzyme made it possible to control the *in vitro* direction in which *recBC* protein travels through a linearized plasmid. As the enzyme travels, it occasionally introduces nicks. When Chi is present in the DNA, there is a high probability a nick will be introduced near Chi, but only when the enzyme is traveling in one of the two possible directions. That direction relative to Chi corresponds to entry of the protein at λ's right end. The location of the Chi-induced cut *in vitro* is a few bases to the right of Chi on the chain that reads (5')-GCTGGTGG.

The demonstration (63) of Chi-induced cutting by purified *recBC* protein proves that *recBC* protein recognizes Chi. That view was previously implied by mutations in the *recB* or *recC* genes that have Rec$^+$ phenotype but fail to respond to Chi (64, 65).

It is tempting to suppose that the single-chain cut at Chi catalyzed by *recBC* protein *in vitro* relates directly to the mechanism of Chi activity, and models in which it plays such a role have been offered (66, 67). Testing those models and some alternatives is an immediate challenge.

VI. Red and RecBC as Models for Meiotic Recombination

Can we use these insights gained from studies on bacteriophage λ as guides to understanding meiotic recombination? As discussed in Section IV,E,3, Hasting's model (39) for fungal recombination supposes that most exchanges are executed as a succession of two single-chain exchanges like those in the model proposed by Holliday (41) and modified by Meselson and Radding (40) (Fig. 8). Recombination mediated by RecBC may be of that sort. Chi, then, would be a model for those recombination-stimulating sequences whose presence is deduced from the changing rates of non-Mendelian segregation along the length of a gene (68). In Hasting's model (39), hybrid DNA created by an Aviemore (Meselson–Radding) exchange (Fig. 8) can be subjected to double-chain cutting if there is a mismatch in the hybrid region. The double-chain cut is repaired when the cut ends invade the homologue and prime DNA synthesis, as shown in Fig. 6. λ's Red pathway and *E. coli*'s RecF pathway appear to be useful models for the double-chain-break repair process. It may even be that *E. coli* houses a model for the proposed double-chain-break reaction provoked by Watson–Crick mismatches. Methylation normally blocks mismatch-provoked excision of parental chains in the regions of DNA close behind replication forks. Newly synthesized chains, being nonmethylated, are subject to excisions. This methyl-directed repair reduces mutation rates. In *E. coli,* when methylation is genetically blocked, the mutation rate rises, presumably because the replication errors, instead of being reversed, are often transferred to the parental chain. Mismatch "repair" in such methylation-defective strains makes *E. coli* dependent on its recombination system—methylation-minus, *recA* double-mutants do not survive. Elimination of the excision system restores viability. Such an interaction among loci is neatly explained by the idea that double-chain breaks, introduced in non-methylated DNA by the excision system, can be repaired by the Rec system (69). Presumably, the RecF pathway, which does recombine DNA near double-chain breaks, is responsible for that repair.

Acknowledgments

The efforts of Susan Rosenberg, Alex Kolodkin, Imran Siddiqi, David Thaler, and Elizabeth Cooksey improved this article. Previously unpublished work here described was supported by grants from the National Science Foundation and the National Institutes of Health. The author is American Cancer Society professor of molecular genetics.

References

1. M. Meselson and J. Weigle, *PNAS* **47**, 857 (1961).
2. M. Meselson, *JMB* **9**, 734 (1964).
3. V. E. A. Russo, M. M. Stahl and F. W. Stahl, *PNAS* **65**, 363 (1970).
4. M. M. Stahl and F. W. Stahl, *in* "The Bacteriophage Lambda" (A. D. Hershey, ed.), p. 431. Cold Spring Harbor Laboratory, Cold Spring Harbor, New York, 1971.
5. F. W. Stahl and M. M. Stahl, *in* "The Bacteriophage Lambda" (A. D. Hershey, ed.), p. 443. Cold Spring Harbor Laboratory, Cold Spring Harbor, New York, 1971.
6. K. D. McMilin and V. E. A. Russo, *JMB* **68**, 49 (1972).
7. A. J. Clark and A. Margulies, *PNAS* **53**, 451 (1965).
8. R. Gingery and H. Echols, *PNAS* **58**, 1507 (1976).

9. E. R. Signer, H. Echols, J. Weil, C. M. Radding, M. Schulman, L. Moore and K. Manly, *CSHSQB* **33**, 711 (1969).
10. R. W. Hendrix, J. W. Roberts, F. W. Stahl and R. A. Weisberg (eds.), "Lambda II." Cold Spring Harbor Laboratory, Cold Spring Harbor, New York, 1983.
10a. D. A. Ritchie, C. A. Thomas, Jr., L. A. MacHattie and P. C. Wensink, *JMB* **23**, 365 (1967).
10b. C. A. Thomas, Jr., T. J. Kelly, Jr. and M. Rhoades, *CSHSQB* **33**, 417 (1968).
11. F. W. Stahl, K. D. McMilin, M. M. Stahl, R. E. Malone and Y. Nozu, *JMB* **68**, 57 (1972).
12. D. G. Ross and D. Freifelder, *Virology* **74**, 414 (1976).
13. L. Enquist and A. Skalka, *JMB* **75**, 185 (1973).
14. A. E. Karu, Y. Sakaki, H. Echols and S. Linn, *JBC* **250**, 7377 (1975).
15. J. Zissler, E. R. Signer and F. Schaefer, *in* "The Bacteriophage Lambda" (A. D. Hershey, ed.), p. 455. Cold Spring Harbor Laboratory, Cold Spring Harbor, New York, 1971.
16. A. Skalka, *in* "Mechanisms in Recombination" (R. F. Grell, ed.), p. 421. Plenum, New York, 1974.
17. W. Fangman and M. Feiss, *JMB* **44**, 103 (1969).
18. K. D. McMilin, M. M. Stahl and F. W. Stahl, *Genetics* **77**, 409 (1974).
19. F. W. Stahl, K. D. McMilin, M. M. Stahl, J. M. Crasemann and S. Lam, *Genetics* **77**, 395 (1974).
20. V. E. A. Ruso, *MGG* **122**, 253 (1973).
21. R. L. White and M. S. Fox, *Genetics* **81**, 33 (1975).
22. F. W. Stahl and M. M. Stahl, *in* "Mechanisms in Recombination" (R. F. Grell, ed.), p. 407. Plenum, New York, 1974.
23. F. W. Stahl, I. Kobayashi and M. M. Stahl, *PNAS* **79**, 6318 (1982).
24. F. W. Stahl, K. D. McMilin, M. M. Stahl and Y. Nozu, *PNAS* **69**, 3598 (1972).
25. F. W. Stahl, I. Kobayashi and M. M. Stahl, *JMB* **181**, 199 (1985).
26. F. W. Stahl and M. M. Stahl, *J. Genet.* **64**, 31 (1985).
27. M. S. Wold, J. B. Mallory, J. D. Roberts, J. H. Lebowitz and R. McMacken, *PNAS* **79**, 6176 (1982).
28. F. W. Stahl and M. M. Stahl, *Genetics* **113**, 1, (1986).
29. A. J. Clark, S. J. Sandler, K. D. Willis, C. C. Chu, M. A. Blanar and S. T. Lovett, *CSHSQB* **49**, 453 (1984).
30. J. R. Gillen and A. J. Clark, *in* "Mechanisms in Recombination" (R. F. Grell, ed.), p. 123. Plenum, New York, 1974.
31. A. H. Doermann, *in* "Genetics Today" (S. J. Geerts, ed.), p. 69. Pergamon, Oxford, 1965.
32. A. H. Doermann and D. H. Parma, *J. Cell. Physiol.* (Suppl. 1) **70**, 147 (1967).
32a. G. Mosig, R. Ehring, W. Schliewen and S. Bock, *MGG* **113**, 51 (1971).
33. M. A. Resnick, *in* "Molecular Mechanisms for Repair of DNA" (P. C. Hanawalt and R. B. Setlow, eds.), p. 459. Plenum, New York, 1975.
34. J. Hicks, A. Hinnen and G. Fink, *CSHSQB* **43**, 1305 (1978).
35. T. L. Orr Weaver, J. W. Szostak and R. J. Rothstein, *PNAS* **79**, 6354 (1981).
36. J. W. Szostak, T. L. Orr-Weaver, R. J. Rothstein and F. W. Stahl, *Cell* **33**, 25 (1983).
37. T. L. Orr-Weaver and J. W. Szostak, *PNAS* **80**, 4417 (1983).
38. J. N. Strathern, A. J. S. Klar, J. B. Hicks, J. A. Abraham, J. M. Ivy, K. A. Nasmyth and C. McGill, *Cell* **31**, 183 (1982).
38a. A. J. S. Klar and J. N. Strathern, *Nature* **310**, 744 (1984).
39. P. J. Hastings, *CSHSQB* **49**, 49 (1984).

40. M. Meselson and C. M. Radding, *PNAS* **72**, 358 (1975).
41. R. Holliday, *Genet. Res.* **5**, 282 (1964).
42. D. G. Catcheside and T. Angel, *Aust. J. Biol. Sci.* **27**, 219 (1974).
43. M. Gutz, *Genetics* **69**, 317 (1971).
43a. F. W. Stahl, M. Lieb and M. M. Stahl, *Genetics* **108**, 795 (1984).
44. D. Henderson and J. Weil, *Genetics* **79**, 143 (1975).
45. F. W. Stahl, J. M. Crasemann and M. M. Stahl, *JMB* **94**, 203 (1975).
46. K. D. McMilin, M. M. Stahl and F. W. Stahl, *Genetics* **77**, 409 (1974).
47. S. T. Lam, M. M. Stahl, K. D. McMilin and F. W. Stahl, *Genetics* **77**, 425 (1974).
48. F. W. Stahl and M. M. Stahl, *Genetics* **86**, 715 (1977).
49. F. W. Stahl and M. M. Stahl, *MGG* **140**, 29 (1975).
50. F. W. Stahl, M. M. Stahl, R. E. Malone and J. M. Crasemann, *Genetics* **94**, 235 (1980).
51. D. Faulds, N. Dower, M. Stahl and F. Stahl, *JMB* **131**, 681 (1979).
52. G. R. Smith, S. M. Kunes, D. W. Schultz, A. Taylor and K. L. Triman, *Cell* **24**, 429 (1981).
53. K. C. Cheng and G. R. Smith, *JMB* **180**, 371 (1984).
54. I. Kobayashi, H. Murialdo, J. M. Crasemann, M. M. Stahl and F. W. Stahl, *PNAS* **79**, 5981 (1982).
55. I. Kobayashi, M. M. Stahl, D. Leach and F. W. Stahl, *Genetics* **104**, 549 (1983).
56. I. Kobayashi, M. M. Stahl and F. W. Stahl, *CSHSQB* **49**, 497 (1984).
57. F. W. Stahl, M. M. Stahl, L. Young and I. Kobayashi, in "Proceedings of the Fifth John Innes Symposium" (D. A. Hopwood, ed.), p. 27. Croom Helm, London, 1983.
58. M. Feiss and A. Becker, in "Lambda II" (R. W. Hendrix, J. W. Roberts, F. W. Stahl and R. A. Weisberg, eds.), p. 305. Cold Spring Harbor Laboratory, Cold Spring Harbor, New York, 1983.
59. M. M. Stahl, I. Kobayashi, F. W. Stahl and S. K. Huntington, *PNAS* **80**, 2310 (1983).
60. E. Yagil and I. Shtromas, *Genet. Res.* **45**, 1 (1985).
61. F. W. Stahl, I. Kobayashi, D. R. Thaler and M. M. Stahl, *Genetics*, **113**, 215 (1986).
62. I. Kobayashi, M. M. Stahl, F. R. Fairfield and F. W. Stahl, *Genetics* **108**, 773 (1984).
63. A. F. Taylor, D. W. Schultz, A. F. Ponticelli and G. R. Smith, *Cell* **41**, 153 (1985).
64. D. W. Schultz, A. F. Taylor and G. R. Smith, *J. Bact.* **155**, 664 (1983).
65. A. M. Chaudhury and G. R. Smith, *PNAS* **81**, (1984).
66. F. W. Stahl, "Genetic Recombination: Thinking about It in Phage and Fungi." Freeman, San Francisco, California, 1979.
67. G. R. Smith, S. K. Amundsen, A. M. Chaudhury, K. C. Cheng, A. S. Ponticelli, C. M. Roberts, D. W. Schultz and A. F. Taylor, *CSHSQB* **49**, 485 (1984).
68. J.-L. Rossignol, A. Nicolas, H. Hamza and T. Langin, *CSHSQB* **49**, 13 (1984).
69. M.-P. Doutriaux, R. Wagner and M. Radman, *PNAS*, **83**, 2576 (1986).

Regulation of Protein Synthesis by Phosphorylation of Ribosomal Protein S6 and Aminoacyl-tRNA Synthetases[1]

J. A. TRAUGH AND
A. M. PENDERGAST

Biochemistry Department
University of California
Riverside, California 92521

Ribosomal protein S6 is a major phosphorylated ribosomal protein in every mammalian cell type and tissue examined. The amount of phosphate associated with S6 varies with the state of the cell. In ribosomes from nongrowing or slowly growing tissues, such as rat liver or quiescent cells in culture, little phosphate is associated with S6, whereas ribosomes from rapidly dividing cells or tissues contain S6 in a highly phosphorylated state. Phosphorylation of S6 is stimulated by hormones and compounds that increase intracellular levels of cAMP, inhibit phosphoprotein phosphatases, stimulate cell division, or are toxic (1).

It has been known for many years that the cAMP-dependent protein kinase (cA kinase, also termed cAMPdPK) incorporates phosphate into S6 (1); however, the first identification of a second protein kinase phosphorylating S6 was made in 1979 with the isolation of protease-activated kinase (PAK) II from reticulocytes (2). The differential activation of protein kinases with the resultant site-specific phosphorylation of S6 was first shown in HeLa cells stimulated with dibutyryl cAMP (Bt_2cAMP) or insulin (3). The early 1980s was a period of considerable research activity on S6, both by researchers in the field of protein synthesis as well as those exploring hormone action at the molecular level.

[1] Abbreviations used: cAMP, cyclic AMP, (adenosine 3′,5′-monophosphate); Bt_2-cAMP, dibutyryl cAMP; cA kinase (cAMPdPK), cyclic AMP-dependent protein kinase; PAK, protease-activated kinase; C kinase, Ca^{2+}, phospholipid-dependent protein kinase; Ap_4A, P^1,P^4-bis(adenosine 5′-)tetraphosphate; eIF, eukaryotic initiation factor; EGF, epidermal growth factor; $PGF_{2\alpha}$, prostaglandin $F_{2\alpha}$; Ab-MuLV, Abelson murine leukemia virus; pp60^{v-src}, gene product of Rous sarcoma virus involved in cellular transformation.

In 1977, it was suggested that phosphorylation/dephosphorylation could play a role in regulating aminoacyl-tRNA synthetase activity as well as other aspects of protein synthesis (4). Presently, seven out of the nine synthetases examined have been shown to be phosphorylated *in vivo*; the consequences of these phosphorylation events are just beginning to be investigated, an exciting period in any field.

This review focuses on phosphorylation of ribosomal protein S6 and the aminoacyl-tRNA synthetases, and the role of these phosphorylation events in the regulation of protein synthesis. This focus was chosen since considerable inroads into our understanding of these processes have been made in the past several years. However, it must be kept in mind that some, if not all, of the protein kinases that phosphorylate S6 and the aminoacyl-tRNA synthetases also phosphorylate other components involved in protein synthesis, including initiation factors and messenger ribonucleoprotein particles.

I. Phosphorylation of Ribosomal Protein S6

A. Phosphorylation of S6 in Response to cAMP

Addition of cAMP or analogs of cAMP stimulates phosphorylation of a single ribosomal protein, S6, in a number of different cell types and tissues (5–12). The peptide hormone, glucagon, stimulates phosphorylation of S6 in rat liver and pancreatic islet tumor cells, and prostaglandin E_1 enhances phosphorylation in thymocytes by elevation of cAMP levels (9, 10, 13, 14).

Ribosomal proteins from the 40-S and 60-S ribosomal subunits have been identified and defined by two-dimensional polyacrylamide gel electrophoresis (15) using the system (and modifications thereof) described by Kaltschmidt and Wittmann (16). Ribosomal protein S6 has a molecular weight of approximately 32,500 and is among the largest of the proteins in the 40-S ribosomal subunit. S6 is found in various phosphorylated derivatives containing up to five phosphates as well as a dephosphorylated form (Fig. 1). The derivatives are easily defined in the two-dimensional electrophoresis system, since addition of each phosphate retards migration in the first dimension, but has little effect on migration in the second. Thus the degree of phosphorylation of S6 is easily monitored.

The most detailed study of phosphorylation of S6 in response to cAMP has been carried out in reticulocytes. In cells incubated in a nutritional medium containing $^{32}P_i$, S6 is the only basic ribosomal protein phosphorylated and is present primarily in nonphosphory-

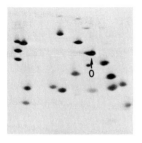 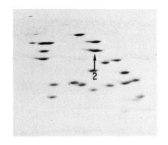 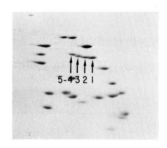

CONTROL   cAMPdPK   PAK II

FIG. 1. Effects of phosphorylation on migration of S6 in two-dimensional polyacrylamide gels. The phosphorylation state of S6 was analyzed by electrophoresis in two-dimensional polyacrylamide gels. Left, 40-S ribosomal subunits incubated in the absence of protein kinase; middle, phosphorylated with cAMP-dependent protein kinase (cA kinase); right, phosphorylated with PAK II (17).

lated and monophosphorylated derivatives. Upon addition of 8-bromo-cAMP (7) or Bt_2cAMP (12) in the presence of 3-isobutyl-1-methylxanthine, phosphorylation of S6 can be stimulated 2.5-fold, to produce the monophosphorylated and diphosphorylated derivatives. Two-dimensional phosphopeptide maps of tryptic digests of S6 from nonstimulated cells show one major and one minor phosphopeptide, identified as A and C, respectively (Fig. 2). Upon addition of 8-bromo-cAMP, another major phosphopeptide, B, is observed (7).

When isolated 40-S ribosomal subunits are incubated with the purified catalytic subunit of cA kinase, only S6 is phosphorylated (7, 18). Occasionally, other researchers have identified additional proteins in the 40-S ribosomal subunit phosphorylated by this protein kinase. However, we have found cA kinase to be specific for S6, and only when 40-S ribosomal subunits are partially denatured is phosphorylation of other proteins observed (18). For instance, removal of magnesium, or freezing and thawing, causes unfolding of ribosomes with the resultant loss in specificity of phosphorylation.

Up to two phosphates can be incorporated into S6 by cA kinase (7, 18). Under these conditions, only the diphosphorylated derivative of S6 is observed (Fig. 1, middle panel) and most, if not all, of the phosphate in S6 is in the B peptide (17, 18). At low concentrations of protein kinase, the monophosphorylated derivative of S6 is predominant. Analysis of these tryptic phosphopeptides shows that most of the phosphate is in the A peptide. From this, it appears that the B peptide and the A peptide are identical except for addition of a phosphate. This has been confirmed by the observation that limited treatment of the B peptide with alkaline phosphatase results in production of the A

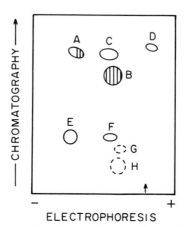

FIG. 2. Two-dimensional phosphopeptide maps of tryptic digests of phosphorylated S6 by cA kinase. Tryptic digests of S6 were analyzed by thin-layer electrophoresis in the first dimension followed by chromatography in the second dimension using the procedure of Traugh et al. (7). Hatched circles are phosphopeptides of S6 phosphorylated by cA kinase; open circles are phosphopeptides from S6 phosphorylated by PAK II; dashed circles indicate phosphopeptides not observed with either purified protein kinase, but phosphorylated in cultured cells in response to insulin or mitogenic hormones/compounds. Letters identify the individual phosphopeptides.

peptide and P_i (19). Thus, the sites phosphorylated in reticulocytes in response to cAMP are identical to the two sites phosphorylated *in vitro* by cA kinase. In similar studies with the cGMP-dependent protein kinase (7, 18), up to 1 mol of phosphate is incorporated per mol of S6; only the A peptide is observed upon phosphopeptide mapping.

Wettenhall et al. (20, 21) have sequenced the region of S6 from rat liver phosphorylated by cA kinase. Under conditions where two phosphates are incorporated by cA kinase, two phosphorylated serines are identified in neighboring positions as shown below.

Arg-Arg-Leu-Ser(P)-Ser(P)-Leu-Arg-Ala-Ser-Thr-Ser-Lys-Ser-Glu-Glu-Ser-Gln-Lys

The first serine is preferentially phosphorylated by cA kinase; under optimal conditions, the second, adjacent, serine also becomes phosphorylated (20). This sequence is located in the COOH-terminal region of the molecule.

B. Phosphorylation of S6 in Response to Growth-Promoting Compounds and Insulin

Hormones and compounds that stimulate mitotic growth stimulate phosphorylation of S6 in quiescent or serum-starved cells (1, 22–32).

TABLE I
EFFECTORS PROMOTING INCORPORATION OF FIVE PHOSPHATES INTO S6

Source	Effectors	Number of phosphates	References
Fibroblasts, chick embryo	Phorbol esters	5	22
	Transformation (Rous)	5	23, 24
Fibroblasts, baby hamster kidney	Infection with pseudorabies virus	5	25
Fibroblasts, 3T3	Epidermal growth factor	5	26
	Prostaglandin $F_{2\alpha}$	5	27
	Platelet-derived growth factor	5	27a
	Insulin	5	27
	Orthovanadate	3–5	28
	Transformation (Ab-MuLV)	1–5	46
3T3-L1	Insulin	5	29
HeLa	Insulin	5	30
Hepatoma, Reuber H35	Insulin	5	31, 32
	Phorbol esters	5	31, 32
Liver	Regeneration	5	33
	Q fever (*Coxiella burnetii*)	3–5	33a
	Cycloheximide	5	34
	Dimethylnitrosamine	1–5	35
	Thioacetamide	2–5	36
Reticulocytes	Slight decrease of pH	5	37
	NaF (5 mM)	5	37
Pancreas	Carbachol	3–5	38
Xenopus oocytes	Progesterone	3–5	39–41
	pp60$^{v\text{-}src}$	4–5	42
	Slight increase of pH	3–5	43
	Ab-MuLV	2–5	46

Increased phosphorylation of S6 has also been observed during liver regeneration (33) and in Q fever (33a). Following transformation of chicken embryo fibroblasts with Rous sarcoma virus, phosphorylation of S6 is elevated and serum starvation does not significantly diminish phosphorylation of S6 (23, 24). Addition of phosphodiesterase inhibitors blocks the insulin and serum-stimulated phosphorylation of S6 (44, 45).

The number of phosphates incorporated into S6 in response to hormones and growth-promoting compounds in various cell types has been determined by monitoring the position of S6 following two-dimensional gel electrophoresis. In all instances examined, phosphorylation is on serine. As shown in Table I, four to five phosphates are incorporated into S6 in a number of different cell types stimulated with insulin, tumor-promoting phorbol esters, epidermal growth factor (EGF), prostaglandin $F_{2\alpha}$ ($PGF_{2\alpha}$), platelet-derived growth factor, and following transformation/viral infection. Thus, compounds and events that stimulate growth and cell division have a similar, if not identical, effect on the amount of phosphate incorporated into S6.

Up to five phosphates have also been shown to be incorporated into S6 in other cells and tissues (Table I). This includes regenerating rat liver, liver in response to Q fever, the pancreas in response to carbachol, and reticulocytes treated with 5 mM NaF or incubated at a pH reduced from 7.6 to 7.2. Fully phosphorylated S6 has also been observed with cycloheximide. In *Xenopus* oocytes, S6 phosphorylation is stimulated in response to progesterone, increased pH, and after injection of the tyrosine protein kinase of $pp60^{v-src}$ or Ab-MuLV.

The time course of phosphorylation of S6 in quiescent cells in response to serum has been studied. Stimulation of phosphorylation of S6 in Swiss 3T3 cells is almost maximal at 30 minutes and remains at this plateau for 2 hours following addition of serum to quiescent cells (28). A similar time-course for serum stimulation of HeLa cells (47) and for phorbol-ester-stimulated phosphorylation of S6 in Reuber H35 cells (32) has been observed.

Removal of serum from quiescent cells stimulated for 1 hour results in dephosphorylation of S6 as monitored by two-dimensional gel electrophoresis and by a decrease in radioactive phosphate (45). Dephosphorylation of S6 is rapidly initiated in serum-stimulated cells by hypertonic shock (48) or by heat shock (49–51), and in liver upon ATP depletion by ethionine (52). Recovery of S6 phosphorylation is observed upon return to isotonic media, moderate temperatures, or addition of adenine, respectively.

The sites phosphorylated in response to the various growth-promoting compounds have been examined by two-dimensional phosphopeptide mapping. Two different systems have been utilized to probe the question (7, 53). In the two-dimensional system of Thomas (53), 10 to 11 major tryptic phosphopeptides are spread along a diagonal. In the system of Traugh (7), eight tryptic phosphopeptides are observed, seven in response to growth-promoting compounds. These are identified as A and C–H in Fig. 2.

Identical phosphopeptide patterns have been observed with Reuber H35 cells stimulated with insulin and phorbol esters (31, 32), with 3T3-L1 cells stimulated with insulin (29), with Swiss 3T3 cells treated with EGF (26), and in reticulocytes incubated in a slightly acidified medium, pH 7.2 (37). In Reuber H35 cells, the same pattern is observed at optimal and suboptimal levels of insulin (10^{-7}, 10^{-10} μM) and phorbol ester (10^{-6}, 10^{-7} M) (31). The A peptide is identical to that obtained with cA kinase, peptides C–H are seen only with growth-promoting compounds, while the B peptide is unique to cells with elevated cAMP levels. The only exception is in Swiss 3T3 cells stimulated with EGF; then, only a small amount of the B phosphopeptide is present. Similar observations have been made by Martin-Perez *et al.* (*54*) in Swiss 3T3 cells with EGF, $PGF_{2\alpha}$, and insulin. This suggests that Swiss 3T3 cells may contain higher levels of cAMP and/or cA kinase or that the latter may be partially involved in mediating mitogenic action in that cell type. Identical chymotryptic phosphopeptide maps have been obtained for chicken embryo fibroblasts treated with phorbol ester or serum, or transformed with Rous virus (*22*). Thus, it appears that the same sites on S6 are phosphorylated in response to a number of different growth-promoting compounds, including expression of a transforming gene product.

Martin-Perez and Thomas (*53*) conducted phosphopeptide analyses of the individual derivatives of S6 following addition of serum to serum-starved Swiss 3T3 cells. Each derivative of S6 contained peptides present in the less phosphorylated derivatives, plus new peptides. They concluded that phosphate added to S6 is incorporated in an ordered manner and not randomly. Perisic and Traugh (*37*) compared the sites of phosphorylation of S6 *in vitro* by the purified S6 kinase, PAK II, and obtained similar results. The monophosphorylated derivative contains two different species of S6 (peptides A and C). Peptide E appears to be primarily responsible for the diphosphorylated derivative, while peptide F contributes to both the di- and triphosphorylated forms. The early sites (peptides A and C) appear to be randomly phosphorylated, followed by the sequential phosphorylation (Fig. 3).

In both studies, all of the phosphopeptides are observed, even when all S6 is multiply phosphorylated. However, more peptides are observed than the number of phosphates incorporated (as analyzed by gel electrophoresis and quantitative incorporation of $^{32}P_i$). Several explanations are possible: there may be more than five sites, but phosphorylation at one site could block phosphate incorporation into another site, so that the total is never more than five; some of the phosphopeptides may be identical except for the number of phosphates present; multiple phosphates in S6 could block trypsin cleavage sites. It is interesting to note that the sequence obtained by Wettenhall and Morgan (*21*) containing the two sites for cA kinase contained four additional serine residues, some of which may be phosphorylated *in vivo* in response to insulin and mitogenic compounds.

Since phosphorylation of S6 is stimulated almost immediately fol-

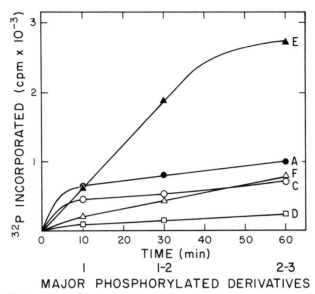

FIG. 3. Time course of incorporation of phosphate into specific sites on S6 by PAK II. Ribosomal subunits were phosphorylated with PAK II and the derivatives analyzed by two-dimensional polyacrylamide gel electrophoresis. The modified sites were identified by two-dimensional phosphopeptide mapping of tryptic digests of S6 (37). Letters identify the individual phosphopeptides, as in Fig. 2.

lowing addition of serum, phorbol ester, EGF, or insulin to serum-starved or quiescent cells, it appears that the protein kinase(s) phosphorylating S6 must be present in an inactive form in the cell, or that the phosphatase activity directed toward S6 is more active than the protein kinase. That a protein kinase phosphorylating S6 is activated in response to growth-promoting compounds was first shown following addition of insulin to serum-starved 3T3-L1 cells (29). Protein kinase activity is assayed following chromatography of the postribosomal supernatant on DEAE-cellulose using 40-S ribosomal subunits as substrate. The inactive form of the S6 kinase is monitored by incubation of the fractions in the presence or absence of trypsin for 30 seconds prior to assay. This brief incubation period is sufficient to activate an inactive form of the enzyme phosphorylating S6. Assays without trypsin pretreatment lead to identification of endogenously activated enzyme. Little active protein kinase is observed in serum-starved cells; following treatment of the cells with insulin for 15 minutes, approximately 50% of the protein kinase activity is activated. After insulin treatment for 1 hour, most of the enzyme has returned to the inactive form (Perisic and Traugh, unpublished results.) This fits well with the time course of phosphorylation of S6 in cultured cells, as described above. Similar results have been obtained in cell extracts following treatment of Swiss 3T3 cells with epidermal growth factor or serum, where up to a 10-fold stimulation of S6 kinase activity is detected (28). Activation is observed as early as 2 minutes after serum stimulation.

The identity of the protein kinase phosphorylating S6 in response to growth-promoting compounds has been elusive. The inactive form of a protein kinase that phosphorylates S6 at the same sites phosphorylated in cultured cells in response to growth-promoting compounds was first purified from reticulocytes (55). The proenzyme is activated in response to limited digestion with trypsin (55, 56) or a Ca^{2+}-stimulated neutral protease from reticulocytes (57), hence the identification as protease-activated kinase II (2). PAK II is obtained in highly purified form from reticulocytes by ion-exchange chromatography and gel filtration (55). The proenzyme form of the protein kinase has a molecular weight of approximately 80,000 as measured by gel filtration, which is diminished to 45,000–55,000 following activation by limited tryptic digestion. This suggests the removal of a regulatory domain upon proteolysis. The enzyme requires Mg^{2+} for activity and is inhibited by other divalent cations and monovalent cations. The enzyme utilizes ATP but not GTP as a phosphate donor. A sulfhydryl reductant is required to maintain activity. Some phosphorylation of S10 is also observed *in vitro*, the degree of which appears to vary with the ribosome preparation. The importance of the structural integrity of the ribosome upon phosphorylation with PAK II has been shown by altering the temperature. A dramatic increase in the number of proteins phosphorylated is observed at temperatures over 43°C, probably due to unfolding of the ribosome.

More recently, PAK II has been shown to be activated in a Ca^{2+}-independent manner by phospholipids and diolein (58). PAK II is separated from C kinase by chromatography on phosphocellulose. The two enzymes have other distinct properties in addition to the differential effect of Ca^{2+} upon activity. PAK II has different kinetics of activation by phospholipids/diolein than C kinase, and has distinct specificity with a number of different substrates (58). Polyclonal antibody prepared to C kinase does not react with PAK II, but does react with C kinase as shown by immunoblotting (C. DeVack and J. A. Traugh, unpublished). The data support the hypothesis that PAK II and C kinase are closely related, but distinct enzymes.

Both the inactive and the activated forms of PAK II have been highly purified from liver by chromatography on DEAE-cellulose and affinity resins (unpublished). The inactive form has an M_r similar to that of PAK II from reticulocytes; the M_r of the endogenously activated form is 60,000 as determined by gel filtration. A number of different protease inhibitors are added at the time of homogenization, so it is unlikely that proteolysis occurs during and following cell disruption. The endogenously activated form of PAK II elutes from DEAE-cellulose at a higher salt concentration than the proenzyme; both forms of the enzyme elute at approximately the same concentrations described previously for PAK II from 3T3-L1 cells (29) and phosphorylate identical sites as the enzyme from reticulocytes, identified previously as PAK II. In addition to S6, PAK II phosphorylates eIF-2 (56, 57), histone 1 (but not other histones) (58), glycogen synthetase (G. M. Hathaway, T. R. Soderling, and J. A. Traugh, unpublished), α_{S1} casein (unpublished), and pp12 from Rous sarcoma virus (58a).

Under optimal conditions, up to three phosphates are incorporated per mol of S6 (17), although all five phosphorylated derivatives are observed (Fig. 1, right panel). In addition, two of the phosphopeptides observed in response to mitogenic compounds are not observed with PAK II (26, 29, 31) as shown in Fig. 2. This suggests the possibility that more than one protein kinase may phosphorylate S6. However, due to the complexity of the substrate, it could also indicate that other components are needed to expose additional sites for phosphorylation by a single protein kinase.

A trypsin-activated protein kinase has been isolated from rat liver using as substrate a peptide analogue of the first eight amino acids of the sequence of S6 identified earlier. The partially purified enzyme phosphorylates at least two sites on S6 (59). One of these is the second serine in the sequence phosphorylated by the cA kinase; the other site is distinct. Donahue and Masaracchia (60) obtained a highly purified protease-activated kinase from lymphosarcoma cells when activated, incorporating up to 3 mol of phosphate per mol of S6 (61). The enzyme is, in many ways, similar to PAK II, but unlike it phosphorylates histone-4. This H4 kinase has been separated from C kinase by chromatography on CM-Sephadex (62). Modulators of C kinase activity have no effect on H4 kinase and NaF is an effective inhibitor of the H4 kinase, but has no effect on C kinase activity.

There is currently a controversy as to whether S6 is phosphorylated directly by C kinase, since C kinase has been reported to be activated by serum (63) and is also the receptor for phorbol esters (64). Several groups have reported C kinase directly phosphorylates S6

(65–67). However, the amount of C kinase used in these studies is 500 to 1000 times the amount of PAK II, based on enzyme units measured with histone 1. At these concentrations of C kinase, two to three phosphates are incorporated into S6 (65, 66). Phosphopeptide maps of S6 phosphorylated by C kinase are similar, but not identical, to those obtained with PAK II (67). As indicated above, both the H4 kinase and PAK II are separable from C kinase. Using purified C kinase from reticulocytes (58), brain, or liver (E. Palen and J. A. Traugh, unpublished), we have been unable to obtain any significant phosphorylation of S6. Phorbol-ester-stimulated phosphorylation of S6 has been dissociated from the S6 phosphorylation stimulated by pp60$^{v\text{-}src}$ or serum in chick embryo fibroblasts. This is shown by the down-regulation of C kinase in response to the continued presence of phorbol ester; this does not eliminate stimulation of S6 kinase activity with serum or upon expression of pp60$^{v\text{-}src}$ (67a).

Rosen and co-workers (68) identified an S6 kinase that is rapidly activated in 3T3-L1 cells in response to insulin or phorbol esters. The activated enzyme has an M_r of 50,000–60,000 with a pH optimum between 8 and 9. It utilizes only ATP as a phosphate donor and is inhibited by Ca^{2+} (0.5 mM) and NaF. The enzyme does not appear to be C kinase since it is soluble and not membrane bound. Previously, Rosen and co-workers (69) had purified an activated form of S6 kinase from 3T3-L1 cells, which they called casein kinase I. However, authentic casein kinase I from reticulocytes (70) does not phosphorylate any of the 40-S ribosomal proteins (P. T. Tuazon and J. A. Traugh, unpublished). Thus, the preparation of S6 kinase from 3T3-L1 cells probably contained at least two protein kinases, casein kinase I and an active S6 kinase.

Recently, an activated S6 kinase has been purified to apparent homogeneity from *Xenopus* oocytes (70). It has an M_r of 92,000 by gel electrophoresis and 70,000–80,000 by gel filtration. As compared to the enzymes described from mammalian cells, the *Xenopus* enzyme has very limited substrate specificity and the only other identified substrate, in addition to S6, is the heptapeptide, Kemptide. The K_m for ATP is 28 μM, and the enzyme is inhibited by numerous compounds including β-glycerophosphate, NaF, potassium phosphate, and heparin. When 40-S ribosomal subnits are phosphorylated with the purified enzyme, from one to five phosphates are observed in S6, and the phosphopeptide maps of proteolytic digests of S6 are identical to those obtained in progesterone-treated oocytes (71).

So far, in every cell type examined, only one inactive and one active form of S6 kinase has been identified by each investigator, in addition to cA kinase. This raises the possibility that there is, in fact, only one enzyme, and the observed differences reported by different investigators are due to the state of activation and degree of purity of the various preparations or to slight differences between cell types. The number of S6 kinases will be resolved when the genes for the individual enzymes are identified and sequenced. It is worthy of note that three isozymes of C kinase have been purified from rat brain, which have different sites of autophosphorylation and are differentially reactive with monoclonal antibodies (71a).

One of the questions of primary interest is the mode of activation of S6 kinase in response to growth-promoting compounds and transformation. Since the activating agents promoting this response are so diverse and the sites on S6 are identical, it appears that the same protein kinase(s) must be activated by different metabolic pathways, or that more than one protein kinase is involved. At this time, there is little evidence to support either hypothesis. Recently, Blenis and Erickson (72) found a single peak of activated S6 kinase in chick embryo fibroblasts in response to expression of $pp60^{v-src}$, to serum, or to phorbol ester. Analysis of the activity by chromatography on anion and cation exchange columns showed that the activated enzyme migrates to an identical position for each of the three treatments, suggesting the protein kinase is the same.

The protein kinase(s) phosphorylating S6 is activated by hormones that bind to receptor-associated "protein-tyrosine kinases" (EC 2.7.1.112) (such as the receptors for insulin and EGF), by the protein-tyrosine kinases of transforming viruses, by phorbol esters (which bind to the serine protein kinase, C kinase), and by changes in pH. There are, of course, a number of possible modes of activation for S6 kinase. These include activation by proteolysis, by the direct phosphorylation of the S6 kinase on tyrosine and/or serine residues, or by production of an activating compound.

Evidence is accumulating that S6 kinase(s) may be activated by phosphorylation of tyrosine and possibly serine residues. Novak-Hofer and Thomas (28) have examined the effects of various phosphatase inhibitors on the activity of S6 kinase. In extracts of EGF-stimulated cells, S6 kinase is most active in the presence of phosphotyrosine, although phosphoserine also has a significant effect. In the absence of phosphatase inhibitors, little S6 kinase activity is observed. Without EGF stimulation, little activated S6 kinase is observed under any conditions. S6 kinase is also activated by incubation of the cells with 1 mM orthovanadate (28), an inhibitor of protein-tyrosine kinases. In addition, expression of Rous virus in chick embryo fibroblasts (23, 24) or injection of Xenopus oocytes with the protein-tyrosine kinases of pp60$^{v\text{-}src}$ (42) or Ab-MuLV (46) enhances phosphorylation of S6.

Utilizing PAK II purified from reticulocytes, Perisic and Traugh investigated the potential phosphorylation of the proenzyme form of PAK II and activation by the tyrosine kinases of the Ab-MuLV, the insulin receptor, and the EGF receptor. Upon addition of the individual tyrosine kinases, a single protein (M_r 70,000 by gel electrophoresis) is phosphorylated on tyrosine in highly purified preparations of PAK II. Coincident with phosphorylation, we have observed up to a 5-fold stimulation of PAK II activity. The same 70,000-Da protein is also phosphorylated by C kinase.

C. Molecular Analysis of Phosphorylation of S6 and Protein Synthesis

One of the early primary responses of quiescent cells in culture to serum or polypeptide growth factors is the stimulation of protein synthesis (73, 74). This stimulation can be due to increased translation of endogenous mRNA and/or enhanced translation of new mRNA. The latter has been shown to be important in the production of ribosomal protein, which effectively results in the stimulation of protein synthesis by increasing the number of ribosomes. Protein synthesis plays an integral role in cell division, as it is required throughout the G_1 phase for initiation of DNA synthesis and during the S and G_2 phase for mitosis and cell division (75–77).

In response to serum stimulation of quiescent cells, there is a shift in the ribosome population from the inactive 80-S storage form to the polysome fraction accompanied by increased movement of endogenous mRNA into polysomes (78, 79). There is a direct correlation between the increase in polysome formation and the level of phosphorylation of S6 in response to EGF, $PGF_{2\alpha}$, and insulin (27). In response to serum stimulation, an increase in polysome formation occurs concomitantly with increased phosphorylation of S6 (80). It has been suggested that this is due to an increased rate of initiation with the highly phosphorylated ribosomes (81, 82). Phosphorylation of S6 does not appear to alter elongation as there is little change in the rate of synthesis at early time points (80); at later times (6 hours), the average transit time decreases considerably, but this coincides with other cellular changes.

To determine whether changes in the synthesis of specific proteins following serum-stimulation of quiescent cells are due to alterations in transcription or translation, cells were incubated with [^{35}S]methionine in the presence and absence of actinomyocin D, and the proteins were analyzed by two-dimensional electrophoresis (83). In the absence of actinomycin D, synthesis of 15 proteins is stimulated; in the presence of the inhibitor, increased synthesis is observed for eight of the proteins. Thus, the increased synthesis of approximately half of the proteins is due to increased translation and recruitment of endogenous mRNA.

The direct effect of phosphorylation of S6 on mRNA binding to 40-S ribosomal subunits has been examined with synthetic and natural mRNA. Binding of AUG and poly(A,U,G) to 40-S ribosomal subunits phosphorylated by cA kinase or PAK II has been compared to that of nonphosphorylated ribosomal subunits (17). Phosphorylation of S6 by cA kinase significantly inhibits binding of both nucleic acids when compared with nonphosphorylated 40-S ribosomal subunits. Phosphorylation of 40-S ribosomal subunits with PAK II increases binding of AUG and poly(A,U,G). The effects of phosphorylation on translation have been monitored using a reconstituted protein synthesizing system containing the different phosphorylated forms of 40-S ribosomal subunits (47). Phosphate on S6 is stable in this system during the entire incubation period. Poly(A,U,G)-directed translation is inhibited with 40-S ribosomal subunits phosphorylated by cA kinase; translation is stimulated with subunits phosphorylated with PAK II.

Recently, Palen and Traugh (84) examined the effects of differential phosphorylation of S6 on the utilization of globin message. Translation of globin mRNA in the reconstituted protein-synthesizing system containing 40-S ribosomal subunits phosphorylated by cA kinase is virtually identical to that observed with nonphosphorylated 40-S ribosomal subunits. No change is observed in total synthesis or in the ratio of synthesis of α- and β-globin. Phosphorylation of S6 with PAK II stimulates translation of globin mRNA. The stimulation is greater than that observed with poly(A,U,G) and is 4-fold greater than that observed with nonphosphorylated subunits. Stimulation of the synthesis of both the α and β chains of globin is observed.

Since the sites phosphorylated by PAK II are identical to those observed *in vivo* in response to insulin and growth-promoting compounds, the data support the hypothesis that enhanced protein synthesis is due to phosphorylation of S6 and that differential phosphorylation of S6 can alter translation of natural mRNA. No effect on globin synthesis is observed following phosphorylation by cA kinase; since stimulation of protein synthesis is not an immediate response to elevated cAMP levels, such a result is expected. The effects of phosphorylation appear to be specific for specific mRNAs since phosphorylation of S6 by either PAK II or the cAMP-dependent protein kinase does not alter collagen synthesis in a reconstituted protein-synthesizing system (85). This differential effect on translation is discussed in detail in Section III of this review.

In the rickettsial infection identified as Q fever, there is multiple phosphorylation of S6, which coincides with increased protein synthesis (33a). The role of phosphorylation of S6 in protein synthesis has been examined using a reconstituted protein-synthesizing system from reticulocyte lysate with ribosomes and mRNA from infected and noninfected liver. The activity of the phosphorylated ribosomes from infected liver is 2-fold greater than that observed with ribosomes from control animals containing low levels of phosphate. Incubation with phosphatase reduces the translational activity of the ribosomes from infected liver to that observed with ribosomes from the control.

Recently, the two phosphorylation sites on the yeast equivalent of S6 were altered by site-specific mutagenesis of the gene coding for this protein (86). Although the mutant protein is not phosphorylated *in vivo*, the cells are able to grow, but at about half the rate of control cells. A comparison of the primary sequence flanking the phosphorylated serine residues from the yeast counterpart of S6 with the sequence for S6 from mammals shows they are almost identical in the region surrounding the first two serines (87). The four serines in the remaining COOH-terminus in mammals are not present in yeast due to an early termination. Thus, the sites on S6, presumed to be modified by growth-promoting compounds, are not present in yeast.

In a few instances, there appears to be little relationship between phosphorylation of S6 and protein synthesis. This occurs under toxic conditions, where S6 is fully phosphorylated (five phosphates), but protein synthesis is inhibited. This has been observed with several toxic compounds, including cycloheximide, dimethylnitrosamine, and thioacetamide (Table I). In these cases, increased phosphorylation of S6 may be considered a result of the toxicity, or the toxicity may override an accompanying trophic response.

In the instances of dephosphorylation of S6 in response to hypotonic or heat shock (48–51), rephosphorylation of S6 does not consistently parallel recovery of protein synthesis. One explanation is that the shock response overpowers the subtle effects of phosphorylation on protein synthesis and overrides the common regulatory system.

II. Phosphorylation of Aminoacyl-tRNA Synthetases

A. Characteristics of Aminoacyl-tRNA Synthetases

The aminoacyl-tRNA synthetases are a family of enzymes that carry out the first step in protein biosynthesis. These enzymes (E) catalyze the covalent attachment of an amino acid (AA) to its specific cognate tRNA according to the two-step reaction

$$E + AA + ATP \rightleftharpoons E \cdot (AA\text{-}AMP) + PP_i \qquad (1)$$

$$E \cdot (AA\text{-}AMP) + tRNA \rightleftharpoons AA\text{-}tRNA + AMP + E \qquad (2)$$

The aminoacyl-tRNA synthetases carry out this reaction with extraordinary precision, as evidenced by the low frequency (<0.03%) (88) with which an amino acid is incorrectly incorporated into proteins *in vivo*. Schimmel (89, 90) has proposed the concept of dual discrimination, one at the level of substrate binding and a second at catalysis. Using stopped-flow and temperature-jump methods (91–93), the first part of the second reaction has been shown to be a diffusion-controlled bimolecular association between activated enzyme and nucleic acid. The second part is a unimolecular conformational change in the complex. The latter occurs in milliseconds, and represents the discrimination step. Thus, with noncognate complexes, the second step is not detected. The K_m and maximal velocity also have an important role in discrimination between tRNA species as shown by steady-state kinetic analysis. The observed differences with cognate and noncognate tRNAs are so great that there is low statistical probability that the noncognate tRNA will be aminoacylated (94, 95). In addition, noncognate complexes may have dissociation constants 100 times those of the cognate complexes (96).

In the past 10 years, extensive research has been conducted on the aminoacyl-tRNA synthetases, particularly those of prokaryotic origin; much less is known about those of the higher eukaryotes. Until recently, progress on the study of the general properties and structural characteristics of the latter has been hampered by the fact that many exist in very large complexes.

These complexes have been described as ranging from fragile supramolecular assemblies comprising almost all of the 20 aminoacyl-tRNA synthetases to those containing a limited number of stably associated enzymes (97). The supramolecular complexes are extremely fragile, and attempts to purify them lead to breakdown. Waller and co-workers (98) established a three-step procedure for the isolation and

TABLE II
The Aminoacyl-tRNA Synthetase Complex from Rabbit

Component (tRNA-synthetase)	M_r	Molar ratio	Isoelectric point[a]	Specific activity[b] (units) 25°C	37°C
Glutamyl-	150,000	1–2	6.80	34	44
Isoleucyl-	139,000	1	6.30	114	20
Leucyl-	129,000	1	6.90	19	43
Methionyl-	108,000	1	6.15	64	92
Glutaminyl-	96,000	1	7.15	36	16
Lysyl-	76,000	3–4	6.30	114	238
Arginyl-	74,000	2–3	6.90	329	496
Aspartyl-	57,000	2	6.60/6.70	N.D.[c]	N.D.
Unidentified band 1	43,000	1	7.80/8.10	—	—
Unidentified band 2	38,000	1	7.20/7.40	—	—

[a] Isoelectric points are given for the synthetase complex isolated from rabbit liver (99).

[b] Specific activities are for the synthetase complex from rabbit reticulocytes. A unit of enzyme activity corresponds to the formation of 1 nmol of aminoacyl-tRNA/min/mg at 25°C (98) or 37°C (107).

[c] N.D., not determined.

purification of a large core complex from a number of different mammalian species. This complex has an apparent M_r of 1.2×10^6 (98, 99) and a sedimentation coefficient of approximately 18 S. The physical association of the aminoacyl-tRNA synthetases in the complex has been demonstrated by a variety of methods including gel filtration (100), ion-exchange chromatography (101, 102), affinity chromatography on lysyldiaminohexyl-Sepharose-B (103), and immunoprecipitation by antibody specific for one of the synthetases (104).

The aminoacyl-tRNA synthetases found in the core complex include the glutamyl-, isoleucyl-, leucyl-, methionyl-, glutaminyl-, lysyl-, arginyl-, and aspartyl-tRNA synthetases. The complex has been purified from sheep liver, sheep spleen, rabbit liver, rabbit reticulocytes (98), rat liver (105), and Chinese hamster ovary cells (106). The same eight aminoacyl-tRNA synthetases are present in each complex. Polyacrylamide gel electrophoresis of the purified complex reveals a similar if not identical polypeptide pattern composed of 10 to 11 protein bands, 8 of which have been assigned to constituent aminoacyl-tRNA synthetases (Table II). Six of the enzymes (the glutamyl-, leucyl-, methionyl , glutaminyl-, lysyl-, and arginyl-tRNA synthetases) were identified by partial reactivation of the isolated polypeptides following extraction from a polyacrylamide gel (108). The assignment of five of the enzymes was confirmed using antibodies raised against each of the electrophoretically separated polypeptide components by immunotitration (*108*) or immunoblotting (*99, 104*). Only later was the identity of the isoleucyl-tRNA synthetase unambiguously established, following isolation of the M_r 139,000 protein component (*109*). Recent studies have also established a correspondence between aspartyl-tRNA synthetase activity and the M_r 57,000 component (cited in *106*). While most of the synthetases are present in a molar ratio of one, the lysyl-, arginyl-, and aspartyl-tRNA synthetases are present in two to four copies per mol of complex. The activities of the synthetases in the complex vary considerably from that of aspartyl-tRNA synthetase, which is not always detectable, to the highly active arginyl-tRNA synthetase (Table II).

In all cases, the purified 18-S complex (*98, 99, 105*) also contains two unidentified proteins of M_r 42,000 and 38,000. Recently a third polypeptide, of M_r 18,000 (*105, 106*), has been identified, and occasionally a fourth, of M_r 235,000 (*98, 106*). These polypeptides do not react with antibody prepared to the aminoacyl-tRNA synthetases, indicating that they are not proteolytic fragments of the synthetases (*105*). It has been suggested that the proteins may be involved in mediating complex formation (*108*), or have a role in modulating expression of the activity of the synthetases in the complex (*110*).

Results from several laboratories indicate that all of the glutamyl-, isoleucyl-, leucyl-, methionyl-, glutaminyl-, and lysyl-tRNA synthetase activities are found exclusively in the core complex (*98, 105, 106, 111–113*). Portions of the arginyl- and aspartyl-tRNA synthetases (up to 30%) appear as free enzyme, except in reticulocytes, where all eight synthetases are in the complex. The free arginyl-tRNA synthetase has been isolated as a monomer of M_r 60,000, substantially smaller than the arginyl-tRNA synthetase (M_r 74,000) in the complex (*105, 112*). The free form of aspartyl-tRNA synthetase has also been purified; the enzyme consists of two identical subunits of M_r 53,000 (*114*), which are slightly smaller than the subunit (M_r 57,000) reported for aspartyl-tRNA synthetase in the core complex (*98, 105, 106*). This suggests that the free arginyl- and aspartyl-tRNA synthetases arose by proteolysis.

Prolyl-tRNA synthetase is found both in the core complex and as a free enzyme, depending on the source from which it is isolated, the presence of protease inhibitors, and the homogenization conditions. Significant levels of prolyl-tRNA synthetase are present in the core complex from Chinese hamster ovary cells (*106*) and rat liver (*105*). A 24-S synthetase complex has been isolated from rat liver (*115*); it contains the enzymes in the 18-S complex plus prolyl-tRNA synthe-

TABLE III
Molecular Weight and Subunit Structure of Free Aminoacyl-tRNA Synthetases from Mammals

Aminoacyl-tRNA synthetase	Source	Subunit structure	Subunit M_r	Reference
Histidyl-	Rabbit reticulocytes	α_2	64,000	118
	Rat liver	α_2	64,000	119
Threonyl-	Rat liver	α_2	85,000	115a
Seryl-	Bovine liver	α_2	87,000	120
	Hen liver	α_2	63,000	120a
Tyrosyl-	Rat liver	α_2	68,000	121
Tryptophanyl-	Human placenta	α_2	58,000	116
	Beef pancreas	α_2	60,000	122
Phenylalanyl-	Rat liver	$\alpha_2\beta_2$	69,000	117
			74,000	
Cysteinyl-	Rat liver	α	120,000	123
		α_2 (major)		
		α_3		

tase, while a 30-S complex also contains threonyl-tRNA synthetase (111). Recentrifugation of the 24-S complex leads to preferential dissociation of prolyl-tRNA synthetase from the core complex (115) indicating that the synthetase is located on the periphery of the complex and can be easily removed during purification. A small portion of the threonyl-tRNA synthetase is also in the core complex, but most of this enzyme is in the free form (106, 111, 115a). Of the remaining 10 synthetases, 9 exhibit M_rs characteristic of free enzymes, ranging from about 100,000 for tryptophanyl-tRNA synthetase (116) to 287,000 for phenylalanyl-tRNA synthetase (117). Valyl-tRNA synthetase is aggregated in a homotypic complex with a native M_r greater than 500,000 (98).

Of the synthetases that occur as free enzymes, two of the better characterized are the threonyl- and histidyl-tRNA synthetases. Threonyl-tRNA synthetase has been purified to near homogeneity from rat liver and is an α_2 dimer with a subunit M_r of 85,000 (115a). Histidyl-tRNA synthetase, purified from rabbit reticulocytes (118) and rat liver (119), has an α_2 structure with a subunit M_r of 64,000. Other free aminoacyl-tRNA synthetases have also been purified from mammalian sources, including seryl-, tryptophanyl-, tyrosyl-, phenylalanyl-, and cysteinyl-tRNA synthetases. The molecular weights and subunit structures of the purified, free aminoacyl-tRNA synthetases are shown in Table III. Most of the enzymes listed have an α_2 structure, with two exceptions: phenylalanyl-tRNA synthetase has an $\alpha_2\beta_2$

structure (117) and cysteinyl-tRNA synthetase has an α and α_3 in addition to the predominant α_2 form (123).

Various studies have established that hydrophobic interactions play a prominent role in the assembly of stable multienzyme complexes. Indirect evidence that this is true for the synthetase complex was first provided by partial destabilization of the complex from rat liver by hydrophobic chromatography (124), and by the synergistic action of neutral detergents and chaotropic salts (125). Direct evidence was obtained recently that leucyl- and lysyl-tRNA synthetases (126), as well as isoleucyl-tRNA synthetase (109), display markedly hydrophobic properties. Additional evidence comes from the observation that fully active, proteolytically cleaved forms of lysyl- (126) and methionyl- (109) tRNA synthetases lose the hydrophobic properties associated with the enzymes in the intact complex. Taken together, these results suggest that assembly of the aminoacyl-tRNA synthetases is mediated by hydrophobic domains present on these enzymes. It has been proposed that the aminoacyl-tRNA synthetase complex contains multiple synthetases with their catalytic centers exposed to the environment, but held together by a core of hydrophobic extensions on the individual proteins, and that lipids or other components may help "glue" the complex together (127). Several reports show that lipids coelute with the synthetase complex (128, $128a$). However, the lipids do not appear to be essential for maintaining the integrity of the complex, as delipidation with detergents does not necessarily lead to breakdown of the complex, although it destabilizes the synthetase activities (125).

Cirakoglu et al. (129) divide the synthetases into two classes with respect to complex formation. Class I includes those synthetases listed in Table I plus prolyl-tRNA synthetase. Class II includes all the other synthetases that are predominantly free in the cytoplasm. Class I synthetases have acquired hydrophobic properties leading to integration in the core complex; class II enzymes appear to lack these properties. This is clearly established for seryl- (120), threonyl- ($115a$), tryptophanyl- (122), tyrosyl- (121), and histidyl-tRNA synthetases (118). They predict the class II synthetases would associate individually with the core complex by electrostatic interactions. The polyanionic character of all of the synthetases would lead to compartmentalization of the enzymes with other components of the cytoplasm, such as the cytoskeletal structure, the rough endoplasmic reticulum (130, 131), and other components of protein synthesis (113, 132–134), through electrostatic associations.

The possibility that association of the synthetases in the complex may alter their kinetic properties as compared to unassociated counterparts has been investigated. For five of the synthetases in the core complex, the apparent Michaelis constants for amino acid, ATP, and tRNA are not significantly different from those obtained with enzymes dissociated by hydrophobic chromatography (*109, 135, 136*). The initial rates of aminoacylation catalyzed by any one of the enzymes in the complex is not affected by the function of the other synthetases (*136*).

Considerable research has been directed toward elucidation of the molecular basis for recognition of tRNA by the synthetases. Three regions of the tRNA molecule have been implicated as contact points with the synthetases in prokaryotes: the anticodon, the 5′-side of the dihydrouridine stem, and the acceptor terminus (*89, 90*). The synthetases span a considerable distance in the crystal structure of yeast tRNA^{Phe}; the acceptor terminus and the anticodon are separated by about 75 Å (*137*). The synthetases appear to bind along the inside of the L-shaped tRNA structure (*138*), and the binding may extend all the way from the 3′ terminus to the anticodon.

Mammalian aminoacyl-tRNA synthetases have other cellular functions in addition to aminoacylation. Lysyl- and phenylalanyl-tRNA synthetases synthesize P^1,P^4-bis(adenosine-5′)tetraphosphate (Ap$_4$A) (*139, 140*). Synthesis of Ap$_4$A is dependent upon the specific amino acid and ATP. The synthetase reacts with ATP and amino acid to form the E·(AA-AMP) which reacts with a second ATP to form Ap$_4$A with the release of amino acid. Zinc stimulates the production of Ap$_4$A *in vitro* up to 100-fold (*141*) while inhibiting aminoacylation. Ap$_4$A is a pleiotropic regulator or signal nucleotide that stimulates DNA synthesis and appears to be involved in the regulation of replication (*142*). The intracellular concentrations of Ap$_4$A increase 50- to 100-fold, to concentrations as high as 1 μM, during the transition from G to S phase in growth stimulated cells (*143*). It is interesting that phenylalanyl-tRNA synthetase is found in the nucleus of rapidly growing cells by immunofluorescence microscopy using antibody specific for the synthetase (*130*). Addition of Ap$_4$A to permeabilized, quiescent cells also stimulates DNA synthesis (*142*). Ap$_4$A binds tightly and specifically to DNA polymerase α (*145*) and appears to serve as a primer for the enzyme *in vitro* (*146*).

The question has been raised as to a regulatory role for the synthetase complex in Ap$_4$A synthesis, as free lysyl-tRNA synthetase shows a 6-fold higher rate of synthesis of Ap$_4$A than the complexed form of the enzyme (*140*). Although the apparent Michaelis constants for ATP and lysine are similar for the two forms of the enzyme, their kinetic mechanisms appear to be distinct (*147*). It has been suggested that the difference in kinetic mechanisms may be caused by an interaction between the hydrophobic domain and the remainder of the synthetase.

B. Phosphorylation of Aminoacyl-tRNA Synthetases

As early as 1977, Berg (4) proposed, on the basis of observations with tissue extracts from mouse uterus and liver, that the mammalian synthetases are regulated by phosphorylation/dephosphorylation. Berg reported that the activities of 12 aminoacyl-tRNA synthetases are decreased 2- to 4-fold by prior incubation with ATP and Mg^{2+}, while the activities of five synthetases are unaltered or increased under these conditions. Treatment with alkaline phosphatase reverses these effects (4).

The activities of the synthetases have also been examined after treatment of mice with 17β-estradiol or Bt_2cAMP (4, 148). In the uterus, the activities of a number of synthetases increase, while the activities of the rest of the synthetases decrease following injection of either compound. In liver, the changes in synthetase activities are similar with Bt_2cAMP, but 17β-estradiol has the opposite effect, as liver is not a target for this hormone. Based on these results, Berg suggests that a phosphoprotein phosphatase is activated in response to elevated levels of cAMP resulting in activation of most of the synthetases. No evidence is provided for the direct phosphorylation or dephosphorylation of any of the enzymes.

Five of the synthetases in the core complex from rat liver are totally inhibited *in vitro* by incubation under phosphorylating conditions and reactivated under conditions where dephosphorylation is favored (149). Inactivation of the leucyl-, methionyl-, lysyl, isoleucyl-, and arginyl-tRNA synthetases required ATP, Mg^{2+}, and the phosphatase inhibitor, sodium fluoride. Upon gel filtration on Sepharose 4B, the synthetases were resolved from an inactivating factor that required ATP and Mg^{2+}, and from a reactivating factor that required Mg^{2+} and was inhibited by sodium fluoride. These factors were alluded to as protein kinases and phosphoprotein phosphatases, respectively (149). The synthetase activities were reactivated by the addition of highly purified protein phosphatases. These studies are marred by the fact that the conditions of the assay were not optimal, including the ATP:Mg^{2+} ratio and the use of *E. coli* tRNA.

Such studies, examining synthetase activities in extracts or partially purified preparations, are fraught with problems. Synthetase activities are dependent on the ATP:Mg^{2+} ratio, the optimum of which differs depending on the synthetase activity. In addition, synthetases are extremely sensitive to changes in ionic strength and buffer conditions; thus, addition of components to reaction mixtures must be carefully controlled. It is not apparent that all of these parameters were carefully considered in all of the studies described above.

In an investigation of the activity of the aminoacyl-tRNA synthetases *in vivo* and in cultured liver cells following hormonal treatment or starvation/refeeding, the five synthetases identified above were largely in the activated state in cells from fed rats, but overnight starvation led to a 50 to 60% inhibition of the activities (*149*). Incubation of hepatocytes from the fed rats with glucagon and the phosphodiesterase inhibitor 3-isobutyl-1-methylxanthine caused a substantial decrease in synthetase activity. Insulin treatment of hepatocytes prepared from starved rats produced the opposite effect, with reactivation of the synthetases after 30 minutes. These results are consistent with the proposal that the aminoacyl-tRNA synthetases are regulated by phosphorylation/dephosphorylation; however, the evidence is indirect, as none of the aminoacyl-tRNA synthetases was shown to be phosphorylated. Thus, no direct correlation between changes in synthetase activity and phosphorylation can be made. In addition, the enzyme activities were measured *in vitro* in the presence and absence of a crude preparation of reactivating factor, the composition of which could vary considerably depending on the hormonal state of the cell. Recently, a reinvestigation of the inactivation/reactivation phenomenon by Schelling and Cohen (personal communication) indicated that the observed changes in synthetase activities could be attributed to the production and cleavage of inorganic pyrophosphate during the inactivation and reactivation reactions, respectively.

Gerken and Arfin (*150*) provided the first direct evidence for phosphorylation of aminoacyl-tRNA synthetases *in vivo* by showing that the threonyl-tRNA synthetase from Chinese hamster ovary cells is phosphorylated on serine following incubation of the cells with $^{32}P_i$ (*150*). For these studies, a mutant cell line that produced 60- to 100-fold more threonyl-tRNA synthetase than the wild type, representing about 1.5% of the total soluble protein as compared to 0.025% in wild-type cells, was utilized (*151*). The synthetase was isolated with antibody prepared in rabbit against purified enzyme from rat liver.

Antibody to aminoacyl-tRNA synthetases has been detected in the blood of patients with autoimmune diseases. In polymyositis patients, antibody directed against a complex of protein and nucleic acid was specific for histidyl-tRNA synthetase complexed with $tRNA^{His}$ (*152, 153*). A second myositis autoantibody immunoprecipitated threonyl-tRNA synthetase (*154*), while other patients had antibody precipitat-

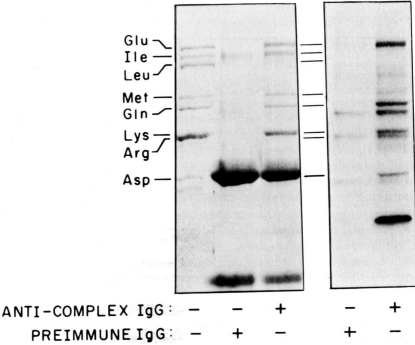

FIG. 4. Identification of phosphorylated aminoacyl-tRNA synthetases in the core complex. Reticulocytes from 5 ml of blood were washed, resuspended in a nutritional medium, and incubated with $^{32}P_i$ for 2 hours. Following lysis and removal of polysomes, the postribosomal supernatant was chromatographed on tRNA-Sepharose. The protein removed by high salt was analyzed by immunoprecipitation with preimmune serum and antiserum prepared to the purified core complex in pigmy goat. Left panel, protein stained with Coomassie blue; right panel, autoradiogram.

ing alanyl-tRNA synthetase (155, 156). Antibodies directed against histidyl-tRNA synthetase were utilized to make the initial observations that this synthetase is phosphorylated in HeLa cells (153); this has been confirmed by the phosphorylation of histidyl-tRNA synthetase on serine in a mutant line of Chinese hamster ovary cells (157).

We find that five of the synthetases in the core complex are phosphorylated upon incubation of reticulocytes with $^{32}P_i$. The complex was partially purified by affinity chromatography of the postribosomal supernatant on tRNA-Sepharose followed by immunoprecipitation with antibody prepared in goat to the purified complex from rabbit. As shown in Fig. 4, the glutamyl-, methionyl-, glutaminyl-, lysyl-, and aspartyl-tRNA synthetases were phosphorylated as well as the 37,000-Da protein, in all instances on serine. The leucyl-tRNA synthetase was removed by detergents during washing of the immunoprecipitate, so it was not possible to determine whether this protein was modified. Previously, an unsuccessful attempt had been made to show that the complex was phosphorylated in Chinese hamster ovary and HeLa cells (106). This negative result may have been caused by the use of NaF as a phosphatase inhibitor; NaF is less effective than the β-glycerophosphate used in the studies described above. The earlier studies (106) were also unable to demonstrate changes in synthetase activities following treatment of cells with a variety of compounds such as Bt_2cAMP, 3-isobutyl-1-methylxanthine, glucagon, insulin, and testosterone in the presence or absence of serum. However, the synthetase activities assayed were those specific for arginyl- and methionyl-tRNA synthetases. We found that the former was not phosphorylated and the latter was only slightly phosphorylated in reticulocytes.

Thus, of the nine synthetases examined (seven in the core complex and two free enzymes), seven can be phosphorylated in cultured cells or reticulocytes. This suggests that phosphorylation of aminoacyl-tRNA synthetases is not a unique event and that most of the synthetases remaining to be examined will probably be phosphorylatable.

C. Effects of Phosphorylation at the Molecular Level: Actual and Hypothetical

As a general rule, every protein phosphorylated *in vivo* is modified by at least two different site-specific protein kinases. The only exception known is phosphorylase. Due to the multiple phosphorylation events occurring *in vivo*, it is difficult to attribute changes in activity to specific phosphorylation events. Thus, initial studies on the effects of phosphorylation at the molecular level have utilized purified synthetase or synthetase complex, and purified protein kinases. Since the mammalian synthetases are present in complex structures and the aminoacylation reaction can be divided into partial reactions, there are a number of activities that can be investigated. These include the potential effects of phosphorylation (1) on the total aminoacylation reaction, (2) on binding of total cognate tRNA, (3) on binding of cognate tRNAs with different anticodons, (4) on Ap_4A synthesis, (5) on association/dissociation of the core complex, (6) on association/dissociation of the fragile synthetase complex, and (7) on association of the complex with other cellular components.

We have examined phosphorylation of the synthetases in the core complex and threonyl-tRNA synthetase, with five different multipotential protein kinases. These include casein kinase I and II, cA ki-

TABLE IV

Specific Phosphorylation of the Aminoacyl-tRNA Synthetases[a]

tRNA synthetase	In vivo	Casein kinase I	cA kinase	C kinase	PAK I
Glutamyl-	+	+	+	+	
Isoleucyl-		+			
Leucyl-	N.D.[b]		+		
Methionyl-	+	+			
Glutaminyl-	+				
Lysyl-	+	+			
Arginyl-					
Aspartyl-	+		+	+	
Threonyl-	+		+	+	+

[a] Data taken from Fig. 4 and *107, 150, 158,* and *159.*
[b] N.D., not determined.

nase, C kinase, and PAK I. As summarized in Table IV, only three of the five protein kinases examined could phosphorylate synthetases in the core complex. Glutamyl-tRNA synthetase was phosphorylated *in vitro* by casein kinase I, by cA kinase, and by C kinase. Only one protein kinase (casein kinase I) phosphorylated methionyl-tRNA synthetase. Lysyl-tRNA synthetase was phosphorylated by casein kinase I and cA kinase, while aspartyl-tRNA synthetase was phosphorylated by cA kinase and C kinase. All four of these synthetases were shown to be phosphorylated in reticulocytes incubated with $^{32}P_i$. None of the protein kinases appeared to phosphorylate the glutaminyl-tRNA synthetase, suggesting that other enzymes, possibly PAK II and the Ca^{2+}, calmodulin-dependent protein kinase, may be also involved.

Threonyl-tRNA synthetase is modified *in vitro* by cA kinase, C kinase, and PAK I (Table IV). Since each of the protein kinases recognizes a different sequence around the phosphorylatable serine, it would be expected that each of the enzymes phosphorylates a different site(s) on the synthetase. This is the case for threonyl-tRNA synthetase as shown by two-dimensional phosphopeptide mapping of tryptic digests of the phosphorylated enzyme.

A study of the effects of phosphorylation of the synthetases in the core complex by casein kinase I has been conducted (*107*). Care was taken to assay for protein kinase activity with casein and histone as substrates during the large-scale purification of the core complex from reticulocytes, and protease inhibitors, but not phosphatase inhibitors, were added. The final purified core complex eluted at 260 to 350 mM NaCl from tRNA-Sepharose and was devoid of protein kinase activity. Upon incubation with purified casein kinase I, up to 0.8 mol of phosphate was added per mol of synthetase. The effect of phosphorylation on tRNA binding was monitored by affinity chromatography on tRNA-Sepharose. Following phosphorylation, the complex was removed from the column at a lower concentration of salt (190 mM NaCl) than the nonphosphorylated control (275 mM NaCl). Phosphorylation appeared to have no effect on the integrity of the complex.

These results are supported by aminoacylation studies (107). Phosphorylation by casein kinase I inhibited aminoacylation of the four phosphorylated synthetases by as much as 38%, compared to the nonphosphorylated controls. No effects of phosphorylation were observed on the activities of other synthetases in the complex, which were not modified by casein kinase I. Similar studies need to be completed with the synthetase complex phosphorylated by cA kinase and C kinase. The effects of phosphorylation on the free synthetases also need to be investigated.

The protein kinase partially copurifying with the synthetase complex on tRNA-Sepharose (107) was identified as casein kinase I (Pendergast and Traugh, unpublished) by phosphorylation of casein and the lack of inhibition by heparin and GTP, which differentiate between casein kinase I and II (160). Casein kinase I is a monomer (M_r 37,000) that is autophosphorylated (160). This led us to investigate the identity of the 37,000-Da protein in the core complex. Phosphopeptide maps of the protein in the complex phosphorylated by casein kinase I and of autophosphorylated casein kinase I were identical (107). This led us to conclude that the protein (M_r 37,000) in the purified complex is an inactive form of casein kinase I.

There are numerous points at which phosphorylation of the synthetases could affect activity. Phosphorylation could alter the activity of individual synthetases by enhancing or diminishing binding of tRNA, amino acid and/or ATP and Mg^{2+}. This could occur by altering the conformation of synthetase and/or the charge at or near the substrate binding site resulting in a change in K_m or V_{max} of the enzyme for the substrate(s). Phosphorylation by the individual protein kinases is selective; not all synthetases are phosphorylated by the same enzymes. Thus, the activity of specific synthetases could be affected by specific growth conditions. This could result in diminished (or increased) levels of specific aminoacylated-tRNAs, resulting in preferential translation of proteins low (or high) in that amino acid.

Alternatively, phosphorylation could preferentially enhance or diminish binding of a specific cognate tRNA. For instance, phosphorylation of glutamyl-tRNA synthetase could alter conformation of the synthetase sufficiently so that the tRNA species recognizing the codon

GAA would bind preferentially to the synthetase, whereas binding of species recognizing GAA would be unaltered or decreased. This could result in a preferential translation of proteins containing specific classes of codons. This hypothesis gains support from the fact that overall levels of individual tRNAs vary with cell type (and possibly hormonal status). This would be an excellent mechanism for enhancing translation of specific classes of messages.

To date, only two mammalian synthetases have been shown to produce significant amounts of Ap$_4$A (*139, 140*). One, lysyl-tRNA synthetase, has been shown to be phosphorylated. Since both cA kinase and C kinase are activated in response to hormones, these phosphorylation events could be involved in regulating levels of Ap$_4$A. Synthesis of Ap$_4$A increases dramatically in the transition from G to S phase (*143*) and appears to be involved in stimulation of DNA synthesis. One would also expect the phenylalanyl-tRNA synthetase to be similarly phosphorylated.

Our initial published (*107*) and unpublished studies offer no evidence for an effect of phosphorylation on association or dissociation of the core complex. This is supported by the observation that all eight of the synthetases in the core complex are tightly associated, with the exception of a small percentage of the arginyl- and aspartyl-tRNA synthetases that were proteolyzed (*105, 106*). Thus, at this time, there is no evidence for dissociation of the eight synthetases to form the core complex upon phosphorylation or dephosphorylation.

Of great potential interest, with regard to phosphorylation, is the association of other synthetases with the core complex. Up to 18 synthetases are associated with a large fragile complex that is readily dissociated to form the core complex upon purification. Two of the last enzymes to remain associated with the core complex are prolyl- and threonyl-tRNA synthetases (*111, 115a*). It has been postulated that the core complex is held together by hydrophobic interactions and the remainder of the synthetases are associated with the core by electrostatic interactions (*127, 129*). Phosphorylation could play an important role in the latter by altering the conformation of specific regions of the synthetases and/or by addition of a phosphate. Since phosphorylation of the core complex and the free enzymes has been observed, phosphorylation could be important in stabilizing or destabilizing synthetase interactions.

Similarly, phosphorylation could position the synthetases for active participation in protein synthesis by altering binding to protein-synthesizing complexes and the cytoskeleton. Protein synthesis could be diminished by dissociating the synthetases from these complexes.

III. Coordinate Regulation of Protein Synthesis

There are several mechanisms by which translation can be altered: by a switching mechanism, by altering the rate of total protein synthesis, or by changing the rate of synthesis of specific proteins. The switching mechanism is utilized in hemin control of protein synthesis in red blood cells, where iron depletion effectively inhibits total protein synthesis via the specific phosphorylation of eIF-2 (*1*). Viral infection of interferon-treated cells has a similar effect (*1*). Since protein synthesis is essential for cell maintenance, it would be expected that only under extreme conditions would protein synthesis be shut down. Thus, normal hormonal changes should result in more moderate alterations of protein synthesis.

Metabolism is coordinately regulated by hormones and compounds that alter levels of cAMP, cGMP, Ca^{2+}, and diacylglycerol, in part by regulating the activity of serine protein kinases. Other hormones and compounds alter metabolism via activation of tyrosine protein kinases. Many of the effects of these compounds on metabolism can be observed immediately; alterations in transcription with the resultant changes in translation require time for processing and transport of new mRNA.

Rates of protein synthesis in mammalian tissues fluctuate in response to dietary and hormonal changes. In skeletal muscle and liver, starvation, diabetes, or treatment with glucocorticoids (*161–165*) lead to decreased rates of protein synthesis. These changes can be reversed by refeeding or by administration of the appropriate hormone. Ca^{2+} is an important requirement for optimal protein synthesis in both normal and tumor cells, including reticulocytes (*166, 167*). In pituitary cells, Ca^{2+} enhances the stimulation of protein synthesis with EGF or phorbol esters over that observed in Ca^{2+}-depleted cells (*167*). Ca^{2+} appears to act at the translational level directly on the protein synthesizing system, possibly through activation of the Ca^{2+}, calmodulin-dependent protein kinase or the Ca^{2+}, phospholipid-dependent protein kinase (C kinase).

Phosphorylation is an ideal way in which to coordinate protein synthesis. For instance, cA kinase phosphorylates S6 and specific aminoacyl-tRNA synthetases as well as one of the subunits of eIF-3 (*168*). Although other protein kinases have not been investigated in as much detail, PAK II and C kinase are activated in response to some of the same compounds (e.g., phorbol esters) and phosphorylate S6 and aminoacyl-tRNA synthetases, respectively.

Evidence has been accumulating in recent years implicating the cytoskeleton as the structural support for organization of the complexes of proteins and nucleic acids comprising the translational machinery. Polysomes (*169–171*), initiation factors (*172*), and tRNA synthetase complexes (*130, 131*) are associated with the cytoskeletal framework. This enhances the proximity of the major components of protein synthesis and simplifies the potential for coordinate regulation of the various components by phosphorylation.

Ribosomal protein S6 is situated near the cleft of the 40-S ribosomal subunit (*173*) and can be cross-linked with synthetic and natural mRNA (*174–176*). Several initiation factors are also cross-linked with ribosomal proteins associated with the site of initiation, including S6 (*177, 178*). Thus, S6 is in a position to modulate translation of mRNA by subtle conformational changes and/or changes in charge. Changes in conformation may occur via phosphorylation of S6, as shown by differences in reductive methylation following intoxication with ethionine and reversal with adenine (*179*). Several ribosomal proteins in the 40-S ribosomal subunit (S3, S4, S7, and S23/24) and seven proteins in the large subunit appear to be involved in the conformational change. The proteins in the 40-S ribosomal subunit are situated near S6 in the ribosome and several have been shown to be involved in initiation in the cross-linking studies cited above.

We propose that phosphorylation of S6 could alter binding or positioning of the message on the 40-S ribosomal subunit, thereby affecting translation. We postulate that there are at least three classes of mRNA that are recognized differentially by 40-S ribosomal subunits.

1. Proteins involved in growth: mRNA is preferentially translated in response to growth-promoting compounds when S6 is multiply phosphorylated by the mitogen-stimulated S6 kinase.

2. Housekeeping proteins: mRNA is translated under all conditions; phosphorylation of S6 has no effect.

3. Proteins involved in catabolism: mRNA is preferentially translated in response to elevated levels of cAMP where S6 is phosphorylated by cA kinase.

The various classes of mRNA would be differentially recognized by the different phosphorylated states of the ribosome. This differential recognition of the mRNA could be mediated by the primary sequence and conformation of the leader region. The leader sequence is responsive to changes in metabolism as shown by studies in which the controlling elements of the heat-shock protein-70 gene were fused to the gene for alcohol dehydrogenase to replace the normal leader sequence of the enzyme (*180*). Message from the native gene, although present in significant concentration, was not translated. Since dephosphorylation of S6 is an immediate result of heat shock, the information in the leader sequence could play an important role in selective translation of mRNA mediated by phosphorylation of S6, initiation factors, and aminoacyl-tRNA synthetases. It is important to note that binding of mRNA to ribosomes is mediated by initiation factors, a number of which have been shown to be phosphorylated *in vivo* and in reticulocyte lysates (*181, 182*). These include initiation factors 2, 3, and 4B. Serum stimulation of quiescent cells stimulates phosphorylation of eIF-4B, one of the factors involved in binding of mRNA to the 40-S ribosomal subunit (*183*). Coordinate regulation of phosphorylation of S6 and initiation factors could concur to enhance binding or positioning of the mRNA.

The specificity in phosphorylation of the aminoacyl-tRNA synthetases could also promote selective translation of mRNA by enhancing (or decreasing) aminoacylation of tRNAs that recognize specific codons. Messages preferentially translated in response to elevated levels of cAMP could contain higher levels of tRNA specific for one or more codons found in low concentrations in the messages for housekeeping proteins. Phosphorylation of the synthetases could result in significantly higher levels of aminoacylation of these species of tRNA, enhancing the rate of elongation of endogenous message. Thus, phosphorylation of S6, aminoacyl-tRNA synthetases, initiation factors, and the proteins associated with mRNA would act in concert to regulate translation of different classes of mRNA in response to hormones and other compounds.

Acknowledgments

Research conducted in our laboratory was sponsored by Grant GM 21424 from the U.S. Public Health Service. We thank Olga Perisic, Emilia Palen, Polygena Tuazon, Chuan Dang, and Craig Byus for many helpful discussions.

References

1. J. A. Traugh, in "Biochemical Actions of Hormones" (G. Litwack, ed.), Vol. 8, p. 167. Academic Press, New York, 1981.
2. J. A. Traugh, G. M. Hathaway, P. T. Tuazon, S. M. Tahara, G. A. Floyd, R. W. Del Grande and T. S. Lundak, *ICN-UCLA Symp. Mol. Cell. Biol.* **13**, 233 (1979).
3. S. M. Lastick and E. H. McConkey, *JBC* **256**, 583 (1981).
4. B. H. Berg, *BBA* **479**, 152 (1977).
5. N. Barden and F. Labrie, *Bchem* **12**, 3096 (1973).
6. M. L. Cawthon, L. F. Bitte, A. Krystosek and D. Kabat, *JBC* **249**, 275 (1974).
7. J. A. Traugh, R. W. Del Grande and P. T. Tuazon, *Cold Spring Harbor Conf. Cell Proliferation* **8**, 999 (1981).

8. A. M. Gressner and I. G. Wool, *JBC* **251**, 1500 (1976).
9. U. K. Schubart, S. Shapiro, N. Fleischer and O. M. Rosen, *JBC* **252**, 92 (1977).
10. R. E. H. Wettenhall and G. J. Howlett, *JBC* **254**, 9317 (1979).
11. D. P. Leader and A. A. Coia, *BBA* **519**, 213 (1978).
12. G. A. Floyd and J. A. Traugh, *EJB* **117**, 257 (1981).
13. C. Blat and J. E. Loeb, *FEBS Lett.* **18**, 124 (1971).
14. A. M. Gressner and I. G. Wool, *Nature* **259**, 148 (1976).
15. E. H. McConkey, H. Bielka, J. Gordon, S. M. Lastick, A. Lin, K. Ogata, J.-P. Reboud, J. A. Traugh, R. R. Traut, J. R. Warner, H. Welfle and I. G. Wool, *MGG* **169**, 1 (1979).
16. E. Kaltschmidt and H. G. Wittmann, *Anal. Biochem.* **36**, 401 (1970).
17. S. J. Burkhard and J. A. Traugh, *JBC* **258**, 14003 (1983).
18. R. W. Del Grande and J. A. Traugh, *EJB* **123**, 421 (1982).
19. R. W. Del Grande, Ph.D. Thesis, University of California, Riverside, 1981.
20. R. E. H. Wettenhall and P. Cohen, *FEBS Lett.* **140**, 263 (1982).
21. R. E. H. Wettenhall and F. J. Morgan, *JBC* **259**, 2084 (1984).
22. J. Blenis, J. G. Spivack and R. L. Erikson, *PNAS* **81**, 6408 (1984).
23. J. Blenis and R. L. Erikson, *J. Virol.* **50**, 966 (1984).
24. S. Decker, *PNAS* **78**, 4112 (1981).
25. I. M. Kennedy, W. S. Stevely and D. P. Leader, *J. Virol.* **39**, 359 (1981).
26. O. Perisic and J. A. Traugh, *FEBS Lett.* **183**, 215 (1985).
27. G. Thomas, J. Martin-Perez, M. Siegmann and A. M. Otto, *Cell* **30**, 235 (1982).
27a. J. Nishimura and T. F. Deuel, *FEBS Lett.* **156**, 130 (1983).
28. I. Novak-Hofer and G. Thomas, *JBC* **259**, 5995 (1984).
28a. J. L. Maller, J. G. Foulkes, E. Erikson and D. Baltimore, *PNAS* **82**, 272 (1985).
29. O. Perisic and J. A. Traugh, *JBC* **258**, 9589 (1983).
30. S. M. Lastick and E. H. McConkey, *BBRC* **95**, 917 (1980).
31. J. M. Trevillyan, O. Perisic, J. A. Traugh and C. V. Byus, *JBC* **260**, 3041 (1985).
32. J. M. Trevillyan, R. K. Kulkarni and C. V. Byus, *JBC* **259**, 897 (1984).
33. A. M. Gressner and I. G. Wool, *JBC* **249**, 6917 (1974).
33a. M. J. Hickey, F. R. Gonzales and D. Paretsky, *Infect. Immun.* **48**, 690 (1985).
34. A. M. Gressner and I. G. Wool, *BBRC* **60**, 1482 (1974).
35. A. M. Gressner and H. Greiling, *Biochem. Pharmacol.* **24**, 2495 (1978).
36. A. M. Gressner and H. Greiling, *Exp. Mol. Pathol.* **28**, 39 (1978).
37. O. Perisic and J. A. Traugh, *JBC* **258**, 13998 (1983).
38. R. Jahn and H. D. Söling, *FEBS Lett.* **153**, 71 (1983).
39. P. J. Nielsen, G. Thomas and J. L. Maller, *PNAS* **79**, 2937 (1982).
40. J. Hanocq-Quertier and E. Baltus, *EJB* **120**, 351 (1981).
41. J. Kruppa, D. Darmer, H. Kalthoff and D. Richter, *EJB* **129**, 539 (1983).
42. J. G. Spivack, R. L. Erikson and J. L. Maller, *MCBiol* **4**, 1631 (1984).
43. W. J. Wasserman and J. G. Houle, *Dev. Biol.* **101**, 436 (1984).
44. S. M. Lastick and E. H. McConkey, *ICN-UCLA Symp. Mol. Cell. Biol.* **12**, 61 (1978).
45. G. Thomas, M. Siegmann, A. M. Kubler, J. Gordon and L. Jimenez de Asua, *Cell* **19**, 1015 (1980).
46. J. L. Maller, J. G. Foulkes, E. Erikson and D. Baltimore, *PNAS* **82**, 272 (1985).
47. S. M. Lastick, P. J. Nielsen and E. H. McConkey, *MGG* **152**, 223 (1977).
48. J. Kruppa and M. J. Clemens, *EMBO J.* **3**, 95 (1984).
49. A. E. Olsen, D. F. Triemer and M. M. Sanders, *MCBiol* **3**, 2017 (1983).
50. C. V. C. Glover, *PNAS* **79**, 1781 (1982).
50a. I. M. Kennedy, R. H. Burdon and D. P. Leader, *FEBS Lett.* **169**, 267 (1984).

51. K.-D. Saharf and L. Nover, *Cell* **30**, 427 (1982).
52. M. A. Treloar, M. E. Treloar and R. Kisilevsky, *JBC* **252**, 6217 (1977).
53. J. Martin-Perez and G. Thomas, *PNAS* **80**, 926 (1983).
54. J. Martin-Perez, M. Siegmann and G. Thomas, *Cell* **36**, 287 (1984).
55. T. H. Lubben and J. A. Traugh, *JBC* **258**, 13992 (1983).
56. P. T. Tuazon, W. C. Merrick and J. A. Traugh, *JBC* **255**, 10954 (1980).
57. S. M. Tahara and J. A. Traugh, *EJB* **126**, 395 (1982).
58. M. I. Gonzatti-Haces and J. A. Traugh, *JBC*, in press.
58a. X. Fu, N. Phillips, J. Jentoft, P. T. Tuazon, J. A. Traugh and J. Leis, *JBC* **260**, 9941 (1985).
59. B. Gabrielli, R. E. H. Wettenhall, B. E. Kemp, M. Quinn and L. Bizonova, *FEBS Lett.* **175**, 219 (1984).
60. M. J. Donahue and R. A. Masaracchia, *JBC* **259**, 435 (1984).
61. B. A. de la Houssaye, T. K. Echols and R. A. Masaracchia, *JBC* **258**, 4274 (1983).
62. P. E. Magnino, B. A. de la Houssaye and R. A. Masaraccia, *BBRC* **116**, 675 (1983).
63. A. Rodriguez-Pena and E. Rozengurt, *EMBO J.* **4**, 71 (1985).
64. Y. Nishizuka, *TIBS* **9**, 163 (1984).
65. C. J. Le Peuch, R. Ballester and O. M. Rosen, *PNAS* **80** 6858 (1983).
66. P. J. Parker, M. Katan, M. D. Waterfield and D. P. Leader, *EJB* **148**, 579 (1985).
67. U. Padel and H.-D. Söling, *EJB* **151**, 1 (1985).
67a. F. Vara and E. Rozengurt, *BBRC* **130**, 646 (1985).
68. D. Tabarini, J. Heinrich and O. M. Rosen, *PNAS* **82**, 4369 (1985).
69. M. H. Cobb and O. M. Rosen, *JBC* **258**, 12472 (1983).
70. E. Erikson and J. L. Maller, *JBC* **261**, 350 (1986).
71. E. Erikson and J. L. Maller, *PNAS* **82**, 742 (1985).
71a. K.-P. Huang and F. L. Huang, *FP* **45**, 1864 (1986).
72. J. Blenis and R. L. Erickson, *PNAS* **82**, 7621 (1985).
73. C. P. Stanners and H. Becker, *J. Cell. Physiol.* **77**, 31 (1971).
74. P. S. Rudland, *PNAS* **71**, 750 (1974).
75. R. F. Brooks, *Nature* **260**, 248 (1976).
76. R. F. Brooks, *Cell* **12**, 311 (1977).
77. R. A. Tobey, D. F. Petersen and E. C. Anderson, in "Cell Cycle and Cancer" (R. Baserga, ed.), p. 409. Dekker, New York, 1971.
78. E. Bandman and T. Gurney, Jr., *Exp. Cell Res.* **90**, 159 (1975).
79. P. S. Rudland, S. Weil and A. R. Hunter, *JMB* **96**, 745 (1975).
80. P. J. Nielsen, R. Duncan and E. H. McConkey, *EJB* **120**, 523 (1981).
81. R. Duncan and E. H. McConkey, *EJB* **123**, 535 (1982).
82. R. Duncan and E. H. McConkey, *EJB* **123**, 539 (1982).
83. G. Thomas, G. Thomas and H. Luther, *PNAS* **78**, 5712 (1981).
84. E. Palen and J. A. Traugh, (unpublished).
85. S. J. Burkhard and J. A. Traugh, *FP* **43**, 1977 (1984).
86. C. Kruse, S. P. Johnson and J. R. Warner, *PNAS* **82**, 7515 (1985).
87. R. J. Leer, M. C. Van Raamsdonk-Duin, M. T. Molenaar, L. H. Cohen, W. H. Mager and R. J. Planta, *NARes* **10**, 5869 (1982).
88. R. B. Loftfield and D. Vanderjagt, *BJ* **128**, 1353 (1972).
89. P. R. Schimmel and D. Soll, *ARB* **48**, 601 (1979).
90. P. R. Schimmel, *CRC Crit. Rev. Biochem.* **9**, 207 (1980).
91. G. Krauss, D. Riesner and G. Maass, *EJB* **68**, 81 (1976).
92. R. Rigler, U. Pachmann, R. Hirsch and H. G. Zachau, *EJB* **65**, 307 (1976).
93. D. Riesner, A. Pingoud, D. Boehme, F. Peters and G. Maass, *EJB* **68**, 71 (1976).

94. J. P. Ebel, R. Giege, J. Bonnet, D. Kern, N. Befort, C. Bollack, F. Fasiolo, J. Gangloff and G. Dirheimer, *Biochimie* **55**, 547 (1973).
95. B. Roe, M. Siroven and B. Dudock, *Bchem* **12**, 4146 (1973).
96. S. S. M. Lam and P. R. Schimmel, *Bchem* **14**, 2775 (1975).
97. C. V. Dang and C. V. Dang, *Biosci. Rep.* **3**, 527 (1983).
98. O. Kellermann, H. Tonetti, A. Brevet, M. Mirande, J. P. Pailliez and J.-P. Waller, *JBC* **257**, 11041 (1982).
99. M. Mirande, O. Kellermann and J.-P. Waller, *JBC* **257**, 11049 (1982).
100. O. Kellermann, A. Brevet, H. Tonetti and J.-P. Waller, *EJB* **99**, 541 (1979).
101. C. Vennegoor and H. Bloemendal, *EJB* **26**, 462 (1972).
102. K. Som and B. Hardesty, *ABB* **166**, 507 (1975).
103. C. V. Dang and D. C. H. Yang, *BBRC* **80**, 709 (1978).
104. M. Mirande, Y. Gache, D. Le Corre and J.-P. Waller, *EMBO J.* **1**, 733 (1982).
105. B. Cirakoglu and J.-P. Waller, *BBA* **829**, 173 (1985).
106. M. Mirande, D. Le Corre and J.-P. Waller, *EJB* **147**, 281 (1985).
107. A. M. Pendergast and J. A. Traugh, *JBC* **260**, 11769 (1985).
108. M. Mirande, B. Cirakoglu and J.-P. Waller, *JBC* **257**, 11056 (1982).
109. M. Lazard, M. Mirande and J.-P. Waller, *Bchem* **24**, 5099 (1985).
110. A. Brevet, C. Geffrotin and O. Kellermann, *EJB* **124**, 483 (1982).
111. P. O. Ritter, M. D. Enger and A. E. Hampel, *BBA* **562**, 377 (1979).
112. M. P. Deutscher and R. C. Ni, *JBC* **257**, 6003 (1982).
113. M. A. Ussery, W. K. Tanaka and B. Hardesty, *EJB* **72**, 491 (1977).
114. G. J. Vellekamp, C. L. Coyle and F. J. Kull, *JBC* **258**, 8195 (1983).
115. C. V. Dang and D. C. H. Yang, *JBC* **254**, 5350 (1979).
115a. J. D. Dignam, D. G. Rhodes and M. P. Deutscher, *Bchem* **19**, 4978 (1980).
116. N. S. Penneys and K. H. Muench, *Bchem* **13**, 560 (1974).
117. J. S. Tscherne, K. W. Lanks, P. D. Salim, D. Grunberger, C. R. Cantor and I. B. Weinstein, *JBC* **248**, 4052 (1973).
118. S. M. Kane, C. Vugrincic, D. S. Finbloom and D. W. E. Smith, *Bchem* **17**, 1509 (1978).
119. D. C. H. Yang, C. V. Dang and F. C. Arnett, *BBRC* **120**, 15 (1984).
120. T. Mizutani, T. Narihara and A. Hashimoto, *EJB* **143**, 9 (1984).
120a. M. A. Le Meur, P. Gerlinger, J. Clavert and J.-P. Ebel, *Biochimie* **54**, 1391 (1972).
121. F. Deak and G. Denes, *BBA* **526**, 626 (1978).
122. O. O. Favorova, I. A. Madoyan and L. L. Kisselev, *EJB* **86**, 193 (1978).
123. F. Pan, H. H. Lee, S. H. Pai, T. C. Yu, J. Y. Guoo and G. M. Duh, *BBA* **452**, 271 (1976).
124. D. L. Johnson, C. V. Dang and D. C. H. Yang, *JBC* **255**, 4362 (1980).
125. R. K. Sihag and M. P. Deutscher, *JBC* **258**, 11846 (1983).
126. B. Cirakoglu and J.-P. Waller, *EJB* **151**, 101 (1985).
127. M. P. Deutscher, *J. Cell Biol.* **99**, 373 (1984).
128. A. K. Bandyopadhyay and M. P. Deutscher, *JMB* **74**, 257 (1973).
128a. C. V. Dang, T. P. Mawhinney and R. H. Hilderman, *Bchem* **21**, 4891 (1982).
129. B. Cirakoglu, M. Mirande and J.-P. Waller, *FEBS Lett.* **183**, 185 (1985).
130. M. Mirande, D. Le Corre, D. Louvard, H. Reggio, J. P. Pailliez and J.-P. Waller, *Exp. Cell Res.* **156**, 91 (1985).
131. C. V. Dang, D. C. H. Yang and T. D. Pollard, *J. Cell Biol.* **96**, 1138 (1983).
132. W. K. Tanaka, K. Som and B. Hardesty, *ABB* **172**, 252 (1976).
133. J. Hradec and Z. Dusek, *Mol. Biol. Rep.* **6**, 245 (1980).
134. M. Smulson, C. S. Lin and J. G. Chirikjian, *ABB* **167**, 458 (1975).

135. C. V. Dang, B. Ferguson, D. J. Burke, V. Garcia and D. C. H. Yang, *BBA* **829**, 319 (1985).
136. M. Mirande, B. Cirakoglu and J.-P. Waller, *EJB* **131**, 163 (1983).
137. E. G. Malygin and L. L. Kisselev, *Mol. Biol.* **18**, 1026 (1984).
138. A. Rich and P. R. Schimmel, *NARes* **4**, 1649 (1977).
139. S. Blanquet, P. Plateau and A. Brevet, *MCBchem* **52**, 3 (1983).
140. S. Z. Wahab and D. C. H. Yang, *JBC* **260**, 5286 (1985).
141. A. Brevet, P. Plateau, B. Cirakoglu, J. P. Pailliez and S. Blanquet, *JBC* **257**, 14613 (1982).
142. F. Grummt, *PNAS* **75**, 371 (1978).
143. E. Rapaport and P. C. Zamecnik, *PNAS* **73**, 3984 (1976).
145. F. Grummt, G. Waltl, H. M. Jantzen, K. Hamprecht, V. Huebscher and C. C. Kuenzyle, *PNAS* **76**, 6081 (1979).
146. P. C. Zamecnik, E. Rapaport and E. F. Baril, *PNAS* **79**, 1791 (1982).
147. S. Z. Wahab and D. C. H. Yang, *JBC* **260**, 12735 (1985).
148. B. H. Berg, *BBA* **521**, 274 (1978).
149. Z. Damuni, F. G. Caudwell and P. Cohen, *EJB* **129**, 57 (1982).
150. S. C. Gerken and S. M. Arfin, *JBC* **259**, 11160 (1984).
151. S. C. Gerken and S. M. Arfin, *JBC* **259**, 9202 (1984).
152. M. D. Rosa, J. P. Hendrick, Jr., M. R. Lerner, J. A. Steitz and M. Reichlin, *NARes* **11**, 853 (1983).
153. M. B. Mathews and R. M. Bernstein, *Nature* **304**, 177 (1983).
154. M. B. Mathews, M. Reichlin, G. R. V. Hughes and R. M. Bernstein, *J. Exp. Med.* **160**, 420 (1984).
155. R. M. Bernstein, M. Reichlin, G. R. V. Hughes and M. B. Mathews, *Arthritis Rheum.* **27**, S66 (1984).
156. R. M. Bernstein, C. C. Bunn and M. B. Mathews, *Br. J. Rheum.* **24**, 101 (1985).
157. S. C. Gerken, I. L. Andrulis and S. M. Arfin, *BBA* **869**, 215 (1986).
158. A. M. Pendergast, Ph.D. Thesis, University of California, Riverside, 1986.
159. A. M. Pendergast and J. A. Traugh, *FP* **45**, 1826 (1986).
160. G. M. Hathaway and J. A. Traugh, *Curr. Top. Cell. Regul.* **21**, 101 (1982).
161. L. S. Jefferson, D. E. Rannels, B. L. Munger and H. E. Morgan, *FP* **33**, 1098 (1974).
162. M. A. McNurlan, A. M. Tomkins and P. J. Gorlick, *BJ* **178**, 373 (1979).
163. D. E. Peavy, J. M. Taylor and L. S. Jefferson, *PNAS* **75**, 5879 (1978).
164. H. E. Morgan, L. S. Jefferson, E. B. Wolpert and D. E. Rannels, *JBC* **246**, 2163 (1971).
165. C. S. Harmon, C. G. Proud and V. M. Pain, *BJ* **223**, 687 (1984).
166. C. O. Brostrom, S. B. Bocckino and M. A. Brostrom, *JBC* **258**, 14390 (1983).
167. M. A. Brostrom, C. O. Brostrom, S. B. Bocckino and S. S. Green, *J. Cell. Physiol.* **121**, 391 (1984).
168. J. A. Traugh and T. S. Lundak, *BBRC* **83**, 379 (1978).
169. M. Cervera, G. Dreyfuss and S. Penman, *Cell* **23**, 113 (1981).
170. R. Lenk, L. Ransom, Y. Kaufmann and S. Penman, *Cell* **10**, 67 (1977).
171. W. J. van Verrooij, P. T. G. Sillekens, C. A. G. van Eckelen and R. J. Reinders, *Exp. Cell Res.* **135**, 79 (1981).
172. J. G. Howe and J. W. B. Hershey, *Cell* **37**, 85 (1984).
173. H. Bielka, *This Series* **32**, 267 (1985).
174. K. Terao and K. Ogata, *J. Biochem.* **86**, 579 (1979).

175. K. Terao and K. Ogata, *J. Biochem.* **86**, 605 (1979).
176. Y. Takahashi and K. Ogata, *J. Biochem.* **90**, 1549 (1981).
177. P. Westermann and O. Nygard, *NARes* **12**, 8887 (1984).
178. D. R. Tolan, J. W. B. Hershey and R. R. Traut, *Biochimie* **65**, 427 (1983).
179. R. Kisilevsky, M. A. Treloar and L. Weiler, *JBC* **259**, 1351 (1984).
180. R. Klemenz, D. Hultmark and W. J. Gehring, *EMBO J.* **4**, 2053 (1985).
181. G. A. Floyd and J. A. Traugh, *EJB* **106**, 269 (1980).
182. R. Benne, J. Edman, R. R. Traut and J. W. B. Hershey, *PNAS* **75**, 108 (1978).
183. R. Duncan and J. W. B. Hershey, *JBC* **260**, 5493 (1985).

The Primary DNA Sequence Determines *in Vitro* Methylation by Mammalian DNA Methyltransferases

ARTHUR H. BOLDEN,
CHERYL A. WARD,
CARLO M. NALIN, AND
ARTHUR WEISSBACH

*Roche Institute of Molecular Biology
Roche Research Center
Nutley, New Jersey 07110*

Mammalian DNA methyltransferases (DNA methylases)[1] modify the deoxycytidine residue of the dinucleotide sequence dC–dG of DNA by enzymatic transfer of the methyl group of S-adenosylmethionine (AdoMet) to carbon-5 of cytosine (1, 2). In naturally occurring DNA, the genomic distribution of 5-methylcytosine is nonrandom (3, 4) with only 40–70% of the dC–dG sites being so methylated (5). The pattern of methylation is heritable from generation to generation in a tissue- and species-specific manner (6–8).

Interest in DNA methylation has been stimulated by evidence suggesting that the level of methylation of DNA may exert control over gene expression. During chemically induced differentiation, murine erythroleukemia cells accumulate globin mRNAs at a rate 10 times that of uninduced cells (9), while an overall decrease in the level of DNA methylation (10, 11) and changes in chromatin structure (12, 13) are also observed. Undermethylation of a variety of active gene sequences, including rat albumin genes (14), *Xenopus* tRNA genes (15), rabbit β-globin-like genes (16), chicken ovalbumin genes (8), and chicken γ-crystallin genes (17), has been demonstrated. However, only a few dC–dG sites, most of which are located near the 5' end of these genes, show a dramatic difference in the level of methylation between nonexpressing and expressing tissue (18–20). In addition, viruses (23, 24) and foreign genes introduced into somatic cells (25, 26) are hypomethylated in cells that express them.

---

[1] DNA (cytosine-5-)methyltransferase, EC 2.1.1.37 [Eds.].

DNA methyltransferases from a number of eukaryotic sources have been purified (2, 25–34). *In vitro*, these nuclear (25, 29) enzymes are capable of methylating single-stranded DNA or double-stranded DNA at previously unmethylated dC–dG sites in a *de novo* reaction (25, 26, 27). The preferred *in vitro* substrate for the known eukaryotic methyltransferases is the unmethylated strand of hemimethylated[2] DNA (28, 34, 35). Methylation of hemimethylated sites proceeds at a rate 10–30 times that of *de novo* methylation (35, 36) and is considered to be important in maintaining the genomic methylation patterns in newly replicated DNA (37).

Using partially purified DNA methyltransferases from HeLa or murine erythroleukemia cells, we have studied the relationship between "*de novo*" and "maintenance" modes of methylation. This work has led to the identification of RNA as well as certain synthetic polyribo- and polydeoxyribonucleotides as inhibitors of DNA methyltransferase activity (36). In addition, we are attempting to elucidate the mechanism by which specific cytosine residues are targeted for methylation by the enzyme. This "targeting" could be accomplished by changes in chromatin structure affecting the exposure of dC–dG sites to the DNA methyltransferase, by a protein or other factor binding to the DNA and thus targeting the site, or by the primary DNA sequence itself. Preparation of a series of synthetic oligodeoxynucleotides of known sequence and containing one or more dC–dG sites has allowed us to examine the influence of the primary DNA sequence on methylation.

I. Characterization of DNA Methyltransferases

A. Purification

The DNA methyltransferases used in these studies were purified 800-fold from the nuclear extracts of HeLa S3 cells (Table I) and 115-fold from extracts of mouse erythroleukemia (MEL) cell nuclei. The strategy of the purification procedure was designed to produce DNA methyltransferase preparations virtually free of deoxyribonuclease activity. DNA methyltransferase activity was extracted from purified

---

[2] Hemimethylated DNA was synthesized *in vitro* by a modification of the procedure of Gruenbaum *et al.* (35) using primed repair synthesis with the (+) strand of ϕX174 DNA as substrate. DNA ligase was omitted from the reaction. The product of the reaction consists of a mixture of 70% nicked (opened) circular and 30% linear DNA molecules in which 85–100% of the cytosine residues of the complementary strand are methylated.

TABLE I
PURIFICATION OF HeLa CELL DNA METHYLTRANSFERASE

Stage of purification	Total protein (mg)	Total activity[a] (units × 10^{-3})	Specific activity (units/mg protein × 10^{-3})	Rate of methylation of HM-DNA: M. luteus
Crude extract	273.0	7.3	0.03	24:1
DEAE-cellulose	27.3	98.3	3.6	25:1
Phosphocellulose	2.5	54.5	22.0	20:1
DNA-agarose[b]	0.4	9.9	25.6	25:1

[a] One unit of DNA methyltransferase catalyzes the transfer of 1 pmol of methyl groups from AdoMet to hemimethylated DNA in 30 minutes under standard assay conditions (36).

[b] Forty-five percent of the phosphocellulose-purified DNA-methyltransferase activity was purified further on DNA-agarose.

nuclei (36) with buffer solutions containing 2 M NaCl and the protease inhibitors α-toluenesulfonyl[3] fluoride (αTosF) and p (or 4)-toluenesulfonyl fluoride (pTosF). Purification was achieved by highspeed centrifugation, and by DEAE-cellulose, phosphocellulose, and single-stranded DNA-agarose chromatography. DNA methyltransferase prepared in this manner contains a low level of ribonuclease activity and no detectable deoxyribonuclease activity. The assay for the latter would detect the solubilization of 1% (1 ng) of the DNA substrate used in the assay. Based on assays employing hemimethylated[2] ϕX174 DNA (35) as a substrate, the specific activities of the partially purified preparations were 50,000 units/mg (HeLa) and 42,500 units/mg (MEL).

B. Biochemical Characterizations

The HeLa DNA methyltransferase exhibits a broad pH optimum with maximum activity observed at pH 7.5–8.0. The methylation reaction does not show a marked requirement for a sulfhydryl reductant although methyltransferase activity is stimulated up to 25% by 1–5 mM dithiothreitol. The enzyme does not require the presence of a divalent cation for catalytic activity and, in fact, is inhibited by 1–3 mM $MgCl_2$ or $MnCl_2$. EDTA (1 mM) doubles the activity. The enzyme is sensitive to high salt concentrations; in 50 mM NaCl, only 30% activity is observed, while NaCl concentrations above 100 mM totally inhibit the reaction. Similarly, in 20 and 50 mM potassium phosphate (pH 7.5) only 50 and 1%, respectively, of the methyltrans-

[3] Often called phenylmethylsulfonyl [Eds.].

ferase activity is observed. The K_m for AdoMet in the presence of hemimethylated DNA is 8 µM (36).

C. Substrate Concentration Dependence

The effect of substrate DNA concentration on the methylation reaction *in vitro* was determined using hemimethylated φX174 DNA, single-stranded and double-stranded *M. luteus* DNAs, poly(dC–dG), and an oligodeoxynucleotide, 27mer-F[4] (d-TCGACCCCCCCCCC-CCCCGGGTCTAG). At 44 units of HeLa DNA methyltransferase per ml, substrate saturation occurs at a DNA concentration of 2 µg/ml with hemimethylated DNA, double-stranded *M. luteus* DNA, or poly(dC–dG). Using 160 units of enzyme per ml, 27mer-F reached saturation at a concentration of 1–2 µg/ml compared to the 2–3 µg/ml required of single-stranded *M. luteus* DNA. Higher substrate concentrations have no effect on enzyme activity.

D. Substrate Specificity

The mechanism of inheritance of methylation patterns from generation to generation is thought to involve maintenance methylation at hemimethylated sites that arise during semiconservative replication of DNA (38). This model suggests that hemimethylated DNA is the preferred substrate for the enzyme. In fact, the known eukaryotic DNA methyltransferases can efficiently methylate *in vitro* the unmethylated strand of hemimethylated duplex DNA (34, 35). In addition, mammalian DNA methyltransferase can methylate a totally unmethylated natural or synthetic DNA in a *de novo* reaction (25, 27). *De novo* methylation *in vivo* occurs when retroviral DNAs are integrated into the genomes of preimplantation mouse embryos, repressing the activity of the proviral genome (38). There is no evidence that the two modes of methylation are catalyzed by two separate enzyme(s), and most partially purified eukaryotic DNA methyltransferase preparations catalyze both modes of methylation *in vitro* (25, 27, 36).

The partially purified HeLa and MEL DNA methyltransferases were assayed with various natural and synthetic DNA substrates (Table II). As expected, hemimethylated φX174 DNA is an excellent substrate, being methylated at a rate 30 times that of double-stranded *M. luteus* DNA by the HeLa enzyme and 8 times that of double-

---

[4] The digits refer to the number of deoxynucleoside residues in the compound. Letters designate a specific oligodeoxynucleotide in the 27mer series.

TABLE II
SUBSTRATE SPECIFICITY OF THE HeLa AND MEL METHYLTRANSFERASE[a]

Substrate	HeLa DNA methyltransferase [relative activity (%)]	MEL DNA methyltransferase [relative activity (%)]
Hemimethylated ϕX174 DNA	100	100
Hemimethylated ϕX174 DNA (denatured)	2	0
M. luteus DNA (double-stranded)	3	12
M. luteus DNA (single-stranded)	10	6
Salmon-sperm DNA (double-stranded)	1	n.d.
Poly(dC–dG)	3	12
Poly(dC) · poly(dG)	0	0
SV40 DNA	1	n.d.
Reovirus RNA (double-stranded)	0	0

[a] The assay mixtures (50 μl) with hemimethylated DNA contained 2.7 μg of DNA per ml. All other assays (200 μl) contained 10 μg/ml of indicated substrate. Each assay mixture contained 50 units of DNA methyltransferase per ml. The 100% values, 2.18 pmol (HeLa), 2.3 pmol (MEL), represents the incorporation of [^{3}H]methyl into hemimethylated ϕX174 DNA.

stranded *M. luteus* DNA by the MEL enzyme. Heat-denatured hemimethylated DNA is methylated at only 2% the rate of the hemimethylated duplex, indicating that the duplex structure is necessary for maintenance methylation. Salmon-sperm DNA [40% (G + C)] and double-stranded *M. luteus* DNA [70% (G + C)] were methylated by the HeLa methyltransferase at approximately the same rate (1–3% that of the hemimethylated duplex DNA), while single-stranded *M. luteus* DNA was methylated at a rate 10% that of the hemimethylated duplex at the enzyme levels used in these experiments. These results agree with earlier reports indicating that a duplex DNA structure is not necessary for *de novo* methylation (25, 39). While the HeLa enzyme appears to methylate single-stranded *M. luteus* DNA more rapidly than double-stranded *M. luteus* DNA, the MEL methyltransferase shows a preference for double-stranded DNA. The alternating copolymer poly(dC–dG) was methylated by both enzymes at a rate comparable to that of double-stranded *M. luteus* DNA. With the HeLa enzyme, 3.5% of the cytosine residues in poly(dC–dG) can be methylated in 6 hours, and if incubation is continued for 24 hours, 12% of the cytosine residues are methylated. The associated homopolymers poly(dC) · poly(dG), poly(dC) · poly(dI), and poly(dA) · poly(dI) are not methylated by the HeLa methyltransferase (40).

II. Methylation of Oligodeoxynucleotides

While it has been possible to study some aspects of the DNA methyltransferase reaction *in vitro* using purified eukaryotic or bacterial DNAs as substrates, the availability of synthetic oligonucleotides of known sequence has allowed significant new progress to be made in understanding the role of the primary DNA sequence. This approach has been used to study the substrate specificity of the enzyme with both single-stranded and double-stranded synthetic DNA molecules and permits one to determine the effect that sequences surrounding a dC–dG site may have on the methylation of that site. Due to the relatively slow rate of *de novo* methylation of these oligomers, the reaction, which is linear for 6 hours, was carried out for 5 hours with 3- to 4-fold more DNA methyltransferase than is used for "maintenance" methylation of hemimethylated DNA. The size of the synthetic oligomers that can act as methyl acceptors is relatively small, as we find that oligomers as short as 12–18 bases in length, e.g., (dC–dG)$_{6-9}$, are substrates for the enzyme. In addition, two dodecamers, d-GGGGGCGCCCCC and d-AAAAACGTTTTT, are methylated at low but reproducible rates (Table III).

The rates of methylation of some oligomers tested are similar to the rates observed with natural DNA molecules. In fact, the rate at which 26mer-C (d-CCGGCCATTACGGATCCGTCCTGGGC) accepts methyl groups is 2.5 times that of single-stranded *M. luteus* DNA. Methylation of 27mer-F proceeds at the same rate as single-stranded *M. luteus* DNA.

From a survey of approximately 50 synthetic oligodeoxynucleotides, it is clear that oligomers containing more than one dC–dG dinucleotide are far more efficient substrates for the enzyme than are those containing only a single such sequence (*40, 41;* Table III). In fact, most oligodeoxynucleotides that contain a single dC–dG dinucleotide were ineffective substrates. In the few cases where an oligodeoxynucleotide containing a single dC–dG dinucleotide could be methylated, the rates were a tenth or less those for substrates having more than one such site. The relevance of this important observation is discussed in Section III.

A. Effect of the Primary DNA Sequence

It has been reported that more than one-half of the 5-methylcytosine residues in bovine satellite DNA occur in palindromic regions containing highly methylated CCGG sequences whereas TCGA sites are methylated to a lesser extent (*20*). In addition, 5-methylcytosine is

TABLE III
Methylation of Synthetic Oligodeoxynucleotides[a]

Compound number			Relative rate of methylation (%)
1	TCGACCCCCCCCCCCCCCCGGGTCTAG	(27mer-F)	100
2	CTAGACCCGGGGGGGGGGGGGGGTCGA	(27mer-F')	0
3	TCGACCCCCCCCCCCCCCCAGGTCTAG		0
4	TCAACCCCCCCCCCCCCCCGGGTCTAG		12
5	TGTCGACCCCCCCCCCCCCCCGGGTCTAG	(29mer-F)	47
6	TCGACCCCCCCCCC		0
7	CCCGGGTCTAG		0
8	CCCCCCCCCCGGGTCTAG		0
9	CCCCCCCCCCCCCCCGGGTCTAG		6
10	CCGGCCATTACGGATCCGTCCTGGGC	(26mer-C)	251
11	GCCCAGGACGGATCCGTAATGGCCGGA	(26mer-C')	50
12	GGAGAGCGTCCATGGACGCGAGAG	(24mer-L)	41
13	CTCTCGCGTCCATGGACGCTCTCC	(24mer-L')	7
14	GGGGGCGCCCCC		10
15	CCCCCCGGGGGG		0
16	CCGGAATTCCGG		0
17	AAAAACGTTTTT		10
18	TTTTTCGAAAAA		0
19	CAGAATTCATGACTGTTGCGCTAC		0
20	AATTAATATGAATGAATTCGGATCCATCGATA		0
21	CGATATTTGGAGGTCAGCACGGTGCTCACG		0
22	TCGATATTTGGAGGTCAGCCCGGTGCTCACG		8
23	(dC-dG)$_{6-9}$		20
24	poly(dC-dG)		33
25	*M. luteus* DNA (single-stranded)		98

[a] Assays (50 μl) were carried out in the presence of 1.4 μg of the indicated oligonucleotide per ml and 55 units of HeLa DNA methyltransferase per ml. *M. luteus* DNA was present at 3 μg/ml. Incubation was at 37°C for 5 hours. The 100% value represents the incorporation of 2.80 pmol of [^{3}H]methyl into 27mer-F.

often found in (G + C)-rich regions (42), and dC–dG dinucleotides are frequently found clustered in the 5'-flanking regions of genes (42–44) where DNA methylation sites affecting gene transcription have been detected (18–20). We have examined the influence of flanking sequences on *in vitro* methylation by preparing a series of synthetic oligodeoxynucleotides, some of which are closely related to 27mer-F, and determined their abilities to act as methyl acceptors (Table III).

The 27mer-F, containing two dC–dG dinucleotides is an efficient substrate for both HeLa and MEL methyltransferases, as has already been stated. However, only one of the two potential methylation sites, the 3′ one at position 19–20 is actually methylated in this oligomer (22). The site of methylation of 27mer-F is discussed in Section III and Fig. 3. If the sequence of 27mer-F is altered by substitution of an adenine for the guanine at position 3 (Table III, compound 4), methylation of the compound is virtually eliminated. Moreover, the addition of two deoxynucleotides, dT and dG, to the 5′ end of 27mer-F (compound 5) decreases its methyl acceptor ability by 50%, indicating the importance of the 5′ flanking sequence on methylation. As observed with other oligomers containing a single dC–dG site, no methyl acceptor activity is detectable with fragments of 27mer-F containing either the 5′ dC–dG (compound 6) or a series of fragments containing the 3′ dC–dG [(compounds 7–9) Table III].

The importance of the flanking sequences on the specificity of methylation is also illustrated by the observed differences in rates and patterns of methylation of three oligodeoxynucleotides (27mer-F, 26mer-C, 24mer-L) and their respective complementary strands (27mer-F′, 26mer-C′, and 24mer-L′) (Table IV). Within these complementary oligodeoxynucleotide pairs, the number of dC–dG pairs and the spacings between them are equivalent; also, the relative percentages of dG + dC are equivalent (81% in 27mer-F and 27mer-F′, 70% in 26mer-C and 26mer-C′, 67% in 24mer-L and 24mer-L′). Despite the similarities between the complementary oligodeoxynucleotide strands, 27mer-F′ has no activity as a methyl acceptor, while 26mer-C′ and 24mer-L′ are only 15–20% as efficient as their complements. Thus, while the number of dC–dG sites, the spacing between them, and the (dG + dC) contents of the sequences appear to be important to the methylation of specific dinucleotides, there appear to be additional determinants for methylation within the primary DNA sequence itself.

B. Effect of the Nucleotide Spacing Between dC–dG Sequences

The efficient methylation of 27mer-F, which contains two dC–dG dinucleotides separated by 14 dC residues prompted us to examine further the effect of spacing between these sites on the rate of methylation. We synthesized a series of molecules that resembled 27mer-F, but differed only in the number of dC residues between the two dC–dG dinucleotides. With both the HeLa and MEL methyltransferases, molecules having a spacing of 13–17 nucleotides between the sites were the best substrates for methylation (Fig. 1). Molecules having a shorter or longer spacing were either inactive or showed only marginal activity as substrates. While these results cannot be used to generalize to all DNA molecules, they suggest that the enzyme may

TABLE IV
METHYLATION OF COMPLEMENTARY SINGLE-STRANDED OLIGODEOXYNUCLEOTIDES AND THEIR DUPLEXES BY HeLa AND MEL DNA METHYLTRANSFERASES[a]

Compound (all deoxys)		HeLa DNA methyltransferase [relative activity (%)]	MEL DNA methyltransferase [relative activity (%)]
27mer-F	TCGACCCCCCCCCCCCCCGGGTCTAG	100	100
27mer-F'	CTAGACCCGGGGGGGGCGGGGGGTCGA	0	0
(27mer-F)·(27mer-F')	TCGACCCCCCCCCCCCCCGGGTCTAG AGCTGGGGGGGGGGGGGGCCCAGATC	0	5
26mer-C	CCGGCCATTACGGATCCGTCCTGGGC	100	100
26mer-C'	GCCCAGGACGGATCCGTAATGGCCGGA	20	9
(26mer-C)·(26mer-C')	CCGGCCATTACGGATCCGTCCTGGGC AGGCCGGTAATGCCTAGGCAGGACCCG	12	24
24mer-L	GGAGAGCGTCCATGGACGCGAGAG	100	100
24mer-L'	CTCTCGCGTCCATGGACGCTCTCC	17	39
(24mer-L)·(24mer-L')	GGAGAGCGTCCATGGACGCGAGAG CCTCTCGCAGGTACCTGCGCTCTC	70	106

[a] Methylation of the oligodeoxynucleotides was carried out at 3 μg of DNA per ml. Assay mixtures contained 130 units of HeLa DNA methyltransferase per ml or 180 units of MEL DNA methyltransferase per ml. Incubation was at 37°C for 5 hours. The 100% values for the HeLa enzyme are 6.5 pmol (27mer-F), 15.5 pmol (26mer-C), and 0.4 pmol (24mer-L). The 100% values for the MEL enzyme are 8.2 pmol (27mer-F), 9.0 pmol (26mer-C), and 0.5 pmol (24mer-L). dC–dG sequences are underlined.

prefer DNA sequences having dC–dG sites within appropriate proximity of one another.

III. *De Novo* and Maintenance Methylation Sites

The single-stranded oligodeoxynucleotides, 27mer-F and 26mer-C, are efficiently methylated *de novo* by both HeLa and MEL DNA methyltransferases (Tables III and IV). In order to determine the sites of methylation of these single-stranded oligomers and their complements, the oligonucleotides, as shown in Fig. 2 for 26mer-C, were methylated with the HeLa DNA methyltransferase in the presence of Ado[*methyl*-^{3}H]Met, and annealed to their nonmethylated complements. The duplexed DNA was subjected to double digestion with appropriate restriction endonucleases, and denatured at 100°C yielding fragments of varying size, each containing a potential dC–dG site for methylation. DNA fragments were separated under denaturing

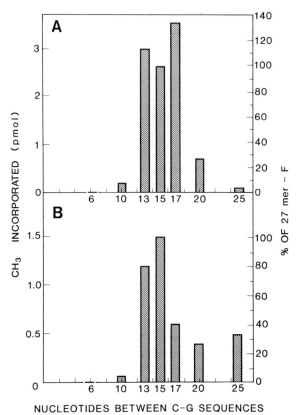

FIG. 1. Methylation of derivatives of 27mer-F. Assays (50 µl), containing 3 µg/ml of oligodeoxynucleotide, and 180 units/ml of either HeLa (A) or MEL (B) DNA methyltransferase, were performed for 3 hours at 37°C. The data are presented as pmol of methyl incorporated per 120 ng of oligodeoxynucleotides. Reproduced from Bolden et al. (45) with permission from the American Society of Biological Chemists, Inc.

conditions on a polyacrylamide gel. The lanes were sliced and fragments containing [^{3}H]methyl groups were located by scintillation counting. In the single-stranded oligomer 27mer-F, which contains two dC–dG sites, methylation was observed only in the one near the 3′ end of the molecule (Fig. 3). Less than 2% of the recovered [^{3}H]methyl group was found in the restriction fragment that contained the dC–dG near the 5′ end of the molecule. The complementary strand of 27mer-F was not methylated at all. As in the case of 27mer-F, the oligodeoxynucleotide 26mer-C, which contains three dC–dG sites, is not methylated at the one nearest the 5′ end of the molecule, but is methylated at the cytosine residues near the 3′ end of the

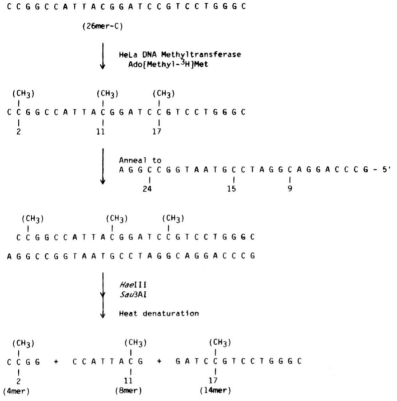

FIG. 2. Scheme for the determination of methylation sites of 26mer-C. (CH$_3$) denotes potential methylation sites.

molecule at positions 11 and 17. Its complement, 26mer-C′, is methylated only at position 15. No methylation of the cytosines at positions 9 and 24 was detected.

It became of interest to know whether the double-stranded forms of these oligonucleotides could be methylated, and if so, how their patterns of *de novo* methylation relate to the patterns of *de novo* methylation established for the single-stranded oligomers. The complementary oligodeoxynucleotides were annealed, and the resulting duplexes (27mer-F) · (27mer-F′), (24mer-L) · (24mer-L′), and (26mer-C) · (26mer-C′) tested as substrates for the HeLa and MEL methyltransferases (Table IV). The 27mer-F duplex is not a substrate for the HeLa methyltransferase but is slowly methylated by the MEL enzyme. Furthermore, a hemimethylated duplex formed by annealing methylated 27mer-F and its nonmethylated complementary strand is a

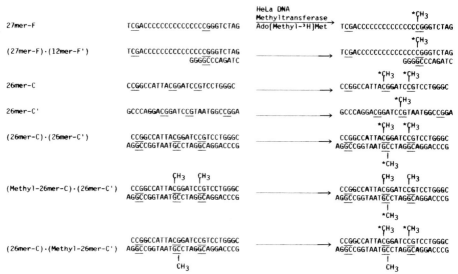

FIG. 3. Methylation patterns of oligodeoxynucleotides. CH₃ denotes observed methylation sites.

poor substrate for either methyltransferase (45). These results indicate that an existing hemimethylated site may not necessarily be an efficient substrate for the methyltransferase.

The poor methyl acceptor ability of the (27mer-F) · (27mer-F') duplex may be due, in part, to the primary structure of the 27mer-F' strand, which contains 15 consecutive dG residues. The oligonucleotide $(dG)_{12-18}$ inhibits the HeLa cell DNA methyltransferase *in vitro* (36), and we find that 27mer-F' (2 µg/ml) inhibits the methylation of hemimethylated ɸX174 DNA by 70% under the conditions used in this study.

To further examine this point, we prepared a partial complement of 27mer-F, 12mer-F' (d-CTAGACCCGGGG). When annealed to 27mer-F, this compound spans the 3' dC–dG dinucleotide of 27mer-F at position 19–20 (Fig. 3). The resulting partial duplex is a more efficient methyl acceptor than the full duplex (27mer-F) (27mer-F') (45). Analysis of the methylated (27mer-F) · (12mer-F') duplex revealed that 80–90% of the [³H]methyl counts incorporated by either the HeLa cell or MEL methyltransferase were located in the 27mer-F strand. This result suggests that the methylation of dC–dG sites in double-stranded DNA may be modulated by the primary DNA sequence surrounding the site.

The rate of methylation of the (26mer-C) · (26mer-C') duplex is intermediate between the rates of methylation of its single-stranded oligodeoxynucleotide components. Incorporated methyl groups were located by digestion with restriction endonucleases (Fig. 2) and polyacrylamide gel analysis as described above. Methyl groups incorporated into the duplexed 26mer-C were found at positions 11 and 17 of the 26mer-C strand and only at position 15 of the 26mer-C' strand (Fig. 3). Thus, the pattern of methylation of the duplex is identical to the pattern of methylation obtained for either the single-stranded 26mer-C or single-stranded 26mer-C'. This pattern of methylation of the 26mer-C duplex generates a hemimethylated site at position 17 of the 26mer-C strand which is not further methylated by the enzyme, as well as one fully unmethylated C–G site at position 2 of 26mer-C and a fully methylated C–G site at position 11 of 26mer-C [(26mer-C) · (26mer-C'), Fig. 3)]. The fact that the unmodified site is at the 5'-end suggests that the enzyme binds to the dC–dG dinucleotide near the 5' end of the molecule and then methylates the next dC–dG located an appropriate distance downstream. The dC–dG dinucleotide near the 5' end of the molecule may serve to position the enzyme so that it can methylate a site that is located 3' to the binding site. This could explain the relatively poor methyl acceptor activity of sequences which contain only one dC–dG dinucleotide.

To examine this point further, we prepared duplexes that contained either (1) an enzymatically methylated 26mer-C strand (methylated at positions 11 and 17) duplexed to a nonmethylated 26mer-C' strand, or (2) a 26mer-C' strand enzymatically methylated at position 15 and duplexed to its nonmethylated complementary 26mer-C strand (Fig. 3). The hemimethylated duplexes were again incubated with the HeLa methyltransferase and the locations of the newly incorporated methyl groups were determined by restriction analysis. The hemimethylated duplex that initially contained methyl groups at positions 11 and 17 of the 26mer-C strand is further methylated only at position 15 of the complementary strand. This observation confirms that position 9 of 26mer-C' cannot be methylated either in a nonmethylated duplex or hemimethylated structure. The converse hemimethylated duplex that initially contains a methyl group at position 15 of the 26mer-C' strand was further methylated at positions 11 and 17 of the 26mer-C strand. Methylation at position 9 of 26mer-C', a hemimethylated site, was not detected. This study demonstrates that *in vitro* methylation of hemimethylated DNA occurs only at dC–dG sites that are also targets for *de novo* methylation, and is consistent with the inability of the methyltransferase to methylate efficiently the hemimethylated site in (27mer-F) · (12mer-F') as discussed earlier.

These results, together with the inability of the enzyme to methylate efficiently the complementary strand of various oligomers (Table IV) and hemimethylated duplexes (Fig. 3), suggest that the pattern of methylation in hemimethylated DNA is determined by the primary sequence of the target strand and not by the pattern of 5-methylcytosine residues in the complementary parent strand.

These observations lead us to hypothesize that the enzyme distinguishes three classes of dC–dG sites in DNA. One class, at the 5' end of a dC–dG cluster, serves primarily as a binding site and may be methylated. The second class is incapable of being methylated when present in either an unmethylated or hemimethylated configuration. The third class can be methylated either in a single- or double-stranded configuration whether the substrate DNA molecules are hemimethylated or unmethylated. The implication that both maintenance methylation and *de novo* methylation are manifestations of the same enzymatic mechanism and share the same sequence specificity suggests that the differences between the two processes is one of rates rather than of substrate recognition. This eliminates the need to postulate two separate enzymes (37, 46) to catalyze these two processes; the known DNA methyltransferase in mammalian cells then suffices for the cellular requirements during replication, development, and differentiation.

IV. Inhibitors of Methyltransferases

A. RNA

During the purification of the DNA methyltransferase from HeLa cells (36), we detected a large apparent increase in total enzyme units at the DEAE-cellulose stage (Table I). The 13-fold increase in activity suggested the presence of an endogenous inhibitor of the enzyme in the crude extract. This inhibitor is insensitive to DNase I (EC 3.1.21.1), is stable at 100°C, and is destroyed by RNase A (pancreatic RNase, EC 3.1.27.5). In crude extracts, the inhibition of DNA methylation by the endogenous inhibitor is prevented or reversed by digestion with RNase A. These observations suggest that the low level of DNA methyltransferase activity observed in crude extracts is caused by inhibition by RNA.

A series of natural ribonucleic acids and synthetic polynucleotides have been tested for their effect on DNA methyltransferase activity.

TABLE V
EFFECT OF HOMOLOGOUS AND HETEROLOGOUS RNA
ON HeLa DNA METHYLTRANSFERASE[a]

Addition	Concentration (µg/ml)	Activity remaining (%)
None		100
HeLa total RNA	10	10
HeLa (16-S) ribosomal RNA	14	14
HeLa (4-S) transfer RNA	10	77
E. coli mRNA	12	0
E. coli transfer RNA	4	5
Yeast transfer RNA	4	36
Reovirus RNA (double-stranded)	4	95

[a] Assays (50 µl) were carried out with 2.7 µg of hemimethylated φX174 DNA per ml. Assay mixtures contained 15 units of DNA methyltransferase per ml and the indicated concentration of RNA. Incubation was at 37°C for 30 minutes. The 100% value of 0.75 pmol represents the incorporation of [^{3}H]methyl into hemimethylated DNA in the absence of any RNA.

Total HeLa RNA as well as messenger and transfer RNA from various heterologous sources are potent inhibitors of the enzyme (Table V). When HeLa RNA is fractionated into ribosomal RNA (16 S) and transfer RNA (4 S) and these species tested as inhibitors, the 16-S species is nearly as effective as total HeLa RNA. Double-stranded reovirus RNA is relatively ineffective as an inhibitor. These results clearly demonstrate that a natural RNA can be a powerful inhibitor of DNA methylation.

B. Polynucleotides and Polydeoxynucleotides

In an attempt to understand further the inhibition of DNA methyltransferase observed with certain natural RNA molecules, a series of synthetic polyribo- and deoxyribonucleotides, oligodeoxynucleotides, and copolymers were tested for their effect on DNA methyltransferase activity (Table VI) using a hemimethylated φX174 DNA substrate. It is interesting that while poly(rG) is a strong inhibitor, the other polyribonucleotides tested, poly(rC), poly(rU), and poly(rA), have relatively little effect (36). The inhibition by poly(rG) of the methylation of hemimethylated DNA (maintenance methylation) and M. luteus DNA (de novo methylation) is almost identical over a range (0.5–5.0 µg/ml) of poly(rG) concentrations (36). Methylation of M. luteus and hemimethylated φX174 DNA is totally inhibited by 2–2.5 µg/ml of

TABLE VI
INHIBITION OF HeLa DNA METHYLTRANSFERASE BY SYNTHETIC POLYNUCLEOTIDES AND OLIGONUCLEOTIDES[a]

Addition	Concentration (μg/ml)	Activity remaining (%)
None		100
Poly(rA)	10	81
Poly(rU)	10	78
Poly(rC)	10	100
Poly(rG)	10	0
Poly(dC)	10	45
Poly(dT)	10	99
Poly(dI)	10	109
Poly(dA)	10	99
Poly(dA–dT)	10	5
Poly(dC) · poly(dG)	10	0
Poly(dC–dG)	10	46
$dG_{(12-18)}$	10	34
$dT_{(12-18)}$	10	110
$dC_{(12-18)}$	10	22
Poly(rU) · poly(rA) · poly(rU)	1	46
	2	4
Poly(dT) · poly(dA) · poly(dT)	1	26
	2	3
Poly(rI) · poly(rA) · poly(rI)	1	45
	2	20
Poly(dA–dC) · poly(dG–dT)	1	100
	2	100
Poly(dG-m^{5}dC)	1	24
	2	7

[a] Assays (50 μl) were carried in the presence of 2.7 μg of hemimethylated ϕX174 DNA per ml, 66 units of HeLa DNA methyltransferase per ml, and the indicated concentration of the compounds tested as inhibitors. Incubation was at 37°C for 30 minutes. Methylation of hemimethylated ϕX174 DNA in the absence of any inhibitor (3.30 pmol) was taken as 100% activity.

this compound. At 1 μg/ml, poly(rG) inhibits methylation of hemimethylated and *M. luteus* DNA by 50 and 30%, respectively. Of the polydeoxyribonucleotides tested poly(dT), poly(dI), and poly(dA) are relatively ineffective as inhibitors. Poly(dC) is slightly inhibitory and small oligomers such as $(dC)_{12-18}$ and $(dG)_{12-18}$, but not $(dT)_{12-18}$, also inhibit the DNA methyltransferase.

The importance of the deoxyribose moieties attached to the cytosine or guanine residues in the binding of DNA to the enzyme has not been determined. The results presented here raise several interesting

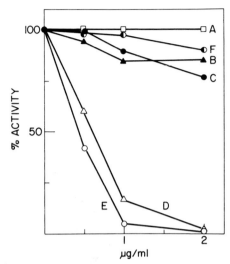

FIG. 4. Effect of polydeoxynucleotides on DNA methyltransferase activity. Assays (50 µl) were carried out with 2.7 µg/ml of hemimethylated φX174 DNA. The assay mixtures contained 31 units of HeLa cell DNA methyltransferase per ml and the indicated concentration of A, double-stranded calf thymus DNA; B, poly(dC–dG); C, poly(dA–dT); D, poly(dC · dG); E, poly(dA · dT); F, poly(dC). The 100% value, 1.55 pmol, was the extent of methylation of hemimethylated φX174 DNA in the absence of any inhibitor. Reproduced from Bolden et al. (36) with permission from the American Society of Biological Chemists, Inc.

points. Since rC–rG in RNA is not methylated, the enzyme must require one or both of the deoxyribose sugars of a dC–dG. However, because poly(rG) is a powerful inhibitor of the enzyme, it may be that only the guanine residue, but not its attached sugar, is recognized. In addition, since poly(rC) is totally without effect whereas poly(dC) shows some inhibition of enzymatic activity, we can conclude that the enzyme recognizes only those cytosines attached to a deoxyribose moiety and the active site may bind to a dC–rG or dC–dG moiety.

The associated homopolymers poly(dA · dT) and poly(dC · dG) show almost complete inhibition of the enzyme at a concentration of 1–2 µg/ml (Fig. 4). In comparison, the alternating copolymers poly(dA–dT) and poly(dC–dG) are much less inhibitory than their homopolymer counterparts. In the presence of 1 µg/ml of poly(dA–dT), 80% of DNA methyltransferase activity is still detected (Fig. 4C). As a control, calf thymus DNA at levels of 1–2 µg/ml has no inhibitory effect on methylation of hemimethylated DNA (Fig. 4A). The dramatic inhibition of DNA methyltransferase by the associated homopolymers poly(dA · dT) but not by the alternating heteropolymer poly(dA–dT)$_n$ is puzzling since neither poly(dA) nor poly(dT) separately inhibits the enzyme and neither A nor T residues are targets for the methylation reaction (25). If the inhibition reflects tight binding of the homopolymer duplexes to the enzyme one can speculate that the presence of stretches of d-AAAAA (d-TTTTT..) near a dC–dG site could affect the stability of DNA–enzyme complexes, with either a positive or negative effect on methylation of that site.

C. Triplexes

Studies using synthetic polymers of deoxyribo- and ribonucleotides show that poly(rA) or poly(dA) can form polynucleotide complexes that do not conform to the Watson–Crick model of double-stranded helical DNA. These complexes have been shown by X-ray diffraction to be triple-stranded, with structural characteristics that differ significantly from those of double-stranded DNA (47). Both deoxyribo- and ribonucleic acids can form these stable triplex structures in high concentrations of inorganic salts (47, 48).

Triplex structures were prepared by annealing the strands in a mixture containing the appropriate ratio of synthetic single-stranded homopolymers (47, 48), and were tested as inhibitors of DNA methylation in an assay using hemimethylated DNA as a substrate. The three putative triplex structures—$dT_n \cdot dA_n \cdot dT_n$, $rU_n \cdot rA_n \cdot rU_n$, and $rI_n \cdot rA_n \cdot rI_n$—inhibit methylation of hemimethylated DNA in our *in vitro* assay (Table VI). Addition of an equivalent concentration of poly(dA), poly(dT), or poly(rU) to the assay did not cause inhibition, and only weak inhibition was observed with poly(rA) or poly(rI) (Table VI). While the physiological significance of this inhibition is not entirely understood, the results may indicate that in addition to recognizing double-stranded DNA, the methyltransferase may be capable of recognizing other structural forms of DNA.

Poly(dC–dG)$_n$ is methylated at a slow rate (Table II). While this copolymer can be a substrate for the enzyme, the fully methylated copolymer poly(dG-m^{5}dC)$_n$ inhibits the methylation of hemimethylated DNA (Table VI). This observation, made with a synthetic copolymer, suggests that the enzyme may also be subject to product inhibition by methylcytosine-containing DNA. Such inhibition would probably reflect competition between methylated and unmethylated DNA for enzyme binding at the catalytic site. Since the poly(dG-m^{5}dC)$_n$ used in this experiment was fully methylated, it is not known if product inhibition can be demonstrated with a partially methylated copolymer or with partially methylated DNA.

V. Summary

The regulation of gene expression in eukaryotes has important implications for differentiation, oncology, genetic diseases, immunology, and virology. Correlations between the level of DNA methylation of specific genomic sequences and expression of certain genes in differentiating tissues suggest that DNA methylation is one of the possible mechanisms for the control of gene expression. If indeed DNA methylation is a primary event in the regulation of gene expression, elucidation of the complex factor(s) that may play important role(s) in DNA methylation would facilitate an understanding of the regulation of gene expression.

In this review, we have presented a summary of some of our experiments that have been directed at understanding the mechanism by which mammalian DNA methyltransferases recognize and modify DNA. The use of defined oligodeoxynucleotides and synthetic polydeoxyribonucleotides as substrates for the enzyme has facilitated the identification of various factors that influence the rate and specificity of DNA methylation *in vitro*.

REFERENCES

1. M. Gold, J. Hurwitz and M. Andres, *PNAS* **50**, 164 (1963).
2. J. F. Turnbull and R. L. P. Adams, *NARes* **3**, 677 (1976).
3. A. Razin and H. Cedar, *PNAS* **74**, 2725 (1977).
4. A. Solage and H. Cedar, *Bchem* **17**, 2934 (1978).
5. Y. Gruenbaum, R. Stein, H. Cedar and A. Razin, *FEBS Lett.* **124**, 67 (1981).
6. C. Waalwizj and R. A. Flavell, *NARes* **5**, 4631 (1978).
7. J. D. McGhee and G. D. Ginder, *Nature* **280**, 419 (1979).
8. J. L. Mandel and P. Chambon, *NARes* **7**, 2081 (1979).
9. R. Gambari, P. A. Marks and R. A. Rifkind, *PNAS* **76**, 4511 (1979).
10. J. K. Christman, P. Price, L. Pedrinan and G. Acs, *EJB* **81**, 53 (1977).
11. J. K. Christman, N. Weich, B. Schoenbrun, N. Schneiderman and G. Acs, *J. Cell Biol.* **86**, 366 (1980).
12. M. Sheffery, R. A. Rifkind and P. A. Marks, *PNAS* **79**, 1180 (1982).
13. J. Yu and R. D. Smith, *JBC* **260**, 3035 (1985).
14. M.-O. Ott, L. Sperling, D. Cassio, J. Levilliers, J. Sala-Trepat and M. Weiss, *Cell* **30**, 825 (1982).
15. A. Bird, J. Taggart and D. Macleod, *Cell* **26**, 381 (1981).
16. C.-K. J. Shen and T. Maniatis, *PNAS* **77**, 6634 (1980).
17. R. M. Grainger, R. M. Hazard-Leonards, F. Samaha, L. M. Hougan, M. R. Lesk and G. H. Thomsen, *Nature* **306**, 88 (1983).
18. M. Busslinger, J. Hurst and R. A. Flavell, *Cell* **34**, 197 (1983).
19. A. Bird, *Nature* **307**, 503 (1984)
20. I. Kruczek and W. Doerfler, *EMBO J.* **1**, 409 (1982).
21. D. Sutter and W. Doerfler, *PNAS* **77**, 253 (1980).
22. M. Groudine, R. Eisenman and H. Weintraub, *Nature* **292**, 311 (1981).

23. Y. Pollack, R. Stein, A. Razin and H. Cedar, *PNAS* **77**, 6463 (1980).
24. M. Wigler, D. Levy and M. Perucha, *Cell* **24**, 33 (1981).
25. P. H. Roy and A. Weissbach, *NARes* **2**, 1669 (1975).
26. D. Simon, F. Grunert, U. V. Acken and H. Kröger, *NARes* **5**, 2153 (1978).
27. G. P. Pfeifer, S. Grünwald, T. L. J. Boehm and D. Drahovsky, *BBA* **740**, 323 (1983).
28. R. L. P. Adams, E. L. McKay, L. M. Craig and R. H. Burdon, *BBA* **561**, 345 (1979).
29. B. Sheid, P. R. Srinivasan and E. Borek, *Bchem* **7**, 280 (1968).
30. F. Kalousek and N. R. Morris, *JBC* **244**, 1157 (1969).
31. N. R. Morris and K. D. Pih, *Cancer Res.* **31**, 433 (1971).
32. R. H. Burdon, M. Qureshi and R. L. P. Adams, *BBA* **825**, 70 (1985).
33. T. W. Sneider, W. M. Teague and L. M. Rogachevsky, *NARes* **2**, 1685 (1975).
34. T. H. Bestor and V. M. Ingram, *PNAS* **80**, 5559 (1983).
35. Y. Gruenbaum, H. Cedar and A. Razin, *Nature* **295**, 620 (1982).
36. A. Bolden, C. Ward, J. A. Siedlecki and A. Weissbach, *JBC* **259**, 12437 (1984).
37. R. Holliday and J. E. Pugh, *Science* **187**, 226 (1975).
38. D. Jähner, H. Stuhlmann, D. L. Steward, K. Harbers, J. Löhler, I. Simon and R. Jaenisch, *Nature* **298**, 623 (1982).
39. D. Drahovsky and N. R. Morris, *JMB* **61**, 343 (1971).
40. A. H. Bolden, C. M. Nalin, C. A. Ward, M. S. Poonian, W. W. McComas and A. Weissbach, *NARes* **13**, 3479 (1985).
41. A. Weissbach, C. M. Nalin, C. A. Ward and A. H. Bolden, *in* "Progress in Clinical and Biological Research. Biochemistry and Biology of DNA Methylation" (A. Razin and G. L. Cantoni, eds.), Vol. 198, p. 79. Liss, New York, 1985.
42. A. Bird, M. Taggert, M. Frommer, O. J. Miller and D. Macleod, *Cell* **40**, 91 (1985).
43. W. Doerfler, *ARB* **52**, 93 (1983).
44. M. L. Tykocinski and E. C. Max, *NARes* **12**, 4385 (1984).
45. A. H. Bolden, C. M. Nalin, C. A. Ward, M. S. Poonian and A. Weissbach, *MCBiol* **6**, 1135 (1986).
46. A. D. Riggs, *Cytogenet. Cell Genet.* **14**, 9 (1975).
47. S. Arnott and E. Selsing, *JMB* **88**, 509 (1974).
48. S. Arnott and P. J. Bond, *Science* **181**, 68 (1973).

The Interferon Genes

> CHARLES WEISSMANN AND
> HANS WEBER
>
> Institut für Molekularbiologie I
> Universität Zürich
> 8093 Zürich, Switzerland

I. Types, Effects, and Properties of Interferons

In 1957, Isaacs and Lindenmann first described interferon (IFN), a factor produced transiently by virus-infected cells that is capable of converting other cells into a virus-resistant state (1). This conversion was later shown to depend on macromolecular synthesis in the target cell (2) and not to be caused by interference with virus adsorption at the cell surface (3).

A. Different Types of Interferons

Many types of cells produce IFN in response to viral infection. IFN produced by human fibroblasts is serologically distinct from that produced by leukocytes, leading to the classifications "fibroblast IFN" and "leukocyte IFN," respectively (4, 5). Fibroblasts usually produce exclusively a species designated IFN-β (but see ref. 6), while leukocytes [in particular monocytes (6a)] produce mainly several closely related IFNs of a type designated α and a small proportion of β (4–8). IFN-α and IFN-β, which have similar biological and physicochemical properties (resistance to acid in particular) are called type I IFNs, and eventually proved to be structurally related (9).

A different kind of IFN, recognized in 1965 (10), and later called type-II, immune, or IFN-γ, is acid-labile; it is produced by T lymphocytes in response to mitogens or enterotoxin, or by specifically sensitized immunocompetent T lymphocytes exposed to the sensitizing antigen (reviewed in 11 and 12).

More recently, evidence has been offered for the existence of an IFN, designated IFN-β_2, that shows little if any structural similarity to any of the IFNs mentioned above, but is neutralized by antibodies directed against IFN-β (12a).

B. The Pleiotropic Effects of Interferons

IFNs have very high specific biological activity, and the antiviral effect of some IFNs can be detected at levels as low as 10^{-12} to 10^{-13} M. Although antiviral activity is the most striking property of IFNs, other activities—such as inhibition of proliferation of certain cell lines (13–15), stimulation of cytotoxic lymphocytes and of natural killer cell activity (16–19), inhibition of monocyte differentiation (20), inhibition of conversion of 3T3-L1 fibroblasts to adipose cells (21), the lowering of myc mRNA level in Daudi cells (22, 22a, 22b), and stimulation of expression of certain cell surface proteins (23)—may be of physiological importance. Type II IFN has most of these (12) and some additional, unique properties, such as macrophage activation activity (23a) and induction of determinant Ia (or HLA-DR) expression (24–26). IFNs induce the formation or increase the level of many proteins in target cells (27–29); in agreement with these biological observations, the proteins induced by type I IFNs are at least in part a subset of those induced by IFN-γ (30, 31).

IFNs show limited species specificity. Some human α IFNs (33–36) are active on mouse cells while others, as well as IFN-β (37, 38), are not. Some, but not all mouse IFNs have very low activity on human cells (39, 40, 40a), and all known human α IFNs are active on certain bovine cell lines (41) and even in cattle *in vivo* (42).

C. Physical Properties of Interferons

The initial isolation and characterization of IFNs was arduous because the starting material was scarce, consisted of different IFN species, and was difficult to purify. Most sources yield IFN at only a few micrograms per liter, and only after years of effort were levels of 100–200 μg/liter achieved with the use of human leukocytes (43). For these reasons, pure IFNs were available only on the microgram scale as recently as 1980 (43a, 44, 44a, 45), and only partial structures could be determined. The complete amino-acid sequence was established only after the cloned cDNAs of the IFNs became available (46–49). In the past few years the availability of monoclonal antibodies (50) has greatly facilitated purification.

The heterogeneity of human IFN preparations very soon became apparent. Human leukocyte IFN was fractionated into multiple components by a variety of techniques (49, 51–54) and IFN-γ into three species (55, 56); IFN-β was obtained as a single species (57). It was not clear whether these multiple fractions represented different gene products or multiple forms of the same primary gene product arising by such posttranslational modifications as proteolytic cleavage or glycation.[1] Ultimately, all these possibilities were found to occur. Some human IFN-α species clearly derive from distinct genes (49, 58), while others represent proteolytic cleavage products (59). Even now many IFN-α species described by Rubinstein, Pestka, and their colleagues have not been correlated with cloned genes (60). As only few (60a), and not the major human α IFNs, are glycated[1] (61), this type of modification cannot contribute to their heterogeneity to a significant extent (60a). In the case of IFN-γ, encoded by a single gene in man, the heterogeneity of the purified protein is ascribed to varying degrees of glycation (55).

Several reviews bearing on the subject matter of this article have appeared in recent years (60, 62–69).

II. Analysis of the Interferon System by Recombinant DNA Technology

The cloning of the IFN genes in 1979 was achieved before any structural information on the proteins was available. This was made possible by two observations, namely, that mRNA from IFN-producing cells injected into *Xenopus* oocytes gave rise to antivirally active IFN (70), and that only induced cells contain IFN mRNA (71–73). These findings provided two approaches for selecting putative IFN cDNA clones. In the "plus–minus" or differential hybridization technique, a cDNA "library" from induced cells is screened with radioactive cDNA made on poly(A)$^+$ RNA from induced and noninduced cells, respectively; the clones hybridizing with the induced, but not the noninduced probes are selected for further analysis (74, 75, 79). In the hybridization–translation assay, the unknown cloned cDNA is irreversibly attached to a membrane, hybridized to poly(A)$^+$ RNA containing IFN mRNA and the filter is washed. If the cDNA (or mixture of cDNAs) contains a sequence complementary to part or all of an IFN mRNA, it will bind the IFN mRNA, which can then be released by denaturation and assayed by the oocyte injection method (76, 77). It should be noted that any cDNA clone with an incidental partial homology to an IFN mRNA will give a positive response in this assay, so that additional verification of the identity of the clone is mandatory.

---

[1] Glycation is proposed by the Nomenclature Committee of IUB (NC-IUB) and the IUPAC-IUB Joint Commission on Biochemical Nomenclature (JCBN) to indicate attachment of a sugar to a protein where it is uncertain whether a glycosyl or a glycoside, or neither, is formed (57a) [Eds.].

Once a cDNA has been identified it can be used to screen cDNA and genomic libraries for further, related sequences.

The human IFN probes were also used to screen DNA libraries of mouse, cattle, rat, and monkey under conditions of reduced hybridization stringency, which led to the isolation of many animal IFN cDNAs and genes. A wide survey of type I IFN genes by "Southern blotting"[2] (78) revealed large IFN-α families in all mammalians tested, and one (or possibly two) IFN-β genes in all vertebrates examined except for ungulates (79, 80), which have several genes.

The mapping of IFN genes to individual chromosomes was achieved by two approaches. (1) A set of DNAs from various human–mouse hybrid cell lines, each of which contained a different combination of human chromosomes, was analyzed for the presence or absence of a particular IFN gene by Southern hybridization, or, less reliably, by determining its capacity to produce IFN in response to induction, and the relevant chromosome was identified by correlation analysis (81–85). (2) A ^{3}H-labeled *IFN* probe of high specific activity was hybridized to chromosome spreads and the location of the gene established by grain counting (86).

A further important contribution of recombinant DNA technology is the expression at high levels of IFN genes in *E. coli* and other microorganisms, allowing the preparation and purification of the different mature IFNs on a large scale and opening the way to the definitive investigation of their properties (87–90). As several IFNs, such as IFN-β (91), IFN-γ (55), and some α (60a, 92) IFNs are glycated, and as *E. coli* cannot provide the necessary glycation, cloned IFN genes have been expressed in animal cells, to yield what are believed to be accurately glycated proteins (93–96).

Finally, recombinant DNA technology has made it possible to modify, by DNA reconstruction and site-directed mutagenesis (96a), natural IFN genes, and explore structure–function relationships at the transcriptional (97–105) and protein level (106–109). In addition, IFN genes encoding either natural or modified amino-acid sequences have been prepared by total organic synthesis (109, 110, 110a).

---

[2] In "Southern analysis," DNA is cleaved with one or more restriction enzymes; the fragments are electrophoretically separated on an agarose gel and transferred (by blotting) from the gel to a filter membrane. The membrane-bound, immobilized DNA is hybridized with a radioactive, specific DNA (for example, a cloned cDNA) to identify the DNA fragments carrying cognate sequences. In "Northern analysis" poly(A) RNA is electrophoresed through an agarose gel and transferred to a membrane. The membrane-bound, immobilized RNA is hybridized with a radioactive DNA to reveal specific RNA species.

III. The IFN-α Genes

A. Human IFN-α Genes

1. GENERAL DESCRIPTION

The first IFN cDNA to be isolated and identified by expression in E. coli was human IFN-α1, using subculture cloning and identification of candidate clones by the hybridization–translation assay and finally by monitoring the bacterial extracts for antiviral activity (77). Using IFN cDNA probes, an additional large number of human IFN-α cDNAs, genes, and pseudogenes were isolated and sequenced (Table I) (48, 111–121a); as set forth below, they are grouped into two subfamilies, namely, subfamily IFN-α_I, comprising at least 15 loci, of which 14 are potentially functional, and subfamily IFN-α_{II} with at least 6 loci, of which all but one are pseudogenes (62, 121a, 122, 123, 136). Two pseudogenes (IFN-α11 and -α12) may represent two additional subfamilies (120). On the basis of their relatedness, the members of subfamily IFN-α_I have been subdivided into two groups: group 1, which encompasses IFN-α1, -α13, -α2, -α5, -α6, and group 2, which includes LeIF C, LeIF F, IFN-α4b, -α7, -α16 (120) (see also Fig. 3). Unavoidably, those who isolated a cDNA or gene used a laboratory-specific designation; in Table I we have indicated the names of sequences that are identical with, or polymorphic variants of, representative sequences of individual loci; it is hoped that the Interferon Nomenclature Committee will soon recommend definitive designations.

All IFN-α (and -β) genes found so far lack introns (112, 120, 124, 125). Their general structure is described in Table II; they encode preIFNs that consists of a 23-residue signal sequence (with the exception of GX-1, which may have only 16 residues) that is cleaved off during maturation, and a mature sequence usually of 166 amino acids; HuIFN-α2 is shorter by one, and HuIFN-$\alpha_{II}1$ longer by six residues.

The amino-acid sequences encoded by known IFN-α loci as well as of some alleles are shown in Fig. 1; a consensus sequence has been derived from all nonallelic sequences. It is evident from Fig. 1 that some regions of the coding sequence are strongly conserved, in particular from position 135 to 151, not only in α IFNs of all other species examined, but also in β IFNs. Erickson et al. (126) have suggested that the IFN-α (and IFN-β) coding regions have a repeat structure, the first half of the molecule being homologous to the second half.

All human α IFNs contain at least four cysteine residues that, at least in the case of IFN-α2, participate in disulfide bonds (Cys^1–Cys^{98};

TABLE I
Compilation of Human IFN-α Genes, cDNAs, and Pseudogenes[a]

A. Genes and cDNAs

	Reference sequences (distinct loci)	Origin	Related sequences (alleles?)	Origin	Differences to reference sequences
Subfamily IFN-α_I					
1.	IFN-$\alpha 1$ (112)	(c, chr) 705–2038	$LeIF$ D (48)	(c) 876–1880	T1341, T1502
2.	IFN-$\alpha 2$ (113)	(c) 950–1651	$LeIF$ A (48)	(c) 870–1879	T870, A880, C881, C882, A1068, C1819, A1860
	$\lambda\alpha 2$ (119)	(chr) 295–2223			
	$p104$ (116)	(c) 1080–1716			
3.	IFN-$\alpha 4b$ (120)	(chr) 1–2231	IFN-$\alpha 4a$ (120)	(chr) 161–1376; 1482–1795	C642, A755, G1151, A1341, T1531
4.	IFN-$\alpha 5$ (120)	(chr) 334–2008	$LeIF$ G (48)	(c) 1100–1880	I: C1800, Δ: 1834
5.	IFN-$\alpha 6$ (120)	(chr) 351–2041			
	$LeIF$ K (222)	(?)			
6.	IFN-$\alpha 7$ (120)	(chr) 863–1827	I-II (111)	(chr) 759–1827	C776, Δ: 793, Δ: 873, I: T879, C1475, A1481
	$LeIF$ J (115)	(chr) 1–2231	$IFLrK$ (226)	(chr) 759–1695	
7.	$LeIF$ B (48)	(c) 901–2004	IFN-$\alpha 8$ (120)	(chr) 884–1517	A901, C904, A920, C953 (Δ/I: 1290–1303)
8.	IFN-$\alpha 13$ (132)	(chr) 564–1989	IFN-$\alpha 14$ (120)	(chr) 616–1828	Δ: 622, Δ: 651, C652, A653, I:G 655, I:G 656, Δ: 800, G922, A925, C1505, G922, A925, T1454, C1505, C1625, T1645, G1656, T1660, T1697, A1826
9.	$\lambda 2h$ (118)	(chr) 399–2207	$LeIF$ H (48)	(c) 875–1919	
	IFN-αN (121)	(chr) 1214–2207[b]	$LeIF$ H_1 (48)	(c) Not given	As $LeIF$ H, in addition A1475, G1481
10.	IFN-$\alpha 16$ (120)	(chr) 321–1580	IFN-$\alpha 17$ (120)	(chr) 164–293; 1152–1453	T212
	IFN-αN(gren) (223)	(c) 925–1922			
	IFN-αWA (224)	(chr) 921–1506			
	IFN-$\alpha 0$ (121)	(chr) 1223–1396[c]			
11.	$\lambda 2c_1$ (118)	(chr) 162–2231	IFN-αT (121)	(chr) 1199–1413	T1292
			IFN-$\alpha 88$[d]	(c) 971–1653	C1453, G1482, T1652
			IFN-$\alpha(Ovch)$ (227)	(c) 986–1499	G1120, C1286, C1352, C1420
12.	$LeIF$ F (48)	(c) 923–1971			
13.	GX-1 (225)	(c) 860–1916			

	LeIF C (48)		
	ΨIFN-α10[e] (120)	(c) 884–1920	T991, G1023, C1265, C1920
	ΨIFLrL[e] (226)	(chr) 745–1577	Δ: 1319
		(chr) 752–1889	A752, T753, Δ: 793, Δ: 869, T871, Δ: 1852, Δ: 1859, T1860, I: A1870, Δ: 1877

Subfamily IFN-α$_{II}$
15. IFN-α$_{II}$1 (122)   (chr) (1544 n.)

B. Human Pseudogenes

Subfamily ΨIFN-α$_I$[e]
1. ΨLEIF E (48)   (c)

Subfamily ΨIFN-α$_{II}$
2. ΨIFN-α$_{II}$2 (23)   (chr)
3. ΨIFN-α$_{II}$3 (123)   (chr)
   ΨIFN-α$_{II}$15 (120)   (chr)
4. ΨIFN-α$_{II}$4 (123)   (chr)
5. ΨIFN-α$_{II}$M (115)   (chr)
   ΨIFN-α$_{II}$8 (120)   (chr)[f]
6. ΨIFN-α$_{II}$9 (120)   (chr)
7. ΨIFN-L130 (136)   (chr)

Other subfamilies
8. ΨIFN-αl1 (120)   (chr)
9. ΨIFN-αl2 (120)   (chr)

IFN-ω(P9A2) (121a)	(c) (877 n.)		G1263, Δ: A1548
IFN-ω(E76E9) (121a)	(c) (875 n.)		G1263, A1602, Δ: A1548

[a] Sequences clearly attributable to different loci are designated as reference sequences. Sequences differing in about 1% or less of their positions from a reference sequence are considered to be allelic to these unless proven otherwise. The origin of the sequences are designated as "c" for cDNA, and "chr" for chromosome-derived, and the numbers mark the beginnings and ends of the sequenced region according to the numbering of Fig. 2 in Henco et al. (120). The differences between the presumed alleles and their reference counterpart are listed in the last column. Δ, deletion; I, insertion. The following correlations between IFN species isolated from natural sources and IFN genes have been made: the leukocyte derived fractions "α1," "α2," "β1," and "β2" are thought to be derived from IFN-α2 (58, 59); the Namalwa cell derived "αA" from IFN-α1 (49); the leukocyte-derived fraction "β3" from IFN-α1 (58). Updated from Henco et al. (120). Namalwa cell derived fraction "αβ3" is a mixture of several components, one of which corresponds to IFN-α1 (49).
[b] The previously (120) described differences from λ2h have been retracted (A. von Gabain, personal communication).
[c] The previously (120) described difference from IFN-α16 have been retracted (A. von Gabain, personal communication).
[d] A. von Gabain (personal communication).
[e] The locus LeIF L is placed among active IFN-α genes because one presumed allele is active.
[f] No: sequenced; identified by position on linkage group and hybridization.

TABLE II
Properties of IFN Genes and Their Products

	Hu α	Bo α	Mu α	Ra α	Hu β	Bo β	Mu β
Number of nonallelic functional genes	>15	>6	>10	>1	1	>3	1
Chromosome location	9		4		9		4
mRNA 5' nontranslated region	67–69	67–69	66–69	72	73–75	76–78[a]	
mRNA 3' nontranslated region	240–440				200		
Number of introns	None	None	None	None	None	None	None
Length signal peptide	23[b]	23	23[c]	23	21	21	21
Length mature protein	166[d]	166[e]	166–167[f]	169	166	165	161
Position of Cys in mature protein[g]	1, 29, 99, 139[h]	1, 29, 99, 139	1, 29, 86, 99, 139	1, 29, 86, 99, 139	17, 31, 141	17	
Position of putative N-glycation sites[g]	None[i]	None	78[j]	167	80	110, 152	29, 69, 76

	Hu γ	Bo γ	Mu γ	Ra γ
Number of nonallelic functional genes	1	1	1	1
Chromosome location	12			
mRNA 5′ nontranslated region	127			110
mRNA 3′ nontranslated region	584		628	
Number of introns	3	3	3	3
Length signal peptide	20[k]	20[k]	19[k]	19[k]
Length mature protein	146[k]	146[k]	135[k]	136[k]
Position of Cys in mature protein[g]	1, 3[k]	None	1, 3, 135	1, 3, 136
Position of putative N-glycation sites[g]	28, 100	19, 86	19, 71	19, 72

[a] By analogy to human IFN-β.
[b] Exception: GX-1, 16.
[c] Exception: MuIFN-α4, 24.
[d] Exceptions: IFN-α2, 165; IFN-α$_{II}$1, 172.
[e] Exception: BoIFN-α$_{II}$1, 172.
[f] Exception: MuIFN-α4, 162.
[g] Position relative to first residue of mature sequence (numbering in Figs. 1 and 4 through 8 are not always applicable because of gapping)
[h] Exceptions: IFN-α1, IFN-α13 additional Cys at pos. 86; LeIF B pos. 100 instead of 99.
[i] Exceptions: GX-1: 2; IFN-α14, LeIF-L, LeIF-H1: 2, 72; IFN-α$_{II}$1: 78.
[j] Exceptions: MuIFN-α6T, MuIFN-αA: no sites.
[k] Note discussion of amino-terminus in text.

	1S	11S	21S	1	11	21
			$	$		$$
consensus sequence	MALSFSLLMA	LVVLSYKSIC	SLG	CDLPQTHSLG	NRRTLMLLAQ	MGRISPFSCL
			#	# # # # #	#	# # ## ##
IFN-α1, IFN-α13	..SP.A...V	.....C..S.	...	....E....D	....	.S....S...
LeIF D	..SP.A...V	.....C..S.	...	....E....D	....	.S....S...
IFN-α2 (λα2)	...T.A..V.	.L...C..S.	.V.	....	.......S..	.R...L....
IFN-α2'	.......V.	.L...C..S.	.V.	........N.	S..	.R...L....
LeIF A	...T.A..V.	.L...C..S.	.V.	....	.......S..	.RK..L....
IFN-α6 (LeIF K)	...P.A....	.....C..S.	..D	....	H...M.....	.R...L....
IFN-α5 (LeIF G)	...P.V....	.....NC...	...	....	.......S..	.....IM...
GX-1	n...p.alM..	.....C..S.	...	.N.S.....N	....	.....IM...
IFN-α14 (λ2h)	...P.A.M..	.....C..S.	...	.N.S.....N	.......M..	.R........
LeIF H	...P.A.M..	.....C..S.	...	.N.S.....N	.......M..	.R........
LeIF H1	...P.A.M..	.....C..S.	...	.N.S.....N	.......M..	.R........
IFN-α8	...T.Y..V.	........FS	...	....	.......A.I	.R........
LeIF B	...T.Y.MV.	........FS	...	....	.......A.I	.R........
IFN-α16	....	VL........	...	....	.......A.I	........H...
LeIF F	....	VL........	...	....	.......A.I	....
IFN (Ovch)	....	VL........	...	....	.......A.I	....
IFN-.4b	....	VL........	...	....	.......A.I	........H...
IFN-α4a	....	VL........	...	....	.......A.I	....
λ2cl (LeIF I)	....	VL........	...	....	.......A.I	....
IFN-α88	....	VL........	...	....	.......A.I	....
IFN-α7 (LeIF J)	..R......V	VL........	...	.......R..	.......A.I	....
IFLrK (I-II)	..R......V	VL........	...	.......R..	.......A.I	....
LeIF C	....	VL........	...	....	.......A.I	.G........
IFN-α_{II}1	...L.F..A.	..MT..SPVG	...	.....N.G.L	S.N..V..H.	.R.....L..
IFN-α_{II}(P9A2)	....	....	..	.....N.G.L	S.N..V..H.	.R.....L..
IFN-α_{II}(E76E9)	....	....	..	.....N.G.L	S.N..V..H.	.R.....L..

	31	41	51	61	71
	$$ $ $	$ $ $$	$	$ $ $$	$$
consensus sequence	KDRHDFGFPQ	EEFDGNQFQK	AQAISVLHEM	IQQTFNLFST	KDSSAAWDET
	## # #	# # ##	##	## # ## #	##
IFN-α1, IFN-α13	M.........	....	.P......L	...I....T.	.........D
LeIF D	M.........	....	.P......L	...I....T.	.........D
IFN-α2 (λα2)	....	.........-	.ET.P.....	...I......	....
IFN-α2'	....	.........-	.ET.P.....	...I......	......
LeIF A	....	....	.ET.F.....	...I......	....
IFN-α6 (LeIF K)	......R...	....	.E......V	....	...V....R
IFN-α5 (LeIF G)	....	....	....	....	....T.....
GX-1	....	....	....	....	.....S....
IFN-α14 (λ2h)	......E...	....	M.........	.N........	....
LeIF H	......E...	....	M.........	.N........	....
LeIF H1	......E...	....	M.........	.N........	....
IFN-α8	......E...	......DK..	....	....	.......L..
LeIF B	......E...	......DK..	....	....	.......L..
IFN-α16	...Y......	.V........	.AF.......	....	....
LeIF F	....	....	....	....	....T.EQS
IFN (Ovch)	....	....	....	....	....T.EQS
IFN-α4b	.........E	.....H....	T.........	....	E.....EQS
IFN-α4a	.........E	.....H....	....	....	E.....EQS
λ2cl (LeIF I)	...P...L..	....	T.........	....	E.....EQS
IFN-α88	....	....	T.........	....	E.....EQS
IFN-α7 (LeIF J)	...E.R..E	.....H....	T.........	....	E.....EQS
IFLrK (I-II)	...E.R..E	.....H....	T.........	....	E.....EQS
LeIF C	.......RI.	....	....	....	....
IFN-α_{II}1	...R..R...	.MVK.S.L..	.HVM......	L..I.S..H.	ER.....NM.
IFN-α_{II}(P9A2)	...R..R...	.MVK.S.L..	.HVM......	L..I.S..H.	ER.....NM.
IFN-α_{II}(E76E9)	...R..R...	.MVK.S.L..	.HVM......	L..I.S..H.	ER.....NM.

FIG. 1. Amino acid sequences of human α interferons. A consensus sequence was derived from the 15 nonallelic gene products (underlined) by determining for each position the amino acid occurring in the largest number of sequences. The amino acids coinciding with the consensus are represented by dots. Numbers preceded by S refer to the signal peptide. Lower case symbols in the signal sequence of *GX-1* represent probably untranslated codons. References are given in Table I. #, Positions conserved

THE INTERFERON GENES

```
                            81         91         101        111        121
                                       $                     $         $  $  $
consensus sequence      LLDKFYTELY QQLNDLEACV IQEVGVEETP LMNEDSILAV RKYFQRITLY
                        ##         ##      #  #          #         ## #  #
IFN-α1, IFN-α13         .....C.... .......... M..ER.G... ...A...... K...R.....
LeIF D                  .....C.... .......... M..ER.G... ...V...... K...R.....
IFN-α2 (λα2)            .......... .......... ..G...T... ..K....... ..........
LeIF A                  .......... .......... ..G...T... ..K....... ..........
IFN-α6 (LeIF K)         ....L..... .......... M...W.GG.. .......... ..........
IFN-α5 (LeIF G)         .......... .......M.. M......D.. ...V....T. ..........
GX-1                    .......... .......M.. M......D.. ...V....T. ..........
IFN-α14 (λ2h)           ..E...I..F ..M....... .......... .......... K.........
LeIF H                  ..E...I..F ..M....... .......... .......... K.........
LeIF H1                 ..E...I..F ..M....... .......... .......... K.........
IFN-α8                  ...E..I..D .......S.. M......I.S ..Y....... ..........
LeIF B                  ...E..I..D ......VLC D.....I.S. ..Y....... ..........
IFN-α16                 ......I..F .......... T.......IA .......... ..........
LeIF F                  ..E..S...N .....M.... .......... ...V...... K.........
IFN (Ovch)              ..E..S...N .......... .......... ...V...... K.........
IFN-α4b                 ..E..S.... .......... .......... ...V...... ..........
IFN-α4a                 ..E..S.... .......... .......... .......... ..........
λ2cl (LeIF I)           ..E..S.... ......N... .....M.... .......... ..........
IFN-α88                 ..E..S.... ......N... .....M.... .......... ..........
IFN-α7 (LeIF J)         ..E..S.... .......... .......... ......F... ..........
IFLrK                   ..E..S.... .......... .......... ......F... ..........
LeIF C                  ..E..S.... .......... .......... .......... ..........
IFN-αII1                ...QLH...H ...QH..T.L L.V..EG.SA GAISSPA.TL .R...G.RV.
IFN-αII(F9A2)           ...QLH.G.H ...QH..T.L L.V..EG.SA GAISSPA.TL .R...G.RV.
IFN-αII(E76E9)          ...QLH.G.H ...QH..T.L L.V..EG.SA EAISSPA.TL .R...G.RV.

                            131        141        151        161        171
                        $       $$ $$ $   $$ $$              $
consensus sequence      LTEKKYSFCA WEVVRAEIMR SFSLSTNLQE RLRRKE
                        #       ### ## ##### ###  #   #       # #
IFN-α1, IFN-α13         .......... .......... .L........ ......
LeIF D                  .......... .......... .L........ ......
IFN-α2 (λα2)            .K........ .......... .......... S..S..
LeIF A                  .K........ .......... .......... S..S..
IFN-α6 (LeIF K)         .......... .......... ...G.R.... ......
IFN-α5 (LeIF G)         .......... .......... ......A... ......
GX-1                    .......... .......... ......A... ......
IFN-α14 (λ2h)           .M........ .......... .L.F.....K .....D
LeIF H                  .M........ .......... ...F.....K .....D
LeIF H1                 .M........ .......... ...F....KK G....D
IFN-α8                  .......S.. .......... ....I...K ..KS..
LeIF B                  .......S.. .......... ....I...K ..KS..
IFN-α16                 .MG....... .......... ...F.....K G....D
LeIF F                  .......... .......... ....KIF... ......
IFN (Ovch)              .......... .......... ....KIF.. ......
IFN-α4b                 .......... .......... .L.F.....K .....D
IFN-α4a                 .......... .......... .......... .....D
λ2cl (LeIF I)           .......... .......... .L.F.....K I....D
IFN-α88                 .......... .......... .L.F.....K .....D
IFN-α7 (LeIF J)         .M........ .......... ...F....KK G....D
IFLrK                   .M........ .......... ...F.....K .....D
LeIF C                  .I.R...... .......... .L.F.....K .....D
IFN-αII1                .K........ .....M...K .LF....M.. ...S.DRDLG SS
IFN-αII(F9A2)           .K.....D.. .....M...K .LF....M.. ...S.DRDLG SS
IFN-αII(E76E9)          .K.....D.. .....M...K .LF....M.. ...S.DRDLG SS
```

in all human IFNs' $, positions conserved in all human α and β IFNs. Since on maturation IFN-β precursors lose 21, and IFN-α precursors 23 amino-terminal amino acids, position n in an IFN-α protein sequence is homologous to position $n + 2$ in an IFN-β sequence. The one-letter (IUPAC-IUB) code used is A, Ala; R, Arg; N, Asn; D, Asp; C, Cys, Q, Gln; E, Glu; C, Gly; H, His; I, Ile; L, Leu; K, Lys; M, Met; F, Phe; P, Pro; S, Ser; T, Thr; W, Trp; Y, Tyr; V, Val.

Cys^{29}–Cys^{138}) (*127*). However, the Cys^1–Cys^{98} linkage is not required for antiviral activity (*127a*); in fact, the four amino-terminal residues can be removed from IFN-α1 without significantly reducing antiviral activity (M. Mishina and C. Weissmann, unpublished results; the conclusion drawn by Zoon and Wetzel (67) from the data of ref. 33, that the first 15 residues of IFN-α2 are not essential for antiviral activity, is not warranted). IFN-α1 and IFN-α13 have an additional, presumably free (*127a*) cysteine residue. Most human IFN-α genes do not encode typical N-glycation[1] sequences of the type Asn-Xaa-Ser (or -Thr) [where Xaa represents an unidentified amino acid (*127b*)], except for GX-1 (position 2), *IFN-α14* (*LeIF H*) (positions 2 and 72), and *IFN-α$_{II}$1* (position 78). Most human α IFNs are not glycated (*58*, *60*), but it has recently been reported that minor IFN species may carry carbohydrate residues (*60a*, *61*).

As shown in Fig. 9, the 5' flanking region shows remarkable conservation in a segment lying between −64 and −109 nucleotides preceding the "cap" site; this region is strikingly rich in GAA and GAAA repeats and contains all the information necessary for the induced transcription of the IFN-α genes (*97–99*). The length of the 3' noncoding region varies from about 240 to 440 nucleotides, as the position and number of polyadenylation sites differ (*48*, *120*). At least one gene, IFN-α1, has three polyadenylation sites, all of which appear to be used (*128*), giving rise to mRNAs differing by about 50 and 90 nucleotides.

2. Linkage and Location of IFN-α Genes

Table I lists independently isolated human IFN-α cDNAs and genomic clones for which complete or significant portions of sequences have been reported. Some sequences are identical; others differ by a few and others by as many as 30% of their nucleotides (*120*, *122*).

In the case of a multigene family, especially when dealing with cDNA-derived sequences, it can be difficult to determine which sequences represent distinct loci and which are allelic variants. Two or more genes are definitely nonallelic when they are found on a single cloned genomic DNA segment or on linkage maps assembled from overlapping genomic clones. The linkage groups I through VI (Fig. 2) have definitively been established; they differ clearly from each other and comprise 18 loci (*120*). Two additional, distinct clones, IX (*119*) and X (*122*), define two further loci. Clone VIII (λHLeIF2, *118*) very likely represents the same segment as V (*120*). Clone VII (*121*) overlaps V (and therefore VIII) with its 5' half, and VI with its 3' third.

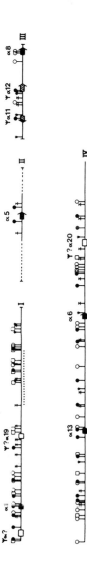
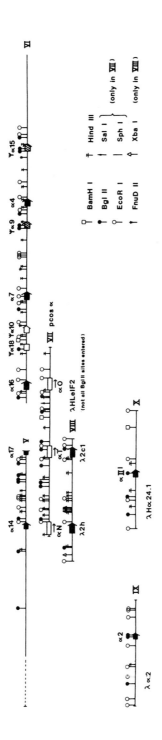

FIG. 2. Restriction maps of linkage groups containing human *IFN-α*-related sequences. The location of genes (black boxes; arrowheads indicate the direction of transcription) and pseudogenes (hatched boxes) were determined by sequencing and/or by Southern blot analyses.[2] Open boxes indicate regions hybridizing to an *IFN-α* probe but not sequenced to a sufficient extent to determine whether they are genes or pseudogenes. Ψ?*IFN-α20* hybridizes weakly to *IFN-α₁* probes but not to Ψ*IFN-α₁9*; it has not yet been further characterized. "Ψα?" designates a region that shows questionable hybridization to *IFN-α₁* probes. The segment underlined with dots was lost on subcloning of the cosmid clone and is not confirmed. From Henco et al. (120).

Therefore, genes *IFN-αN*, *-αT*, and *-αO*, which lie on clone VII, are allelic to *IFN-α14*, *-α17*, and *-α16*, respectively (*120*). The reservations expressed in ref. *120* regarding this assignment have been superseded by the finding that *IFN-αN* and *IFN-α14* have the identical nucleotide sequence, rather than differing in 22 positions as originally claimed (von Gabain, personal communication). The overlap V–VI–VII defines a 100-kilobase segment of DNA, carrying 9 loci. Altogether, the linkage groups of Fig. 2 comprise 20 loci, of which 12 are potentially functional, 6 are not, and 2 remain to be characterized. Within each linkage group the IFN genes examined so far have the same orientation. Most or all IFN-α genes are located on chromosome 9 (*81, 82, 85*), probably close to the IFN-β gene (*129*).

Three cDNA clones (ΨPLeIF E, LeIF F, and GX-1) show 9% or more nucleotide differences relative to the genes defined above. As IFN genes known to be allelic all show less than 1% nucleotide differences, we assume they represent additional loci. However, this conclusion is tentative, as allelic genes with 7.6–10% nucleotide mismatches are known in other gene families (*130, 131*). Seventeen additional sequences, all of which differ by less than 1% from the genomic loci described above, are listed as alleles in Table IA. This similarity is again an unreliable criterion, because two IFN genes, for example, *IFN-α1* and *-α13*, may have completely identical coding sequences and yet be nonallelic (*132*). LeIF C, which encodes a translatable sequence, is considered to be allelic to Ψ*IFN-α10*, which by virtue of one nonsense and one frameshift mutation, is nonfunctional. Three additional pseudogenes have been reported by others (*123, 136*).

In summary, by the criteria applied above, the IFN-α gene family consists of at least 24 loci, of which 14 are potentially functional, 9 are not, and 1 probably has functional and nonfunctional alleles. Two additional loci have been identified but remain to be characterized. At least 17 additional sequences, allelic to the above, have been identified.

3. The Relatedness of the Human IFN-α Genes

The relatedness of two coding sequences can be quantified by aligning them and counting the number of mismatches. It is useful to distinguish *replacement* substitutions, which cause an amino-acid change, from *silent* substitutions, which do not. Divergence between two sequences is calculated separately for replacement and silent substitutions, in essence by dividing the number of actual by the number of possible substitution events. A correction is applied to account for multiple events at one site, as any one position will undergo more than one mutation in the course of time.

The nucleotide sequences of the IFN-α coding regions have been compared pairwise; divergences in silent and in replacement sites, corrected for multiple events, were calculated following Perler et al. (133) and recorded in Henco et al. (120).

All genes except IFN-$α_{II}1$ differ from each other by divergence values of 8.3 ± 2.5% (range 1.6–13%) in the replacement, and 23 ± 8.3% (range 1.8–35%) in the silent sites, while the corresponding values for IFN-$α_{II}1$ relative to all other genes are 30.5 ± 1.4% (28–34%) and 67.5 ± 2.9% (62–72%), respectively (Table III). Capon et al. (122) therefore classified the IFN-α genes into two subfamilies, IFN-$α_I$ and IFN-$α_{II}$. The few genes and pseudogenes assigned to subfamily IFN-$α_{II}$ have been specifically designated with the subscript II; all genes not so designated may be considered as pertaining to subfamily IFN-$α_I$.

As judged by the divergence data, the IFN-$α_I$ genes fall into two major groups (Fig. 3). Group 1 comprises IFN-α1, -α2, -α5, -α6, -α13, and their presumed alleles. Group 2 corresponds to group C proposed by Ullrich et al. (115) and comprises IFN-α4b, IFN-α7, LeIF C, λ2c_1, LeIF F, and IFN-α16. IFN-α8 (LeIF B) appears equally distant from the members of the two groups.

However, not all sequences fit into this classification. Henco et al. (120) examined the relationships between IFN-α sequences by charting "characteristic mismatches" according to Gillespie and Carter (134). This approach showed that IFN-α14 is more closely related to IFN-α5 with its 5', and to IFN-α7 with its 3' moiety. Even more striking is the relationship of GX-1 to IFN-α5 and IFN-α14: the upstream moiety of the GX-1 coding sequence, down to (mature) codon 10 is, with one exception, identical to that of IFN-α14 (λ2h), whereas the 3' part is identical with that of IFN-α5; these similarities extend to the 5' and 3' noncoding regions. GX-1 thus represents a remarkable instance of a natural hybrid gene of very recent origin, which may have arisen by unequal crossing over or gene conversion. IFN-α14 probably originated in a similar fashion from members of groups 1 and 2. Hybrid genes resulting from unequal crossing-over of the human α1- and α2-globin genes or of the human δ- and β-globin genes (Hb Lepore) have been reported (135). The cDNA sequence LeIF H, which we believe is allelic with λ2h and IFN-α14 (Table I), differs from the latter in five positions within a short 3' noncoding segment (position 1625–1697 in Fig. 2; ref. 120); this segment is identical with the corresponding section of IFN-α7 and could have arisen by limited gene conversion.

TABLE III
DIVERGENCE OF IFNs[a]

A. Average divergence between α IFNs

	Hu α_I		Hu α_II		Bo α_I		Bo α_II		Mu α	
	R	S	R	S	R	S	R	S	R	S
Hu α_I [13][b]	8.3 ± 2.5 (1.6–13)	23.0 ± 8.3 (1.8–35)								
Hu α_II [1]	30.5 ± 1.4 (28–34)	67.5 ± 2.9 (62–72)								
Bo α_I [3]	22.4 ± 0.9 (20.9–24.4)	43.5 ± 3.0 (38.2–51.5)	38.7 ± 1.4 (37.4–40.1)	76.2 ± 1.5 (74.9–77.8)						
Bo α_II [1]	32.8 ± 1.1 (31.4–35.4)	59.9 ± 5.1 (51.3–65.6)	22.8	38.5	3.7 ± 0.8 (2.8–4.2)	10.8 ± 1.8 (9.2–12.7)				
Mu α [7][c]	26.0 ± 1.7 (22.7–29.5)	57.4 ± 6.7 (43.5–72.7)	42.3 ± 1.7 (39.5–44.5)	81.0 ± 7.1 (70.0–92.0)	35.2 ± 0.3 (35.0–35.6)	55.3 ± 5.7 (52.5–60.7)	41.9 ± 2.0 (39.6–45.3)	73.1 ± 8.9 (60.6–86.6)	8.5 ± 2.0 (4.7–12.1)	23.1 ± 9.7 (7.3–40.1)
Ra α [1]	26.4 ± 0.7 (25.5–27.4)	54.9 ± 4.1 (47.2–60.0)	42.4	71.2	32.2 ± 1.4 (30.1–34.2)	51.9 ± 7.6 (39.1–69.7)				
					31.7 ± 0.2 (31.5–31.9)	53.2 ± 2.2 (51.9–55.8)	41.8	69.2	11.0 ± 1.3 (8.5–12.1)	24.8 ± 8.4 (16.1–36.9)

B. Divergence of β IFNs

	Hu β		Mu β		Bo β	
	R	S	R	S	R	S
Hu β			36.2	71.1	30.7 ± 3.5 (27.8–34.6)	65.8 ± 4.1 (61.5–69.7)
Mu β					50.1 ± 2.4 (47.6–52.4)	79.7 ± 5.7 (76.0–86.3)
Bo β [3]					7.8 ± 1.7 (6.0–9.3)	12.2 ± 3.5 (8.6–15.6)

C. Divergence of γ IFNs

	Mu γ		Ra γ		Bo γ	
	R	S	R	S	R	S
Hu γ	44.9 ⟨59.3⟩	79.5	47.4 ⟨61.1⟩	91.2	⟨38.6⟩	
Mu γ			7.8	40.6		
Ra γ			⟨16.7⟩		⟨57.3⟩ ⟨59.9⟩	

D. Divergences between selected IFN-α and IFN-β genes

	Hu α6		Bo α2		Mu α6T	
	R	S	R	S	R	S
Hu β	68.0	106.3	67.5	110.0	78.8	130.1
Bo β₁	66.9	119.5	75.1	113.5	75.8	133.6
Mu β	75.3	96.3	81.1	98.6	79.7	102.5

[a] Percentage divergence values between pairs of genes were calculated according to Perler et al. (133) (see Table II in ref. 120) and averaged. The values in parentheses indicate the range. Numbers in square brackets in the first column indicate the number of genes compared. R, Replacement sites; S, silent sites. In cases where only protein sequence data were available, amino-acid differences (%) are given between angular brackets. For comparisons between α and β IFN genes, the most distant individual a and b subtypes of a species were selected as judged by preliminary divergence calculations using visually aligned sequences. The selected genes were aligned using the program "GAP" of the Program Library of the University of Wisconsin Genetics Computer Group (J. Devereux and P. Haeberli, personal communication) and a gap-weight value of 5.0 (number of gaps introduced: 0–6). The Perler algorithm for calculation of divergence in some cases gives rise to logarithms of negative numbers. This was always due to a single nucleotide of replacement site category 2 of the BoIFN-α₁₁ gene. In such cases, the codon in question was disregarded in the calculations. It is generally advisable to disregard replacement sites of category 2 and silent sites of category 1, as they may give rise to anomalous values by the Perler method.

[b] The 13 nonallelic human IFN-α genes with different coding sequences listed in Table IA.

[c] MuIFN-α1, -α2, -α4, -α5(P), -α6(P), -α6(T), -αA(7).

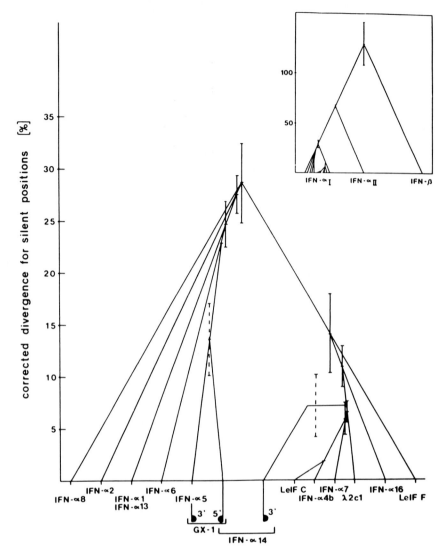

THE SUBFAMILY IFN-α_I

FIG. 3. Relationship graph of human *IFN* coding sequences. The relationships of the coding sequences of 14 distinct human IFN-α_I loci, based on the divergence of the silent sites, are plotted. The inset shows the relationship of *IFN-α_I*, *IFN-α_{II}*, and *IFN-β* coding sequences. Branch points are calculated as for an evolutionary tree on the basis of the divergence data in Henco *et al.* (120), assuming a simple gene duplication process at each branch point (220). The bars indicate the standard error of the mean in the determination of the branch points. The branch points were calculated separately for both the 5'- and 3'-part of *GX-1* and *IFN-α14*. The gene for *GX-1* is believed to have resulted from a very recent gene conversion or unequal crossover event, because its 5' and 3' segments are identical to their counterparts in *IFN-α14* and *IFN-α5*, respectively. From Henco *et al.* (120).

Very extensive gene conversion may have occurred in the case of the nonallelic genes *IFN-α1* and *-α13*, which are identical over their entire coding sequence but show 4.5% nucleotide differences in the 5' flanking and noncoding regions, and 3.8% differences in the 3' noncoding region (*132*).

As gene conversion and/or crossover events are probably frequent in the IFN-α gene family, the representation of Fig. 3 should not be considered as a phylogenetic tree from which dates of gene duplication can be derived, but as a "relationship graph" indicating the degree of similarity of individual sequences or parts thereof.

4. Pseudogenes

Strictly speaking, a pseudogene is a mutated gene that can no longer give rise to a biologically active product, either because no translatable mRNA is generated, or because the translation product fails to arise or is no longer functional. In practice, and especially in the case of the IFN genes, it is not always easy to establish these criteria. In this review we use the term pseudogene to designate coding sequences clearly related to IFN genes in which the reading frame has been shifted by insertions or deletions, or interrupted by a stop codon, unless it is known that a biologically active product is made.

The λ chromosomal clones carrying the IFN-α genes also contain a number of DNA sequences hybridizing very weakly to an *IFN-α1* probe (*117*). These regions were subcloned and in turn used as probes for Southern analysis[2] of the λ chromosomal IFN clones. The sequence now designated as Ψ*IFN-α$_{II}$9* hybridized strongly with Ψ*IFN-α$_{II}$15*, Ψ*IFN-α$_{II}$18*, and Ψ*(?)IFN-α$_{II}$19* (type z); Ψ*IFN-α11* (type y) and Ψ*IFN-α12* (type x) cross-hybridized neither with each other nor with type z sequences (*120*). Ψ*LeIF-α$_{II}$M* (designated *LeIF-M* by Ullrich *et al.* (*115*)) belongs to type z as determined by sequence analysis; it is more similar to the recently discovered *IFN-α$_{II}$1* than to any of the subfamily IFN-α$_I$ genes (*122*). Therefore, all z-type sequences (which differ by about 6% mismatches) are assigned to subfamily *IFN-α$_{II}$*.

Goeddel and his colleagues (*122, 123*) determined the existence of about 6–7 subfamily IFN-α$_{II}$ loci by Southern analysis and sequenced four of them. Ψ*IFN-α$_{II}$2*, *-α$_{II}$3*, and *-α$_{II}$4* were pseudogenes; Ψ*IFN-α$_{II}$3* was identical to Ψ*IFN-α$_{II}$15* (*120*). Feinstein *et al.* (*136*) also identified a set of pseudogenes and sequenced one of them, *L130*, which was in general very similar to Ψ*IFN-α$_{II}$3* (and *-α$_{II}$15*) but showed distinctive differences in the 5' and 3' regions; it is clearly also a member of the IFN-α$_{II}$ family and we designate it Ψ*IFN-α$_{II}$L130*. Because the Ψ*IFN-*

α_{II} pseudogenes are so similar to each other and differ from the *IFN-α* consensus sequence by identical insertions and deletions, it is very likely that they arose from an ancestral pseudogene by duplication events. Both Ψ*IFN-α11* and Ψ*IFN-α12* (*120*), which differ considerably from each other and from the Ψ*IFN-α$_{II}$* pseudogenes (40–50% mismatches), show distinctive deletion and insertion patterns, and represent two independent evolutionary developments. They may represent additional subfamilies that could also contain as yet unidentified functional genes.

Ψ*LeIF E* clearly belongs to subfamily IFN-α_I, but has a point mutation in the 39th codon that introduces a frameshift, as well as many nucleotide changes in positions otherwise conserved. This pseudogene is still transcribed to some extent, as its cDNA was cloned from KG-1 cell RNA (*48*). The pseudogenes Ψ*LeIF-L* and Ψ*IFN-α10* are defective by virtue of one and two point mutations, respectively; they are otherwise very similar to *IFN-α_I* genes.

5. Other IFN-α-Related Loci

Several loci have been identified as *IFN-α1*-related, but have not been characterized sufficiently. The DNA segments designated Ψ*IFN-$\alpha_{II}18$*, Ψ*(?)IFN-$\alpha_{II}19$*, and Ψ*(?)IFN-α20* (Fig. 2) hybridize weakly with an *IFN-α1* probe. Ψ*IFN-$\alpha_{II}18$* is probably allelic with Ψ*LeIF-$\alpha_{II}M$* (*120*). Ψ*(?)IFN-$\alpha_{II}19$* hybridized strongly with the Ψ*IFN-$\alpha_{II}9$* probe and therefore is probably a subfamily IFN-α_{II} member; however, it has not been sequenced, so it is therefore not known whether or not it is a pseudogene. Ψ*(?)IFN-α20* has not been further characterized (*120*).

6. Biological Activity and Expression Level

The multiplicity of IFN-α genes raises the question as to whether the individual species preferentially fulfill different functions. The following general conclusions about the biological properties of the different IFN-α species appear justified. (1) They may show very different patterns of target specificities when tested on cell lines from various species (*33, 35, 36, 137*). This may mean that the IFN-α receptors of different species show different "fits" for the various human IFN-α species. The target specificities of natural and hybrid human α IFNs on murine, bovine, and human cells led to the proposal that the two halves of the IFN molecule contribute to its activity by interacting simultaneously with different sites on the receptor or to different receptor moieties (*33*). (2) Individual human IFN-α species show different relative activities when tested on different *human* cell lines (*35*,

138). This can be explained if there are more than one kind of receptor, and different cell lines have different proportions of these receptors. Alternatively, a cooperative binding model which assumes that the apparent specific activity of an IFN species depends both on the surface density of (a single kind of) receptor and its affinity for the IFN could also explain the findings. (3) The ratios of the different biological activities, such as antiviral activity, natural killer cell activation, antiproliferative activity, and (2'-5')-polyadenylate synthetase induction (as measured on one cell type) may differ for the various IFN-α species (34, 61, 138–141). It has also been claimed that the antiviral spectra of different IFN species, measured on the same cell line, may differ (35). These findings are best accommodated by the assumptions that the different IFN-α species fulfill different functions and that there exist more than one type of receptor; however, the appropriate experiments to test this hypothesis have yet to be reported.

The individual IFN genes are expressed at different levels in different cell types. In normal, virus-induced leukocytes the highest transcript levels were those of *IFN-α1*, *-α13*, *-α2*, and of an IFN-$α_{II}$1-related gene, and to a lesser extent of *LeIF F* (7, 48, 122, 132). Our laboratory had previously reported that *IFN-α4* was expressed at the transcript level, at least in some leukocyte preparations (7, 8); we have recently found that the transcript we had detected was probably not from *IFN-α4* but from an as yet unidentified, closely related gene (M. Streuli, unpublished results). No expression of *IFN-α6* has been reported. In leukemic leukocytes and lymphoblastoid (Namalwa) cells, the ratio of *IFN-α2* to *IFN-α1* transcripts is increased, and in addition there is a striking increase in the level of *IFN-α14* in leukemic myeloblasts (8).

B. IFN-α Genes in Nonhuman Species

1. Bovine IFN-α Genes

Capon *et al.* (*122* and D. Goeddel, personal communication) isolated clones from a bovine genomic library using a human IFN-$α_I$ cDNA probe and determined the sequence of four genes. Three of these, *BoIFN-$α_I$1*, *-$α_I$2*, and *-$α_I$3*, were quite similar to each other and to two distinct bovine genes (cos1 and cos26) isolated by T. Kovacic (Biogen Inc., personal communication), with about 3.7% (corrected) average divergence at replacement sites (less than 10% difference at the amino-acid level) (Table III). As shown in Fig. 4, *BoIFN-$α_I$1*, *-$α_I$2*, and $α_I$3 encode mature proteins of 166 amino acids. However, *BoIFN-$α_{II}$1* showed 35% divergence from the *BoIFN-$α_I$1* sequence (46% dif-

```
                           S1         S11        S21   1         11         21
                           $          $ $        $          $              $$
consensus sequence         MAPAWSLLLA LLLLSCNAIC SLG   CHLPHTHSLA NRRVLTLLRQ LRRVSPSSCL
                           ##    # #  # # #  #         # #   # #  #   # ## ## ## ##
BoIFN-α1                   ..........  ..........  ...  .....S....  K.........  ..........
BoIFN-α2                   ......F...  ..........  ...  ........P.  ..........  ..........
BoIFN-α3                   ..........  ..........  ...  ........I..  ......M..G.  ..........
BoIFN-α_{II}1              ..FVL...M.  .V.V.YGPGG  ...  .D.SPN.V.V  G.QN.R..G.  M..L..RF..

                           31         41         51         61         71
                           $$ $    $  $         $ $$ $  $   $  $             $
consensus sequence         QDRNDFAFPQ EALGGSQLQK AQAISVLHEV TQHTFQLFST EGSAAVWDES
                           ### ######  #         ## #   ##########    #  # ##    # #     #
BoIFN-α1                   ..........  ..........  ..........  ..........  ..........
BoIFN-α2                   ..........  ..........  ..........  ..........  .......Q.
BoIFN-α3                   ..........  ..........  ..........  ..........  ....DHHV...
BoIFN-α_{II}1              ...K......  .MVEV..F.E  .........M  L.QS.N..HK  .R.S.A..TT

                           81         91         101        111        121
                           $   $    $ $                               $          $
consensus sequence         LLDKLRTALD QQLTDLQACL RQEEGLPGAP LLKEDSSLAV RKYFHRLTLY
                           ## #   #   ### ##  ##                #         ##     ##      #
BoIFN-α1                   ...R......  ..........  ..........  ..........  ..........
BoIFN-α2                   ......A...  ..........  ......R...  ......A....  ..........
BoIFN-α3                   ......D...  .......F..  .....E.Q...  ..........  ..........
BoIFN-α_{II}1              ..EQ.L.G.H  ...D..D...  GLLT.EEDSA  .GRTGPT..M  KR..QGIHV.

                           131        141        151        161        171
                           $  $      $$ $ $              $  $        $
consensus sequence         LQEKRHSPCA WEVVRAQVMR AFSSSTNLQE RFRRKD
                           ####    # ##  ## ##    ##    #### ###   # #
BoIFN-α1                   ..........  ..........  ..........  ......
BoIFN-α2                   ..........  ......E...  ..........  K.....
BoIFN-α3                   ..........  ..........  ..........  S.....
BoIFN-α_{II}1              ....GY.D..  ...I..LEI..  SL....S...  .L.MM.GDLK SP
```

FIG. 4. Amino-acid sequences of bovine α interferons. Conventions are as in Fig. 1. In positions where no plurality amino acid could be defined, the consensus shows the residue found in BoIFN-α1. The sequences are from genomic clones (122 and D. Goeddel, personal communication). #, conserved in all bovine α IFNs; $, conserved in all bovine α and β IFNs. Since on maturation IFN-β precursors lose 21, and IFN-α precursors 23 amino-terminal amino acids, position n in an IFN-α protein sequence is homologous to position $n + 2$ in an IFN-β sequence.

ference at the amino-acid level) and encoded a mature protein of 172 residues. This led to the recognition of two subfamilies of *IFN-α*. Capon et al. (122) then identified the two homologous families in man, as described above, and showed that bovine *IFN-α_{II}* was more closely related to human *IFN-α_{II}* than to bovine *IFN-α_{I}*. This clearly shows that the IFN-α_{I} and -α_{II} families diverged prior to the mammalian radiation. Southern analysis of bovine genomic DNA under stringent hybridization conditions revealed the presence of about 10–12 genes related to *BoIFN-α_{I}* and 15–20 genes related to *BoIFN-α_{II}*. The number and distribution of cysteine codons are the same as in the human α IFN genes (with the exception of HuIFN-α1 and -α13); no potential N-glycation[1] sites are present.

2. RODENT IFN-α GENES

Southern analysis suggests the existence of not less than 12 rat (142) and 10 mouse *IFN-α$_I$*-related genes (40). All murine IFN-α genes are probably located on chromosome 4 (83, 143), linked to both the histocompatibility locus H-15 (84) and the MuIFN-β gene (143a), in the 4C3 → C6 region (144, 143a). Murine (40, 145–147) and rat (142) IFN-α sequences were originally isolated from genomic and/or cDNA banks using human *IFN-α* probes. Figure 5 shows the amino-acid sequences of the polypeptides encoded by the nine murine genomic and cDNA sequences. Six genes are definitely nonallelic; *MuIFN-α1*, -α4, -α5, and -α6(P) were isolated from a single 28 kilobase DNA fragment (146), and *MuIFN-α2*, -α6(T) are from a 15 kilobase fragment (147). *MuIFN-αA* and -αB are also most probably distinct alleles, because of the relatively large number of amino-acid replacements; we propose they be designated *MuIFN-α7* and *MuIFN-α8*, respectively. *MuIFN-α5(P)* is probably an allele of *MuIFN-α5(T)*. *MuIFN-α3* was only partially sequenced and is therefore not presented (W. Boll, personal communication; it most closely resembles *MuIFN-α5* and *MuIFN-α6*). Quite recently, a further gene, designated *MuIFN-α9*, has been described (147a).

The general organization of the rodent *IFN-α* genes is similar to that of man and cattle (Table II). Murine and rat *IFN-α* genes, other than most other IFN genes, have five rather than four cysteine residues, in the same location as human *IFN-α1* (Fig. 6). With the exception of *MuIFN-α6(T)* and *MuIFN-αA(-α7)*, all known rodent *IFN-α* sequences encode a potential N-glycation site at codon 78. Natural murine α IFNs are glycated; glycation is, however, not essential for antiviral activity, at least in some cases (92).

All published murine *IFN-α* coding sequences are as closely related to each other as are those of the human IFN-α$_I$ subfamily (Table IIIA), with (corrected) divergences for replacement sites between 4.7 and 12.1%. They are about as closely related to the human IFN-α$_I$ genes (average replacement site divergence, 26 ± 2%) as are the bovine IFN-α$_I$ genes (23 ± 1%) and therefore constitute the homologous murine IFN-α$_I$ subfamily. The individual murine genes have some particularities worth mentioning: *MuIFN-α4* has a five-codon deletion between codons 102 and 108, a region that in general shows little conservation at the amino-acid level; *MuIFN-αB (-α8)* has a single codon deletion at amino acid position 17; and *MuIFN-α2*, -α4, and -αA (-α7) have an additional glutamate codon at the carboxyl terminus.

The antiviral activities of MuIFN-α1, -α2, and -α5 expressed in

```
                          S1           S11            S21     1             11              21
                          $            $                      $             $ $    $                    $
consensus sequence        MARLCAFLMV   LAVMSY-WST     CSLG    CDLPQTHNLR    NKRALTLLVQ      MRRLSPLSCL
                          ### ###              ##  #  ####    #### #   #    ## # #          #### ####
MuIFN-α1   (40,146,147)   ..........   ..........    ...P.   ..........    ..........      ..........
MuIFN-α2   (40,147)       ........VM   .I........I   ....    ....H.Y...    .....KV.A.      ....PF....
MuIFN-α4   (146,147)      .........I   LVM...Y..A    ....    ....H.Y..G    .....V.EE.      .......P..
MuIFN-α5(P) (146)         ..........   .P.L......    .P..    ..........    ..........K     ..........
MuIFN-α6(P) (146)         ..........   ...L......    .P..    ..........    ..........K     ..........
MuIFN-α6(T) (147)         ..........   ..........    ....    .......K..    ........I.      ..........
MuIFN-αA(α7) (145)        ..........T  .L........    ....    ..........    ..........      ..........
MuIFN-αB(α8)  *           ..........   ..........    ....    ..........    ........-.      ..........
MuIFN-α9   (147a)         ...PF.....   .V.I......    ....    ..........    ..KI......A.    ..........

                          31            41            51             61             71
                          $  $$$   $    $                            $              $   $  $
consensus sequence        KDRKDFGFPQ    EKVDAQQIQK    AQAIPVLSEL     TQQILNIFTS     KDSSAAWNAT
                          ### ## ##     ###     #     ## ##         ### # ###  #   ## #  #
MuIFN-α1                  ..........    .........K    ..........    ..........     ..........
MuIFN-α2                  ...Q.....L    ....N.....    .......RD.    ...T..L...     .A........
MuIFN-α4                  .........L    ....N.....    ....L..RD.    ........L.     ..L..T....
MuIFN-α5(P)               ..........    .....G....E   ..........    .......V..     ..........
MuIFN-α5(T) (147)         ..........    ..........    ..........    ...V......     ..........
MuIFN-α6(P)               ..........    .....G....E   .......T..    .......TL.     ..........
MuIFN-α6(T)               ..........    ....TLK...    EK.......V    ..........     .........D..
MuIFN-αA(α7)              ....R.....    ..........    ....N.....    ...Q...V..     ........D.S
MuIFN-αB(α8)              ..........    .........E    .......T..    ..M.TL....     .A........
MuIFN-α9                  ..........    .........E    ..........    .......TL.     ..........

                          81            91             101            111             121
                                        $                                             $    $
consensus sequence        LLDSFCNDLH    QQLNDLQACL     MQQVGVQEPP     LTQEDSLLAV      RKYFHRITVY
                          ### ##        # ##### #       #         #    #### ##       ########
MuIFN-α1                  ..........    .......G..     .......F..     ......A...      ..........
MuIFN-α2                  ..........    .......T..     ..........     ......A...      ..........
MuIFN-α4                  ..........    ......K..V     ..------..     ..........      .T........
MuIFN-α5(P)               .....EV...    ......K..V     .......S..     ..........      ..........
MuIFN-α5(T)               .....EV...    ......K..V     .......S..     ..........      ..........
MuIFN-α6(P)               ..........    .L.....G..     ....EI.AL.     ..........      .T......F
MuIFN-α6(T)               ...T.....Y    ..........     V...RL....     ......V...      ..........
MuIFN-αA(α7)              ..........    ......K..V     ....E.....     .........Y.     ...T......
MuIFN-αB(α8)              ..........    ..........     ..........     ..........      ..........
MuIFN-α9                  ......TG..    .L.....G..     ..L..MK.L.     ......Q..M      K.........

                          131           141            151            161
                          $             $ $ $$$  $
consensus sequence        LREKKHSPCA    WEVVRAEVWR     ALSSSANLLA     RLSEEK
                          ## ####  ##    ### ###### #   # ###  #     ##
MuIFN-α1                  ..........    ..........     ......V.G.     ..R...
MuIFN-α2                  ..........    ..........     ......V...P    ......E
MuIFN-α4                  ..K....L..    ....I.....     .......T..     ......E
MuIFN-α5(P)               ..........    ..........     ......V...     ...K.E
MuIFN-α5(T)               .......L..    ..........     ......V...     ...K.E
MuIFN-α6(P)               ..........    ..........     .........K    ..N.DE
MuIFN-α6(T)               ..........    ..........     ......V.G.     ..R...
MuIFN-αA(α7)              ..........    ..........     .M....K...     ......E
MuIFN-α9                  ..........    ..........     .......V..     ......E
```

FIG. 5. Amino-acid sequences of murine α interferons. Conventions are as in Fig. 1. In positions where no plurality amino acid could be defined, the consensus shows the residue found in *MuIFN-α1*. All genes listed are from genomic clones; MuIFN-α2 was previously isolated from a cDNA clone (40). *MuIFN-α5(P)* and *MuIFN-α5(T)* are very probably allelic. As the laboratories of P.M. Pitha (146) and of J. Trapman (147) used the designations *MuIF-α5* and *-α6* for two different pairs of isolates, we have added the suffix (P) and (T) to differentiate between them. *, S. Pestka, personal communication; #, conserved in all murine α IFNs; $, conserved in all murine α and β IFNs. Since on maturation IFN-β precursors lose 21, and IFN-α precursors lose 23 amino-terminal amino acids, position n in an IFN-α protein sequence is homologous to position $n + 2$ in an IFN-β sequence.

THE INTERFERON GENES

```
                              S1         S11        S21    1           11            21
all IFN-α                                                 * *          *                           **
all IFN-α & -β                                            $                                        $
HuIFN-α consensus             MALSFSLLMA LVVLSYKSIC SLG   CDLPQTHSLG   NRRTLMLLAQ    MGRISPFSCL
conserved in hum.                                         # #    # #   #     #        # # ##   ##

BoIFN-α consensus             ..PAW...L. .LL..CNA.. ...  .H..H....A   ..V.T..R.     LR.V..S...
conserved in bov.             ##  # # # # # #        ###  # #    # #  #  # ##        ## ## ##

MuIFN-α consensus             ..RLCAF..V .A.M..WST. ...  .......N.R   .K.A.T..V.    .R.L..L...
conserved in mur.             ### ###        ### # ###  #### #  #  ## #   #         #### ####

RaIFN-α1 (221)                ..RLCAF..S ...V..W.A. C..  ....H..N.R   .K.VFT....    .R.L..V...


                              31         41         51           61           71
                              * *        *  *       *                *        *
                                         $  $       $                $
HuIFN-α consensus             KDRHDFGFPQ EEFDGNQFQK AQAISVLHEM   IQQTFNLFST   KDSSAAWDET
                              ##   #  #  #     # ##           ##  ## # ## #   ##

BoIFN-α consensus             Q..N..A... .ALG.S.L.. .........V   T.H..Q....   EG.A.V...S
                              ### ######  #     ## #  #########   #  # ##    # #         #

MuIFN-α consensus             ...K...... .KV.AQ.I.. ....P..S.L   T..IL.I.TS   .......NA.
                              ### ## ##  ###       #   ## ##    ### #  ### #  ## #  #

RaIFN-α1                      ...KY....L .KV..Q.I.. ....P....L   T..ILS..TS   .E..T...A.


                              81         91         101          111          121
                              **         *                                    **
                                                                              $
HuIFN-α consensus             LLDKFYTELY QQLNDLEACV IQEVGVEETP   LMNEDSILAV   RKYFQRITLY
                              ##         #  ##      #              #          ## # #

BoIFN-α consensus             ....LR.A.D ...T..Q..L R..E.LPGA.   .LK...S...   ....H.L...
                              ##     #     ### ## ##                #          ##  ##    #

MuIFN-α consensus             ...S.CND.H ......Q..L M.Q...Q.P.   .TQ...L...   ....H...V.
                              ### ##     # ####  #     #          # ####  ##  #######

RaIFN-α1                      ...S.CND.Q ...SG.Q..L M.Q...Q.S.   .TQ...L...   .E..H...V.


                              131        141        151          161
                              *          * **  **                *            .
                              $          $  $
HuIFN-α consensus             LTEKKYSPCA WEVVRAEIMR SFSLSTNLQE   RLRRKE
                              #          ### ## ##### ###  #      #      # #

BoIFN-α consensus             .Q..RH.... ......QV.. A..S......   .F...D
                              ####    # ## ## ##      ##          #### ###   # #

MuIFN-α consensus             .R...H.... ........VW. AL.S.A..LA  ..SEEK
                              ## ####  ## ### ###### # ###        #  ##

RaIFN-α1                      .R.N.H.... ....K..VW. AL.S.A..MG   ...EERNES
```

FIG. 6. Comparison of the amino-acid sequences of all known mammalian α IFNs. Positions in the animal IFN sequences [BoIFN-α consensus, MuIFN-α consensus, rat IFN-α1 (221)] coinciding with the human consensus sequence are indicated by dots. #, Positions conserved in α IFNs within a species; *, positions conserved in α IFNs in all species; $, positions conserved in all α and β IFNs in all species. Since on maturation IFN-β precursors lose 21, and IFN-α precursors lose 23 amino-terminal amino acids, position n in an IFN-α protein sequence is homologous to position $n + 2$ in an IFN-β sequence.

eukaryotic cells or *E. coli* are similar (*40, 147*); that of MuIFN-α4 (which has a five-amino-acid deletion) was about 1% of the others when expressed in *E. coli* (*213*) but equal to that of the others when expressed in eukaryotic cells (J. Trapman, personal communication). As in the case of the human genes, there is evidence for recombination or conversion between the different genes; for example, *MuIFN-α4* resembles *MuIFN-α2* at its 5′, but not at its 3′ half. The highest transcript levels were in *MuIFN-α4*, followed by *MuIFN-α6* and -α2, while little (*40, 147*) or no transcripts (*213*) were found in *MuIFN-α1* and -α5.

IV. The IFN-β Genes

A. The Human IFN-β Gene

A cDNA clone encoding human IFN-β was first isolated by Taniguchi and his colleagues using the differential hybridization and hybridization translation techniques (*148*) and characterized by sequence analysis (*46*) and subsequently by expression in *E. coli* (*149*). IFN-β cDNA clones isolated independently by Derynck et al. (*150*) and Goeddel et al. (*90*) had identical nucleotide sequences. One clone isolated by May and Sehgal (*151*) encoded a single Leu-to-Met change at position 47; this change entailed no loss of antiviral activity. Several IFN-β gene isolates (*124, 152–155*) were identical, with the same coding and noncoding sequences as the first cDNAs isolated. The mature sequence (Fig. 7) encodes three cysteine residues, at positions 17, 31, and 141. One partial cDNA clone (*108*) had a Cys-to-Tyr replacement at position 141, which caused a total loss of antiviral activity, presumably because of the disruption of a putative disulfide bridge between Cys^{31} and Cys^{141}; such a bond has been demonstrated in the case of HuIFN-α2 (*127*). Conversion of Cys^{17} to Ser by site-directed mutagenesis had no effect on biological activity (*107*).

A potential N-glycation site is found at position 80; human IFN-β is known to be glycated (*91, 157, 158*). As in the case of the IFN-α genes, there are no introns. The striking homologies within coding and 5′ flanking regions of the IFN-β and IFN-α genes are discussed in Section VI.

Southern analyses (*79, 122, 159*) identified only a single gene at hybridization stringencies that clearly revealed the multiplicity of the IFN-α genes. Therefore, the variant cDNAs mentioned above could represent alleles or, possibly, cloning artifacts. There is a report suggesting a duplication of the IFN-β gene in two human families (*159*).

```
                            S1         S11        S21 1          11         21
all IFN-β                   *     *    **********  *       *   *       *    ** * **
all IFN-β & -α
HuIFN-β (46,90,150)         MTNKCLLQIA LLLCFSTTAL  S   MSYNLLGFLQ RSSNFQCQKL LWQLNGRLEY
BoIFN-β consensus (172)     ..YR....MV ..........  ..  R..S..R.Q. .R.Lxx.... .G..PSTPQH
conserved in bov.           ## #### ## ########## #     ########## # #  #### # ## ## ##
MuIFN-β (173)               .N.RWI.HA. F.........  .   IN.KQ.QLQE .TNIRK..E. .E....KI--


                            31         41         51         61         71
                            *     *    * *              *    * ** *          *****
                            $          $ $                                   $
HuIFN-β                     CLKDRMNFDI PEEIKQLQQF QKEDAALTIY EMLQNIFAIF RQDSSSTGWN
BoIFN-β consensus           ..EA..D.QM ...M..A... .....I.V.. ....Q..N.L TR.F.....S
                            #### ####  #### # ### ######## # # ## ## ## ##########
MuIFN-β                     N.TY.AD.K. .M.MTE-KM- ..SYT.FA.Q .....V.LV. .NNF......


                            81         91         101        111        121
                            ***   **   *     *    *          *          ** **
                                                                         $
HuIFN-β                     ETIVENLLAN VYHQINHLKT VLEEKLEKED FTRGKLMSSL HLKRYYGRIL
BoIFN-β consensus           ...I.D..xE L.x.M.R.QP IQK.IMQ.QN S.M.DTTV-. ..xK..FNLV
                            ######### # ## ### # # ######## ## # ## # ## ######
MuIFN-β                     ....VR..DE LHQ.TVF... .....Q.-.R L.WEMSSTA. ...S..W.VQ


                            131        141        151        161
                            **    *    ** *    *  *   ** *
                            $          $$
HuIFN-β                     HYLKAKEYSH CAWTIVRVEI LRNFYFINRL TGYLRN
BoIFN-β consensus           Q..ES...NR ....V...Q. .T..S.LM.. .....D
                            ### #### # ###### ### # # ### ## #    #
MuIFN-β                     R...LMK.NS Y..MV..A.. F...LI.R.. .RNFQ.
```

FIG. 7. Comparison of the amino acid sequences of all known mammalian β IFNs. Conventions are as in Fig. 1 and Fig. 6. The bovine sequence is a consensus of *BoIFN-β1, -β2*, and *-β3*; conserved positions are marked by # below the amino acid symbol and positions where all three bovine sequences are different are indicated by x. *, Positions conserved in β IFNs in all species; $, positions conserved in all α and β IFNs in all species. Since on maturation IFN-β precursors lose 21, and IFN-α precursors lose 23 amino-terminal amino acids, position n in an IFN-α protein sequence is homologous to position n + 2 in an IFN-β sequence.

The IFN-β gene has been localized to chromosome 9 by classical somatic cell genetics (*160, 161*) and by Southern hybridization techniques (*81, 82, 86*), and is closely linked to the IFN-α genes (*159*); nevertheless, a physical linkage at the DNA level has not been demonstrated despite the existence of a 46-kilobase map around the IFN-β gene (*64*).

B. Other Human "IFN-β" Genes

The existence of human β IFNs and IFN-β genes other than the ones described above has been claimed on the basis of independent studies at the protein, RNA, and DNA level. Revel and his colleagues found that 10% of the antiviral activity produced by induced human fibroblasts (designated IFN-β_2) do not bind to Blue Sepharose while regular IFN-β (which they call IFN-β_1) do. IFN-β_2 could be neutralized (but not precipitated) by antibodies against "IFN-β_1" (unpublished data, quoted in ref. 63).

A second line of investigation, at the level of mRNA, led to the finding that there are two species of poly(A) RNA in induced human fibroblasts (11 S and 14 S) both of which, when injected into oocytes, give rise to the secretion of antiviral activity that can be neutralized by (polyclonal) antibodies to IFN-β (162). The product of the 14-S poly(A) RNA was equated with the IFN-β_2 described above. The cDNA corresponding to the 14-S RNA when cloned did not hybridize to *IFN-β* DNA (162). Using the IFN-β_2 cDNA as probe, two structurally different genomic clones, *IFA-2* and *IFA-11*, were isolated. *IFA-2*, which contains at least four exons, is believed to correspond to the cloned IFN-β_2 cDNA. The gene, when placed under the control of the SV40 early promoter, gave rise to an antivirally active protein, albeit with a specific activity 30–100 times lower than that of IFN-β (12a).

Although the amino-acid sequence of IFN-β_2 shows only a remote similarity to that of IFN-β, its antiviral activity is neutralized by all antibodies (including monoclonal ones) to IFN-β tested (12a). This remarkable finding suggested that the IFN-β_2 might be an inducer of IFN-β, rather than an antiviral protein in its own right. However, this proved not to be the case (162b). The second genomic clone isolated with the IFN-β_2 cDNA, *IFA-11*, had a 3' exon virtually identical with that of IFA-2, but showed no homology in the 5' proximal exons. When introduced into murine cells, *IFA-11* DNA also gave rise to antiviral activity that could be neutralized with IFN-β antibodies. In view of the lack of similarity at the genetic structural level between *IFN-β* and *IFN-β_2*, it seems to us that the designation "β_2" should be replaced by a different Greek letter.

More complexity is generated by a series of papers by Sehgal and his collaborators. Ruddle and his colleagues had earlier developed a series of human–mouse somatic cell hybrids containing different combinations of human chromosomes 2, 5, and 9; they reported that lines containing any one of these chromosomes could be induced to produce IFN-β and concluded that IFN-β genes are located on these chromosomes (129, 163, 164). Sehgal et al. assayed fractionated poly(A) RNA from several cell lines lacking chromosome 9 (which contains the IFN-β gene) by oocyte injection and found RNAs of a variety of sizes that gave rise to (low levels of) antiviral activity neutralizable by antisera to IFN-β. These mRNAs were assigned the designations IFN-$β_3$ to -$β_5$ and attributed to genes on various chromosomes. Pitha et al. (165) independently found that poly(A) RNA from (the same) somatic cell hybrids lacking chromosome 9 gave rise in oocytes to antiviral activity neutralizable by antiserum to IFN-β, but did not hybridize with IFN-β cDNA. These findings were taken to signify the existence of genes located on chromosomes other than 9, encoding antiviral proteins similar enough to IFN-β to be neutralized by anti-IFN-β antisera, but yet so different at the nucleotide sequence level that they would not cross-hybridize. The relationship between these findings and those of Revel and his colleagues remains to be established.

Sehgal et al. recently cloned two human genomic DNA fragments that hybridize to *IFN-β* probes under very nonstringent conditions (166). Clone λB3, which mapped to chromosome 2 (167), contained scattered regions with homology to various segments of the IFN-β gene (168), and "Northern analysis"[2] of induced fibroblast RNA with a λB3 probe revealed several bands that hybridized about equally well with *IFN-β* DNA. There is no convincing evidence to support the contention that these bands represent specific λB3 transcripts, and even less that λB3 transcripts encode an interferon, as suggested by Sehgal et al. (166). The cross-hybridization of these and other clones with *IFN-β* DNA under relaxed conditions may arise from traces of a common ancestry, residues of IFN pseudogenes, genomic segments that have acquired IFN-β-like sequences by gene conversion or fortuitous (i.e., biologically meaningless) homologies.

It is interesting to note that van Damme et al. (169, 170) reported a 22 kDa protein from Con A-stimulated leukocytes that exercises an antiviral effect and whose activity could be neutralized by antibodies against IFN-$β_1$. This protein, first thought to be an IFN, turned out to be an inducer of IFN-β (171) and has recently been shown to be closely related to IL-1 (172).

C. Nonhuman IFN-β Genes

Southern analysis[2] revealed the presence of a single IFN-β gene in a variety of primates, as well as in the mouse, rabbit, cat, and lion (79, 80). In the case of rabbit and lion, a very weak additional band suggests a second gene distantly related to *IFN-β* (79). Cow, horse, pig, and blackbuck each contains at least five genes related to IFN-β.

Leung *et al.* (80) isolated and sequenced three genomic bovine *IFN-β* clones. The coding sequences of the three genes show 6–9.3% corrected divergence at replacement and 8.6–15.6% at silent sites (Table IIIB), the same range found for the human IFN-$α_1$ genes. They have three cysteine residues in the same positions as the human IFN-β gene and potential glycation[1] sites at positions 110 and 153 (Table IIIB); it is not known whether bovine IFN-β is glycated.

Murine IFN-β cDNA (173) (Table II) encodes a 161-residue mature protein with three potential glycation sites; the cognate IFN, with an apparent molecular weight of 36,000, is a glycoprotein (92, 174). Interestingly, there is only a single cysteine, at position 17, which is dispensable for function in human IFN-β; there are no cysteine residues at positions 31 and 141, which probably form an essential disulfide bond in human IFN-β.

As mentioned above, only a single IFN-β gene has been identified by Southern analysis, apparently devoid of introns; it was mapped to chromosome 4 (173a). Skup *et al.* (175) selected two cDNA clones from virus-induced murine C243 cells, using the hybridization–translation assay; the genomic counterparts, which contained at least one intron each, were also isolated (176) and mapped to chromosomes 12 and X (quoted in 173a). The coding regions of the cDNA clones showed slight homologies to the human IFN-β but not to the murine IFN-β DNA sequence, and so far no proof has been offered that the proteins encoded by these sequences have an antiviral effect. It should be noted that the hybridization–translation procedure can give a false-positive result if a cDNA clone has a region homologous to any part of the IFN-β mRNA, as then IFN-β mRNA will be bound and subsequently give rise to authentic IFN-β.

V. The IFN-γ Genes

A. The Human IFN-γ Gene

Human IFN-γ cDNA was first cloned by Gray *et al.* (177) and Devos *et al.* (178), and both groups subsequently isolated the cognate chromosomal gene (179, 180). There were some differences between the sequences of different isolates: codon 140 was CAA (Glu) in the first cDNA clone characterized (177) but CGA (Arg) in all subsequently isolated cDNA and genomic clones, and codon 9 was CAA

```
                         S1          S11         1            11          21
all IFN-γ                *   *   *   *           *            * *  ***    *
HuIFN-γ  (177-180)       MKYTSYILAF  QLCIVLGSLG  CYCQDPYVKE   AENLKKYFNA  GHSDVADNGT
BoIFN-γ  (68)            ......F..L  L..GL..FS.  S.G.GQFFR.   I....E....  SSP...KG.P
MuIFN-γ  (66,186,187)    .NA.HC...L  ..FLMAV.-.  ...HGTVIES   L.S.NN...S  SGI..-EEKS
RaIFN-γ  (188)           .SA.RRV.VL  ...LMAL.-.  ....GTLIES   L.S..N...S  SSM.AMEGKS


             31          41          51           61          71
             *   *  **   **   *  **  ***   **     **  *  *    *
HuIFN-γ      LFLGILKNWK  EESDRKIMOS  QIVSFYFKLF   KNFKDDQSIQ  KSVETIKEDM
BoIFN-γ      ..SD......  D...K..I..  ..........   E.L..N.V..  R.MDI..Q..
MuIFN-γ      ...D.WR..Q  KDG.M..L..  ..I...LR..   EVL..N.A.S  NNISV.ESHL
RaIFN-γ      .L.D.WR..Q  KDGNT..LE.  ..I...LR..   EVL..N.A.S  NNISV.ESHL


             81          91          101          111         121
             *           *   *       *    *       *  *** *    *  *
HuIFN-γ      NVKFFNSNKK  KRDDFEKLTN  YSVTDLNVQR   KAIHELIQVM  AELSPAAKTG
BoIFN-γ      FQ..L.GSSE  .LE..K..IQ  IP.D..QI..   ...N...K..  ND...KSNLR
MuIFN-γ      ITT..SNS.A  .K.A.MSIAK  FE.NNPQ...   Q.FN...R.V  HQ.L.ESSLR
RaIFN-γ      ITN..SNS.A  .K.A.MSIAK  FE.NNPQI.H   ..VN...R.I  HQ...ESSLR


             131         141
             *****
HuIFN-γ      KRKRSQMLFR  GRRASQ
BoIFN-γ      ......N...  .....M
MuIFN-γ      .....RC
RaIFN-γ      .....RC
```

FIG. 8. Comparison of the amino-acid sequences of all known mammalian γ IFNs. Conventions are as in Fig. 1 and Fig. 6. *, Positions conserved in γ IFNs in all species.

(Gln) in the sequence of Taniguchi *et al.* (*182*) and AAA (Lys) in all other cases.

Cloning and Southern hybridization experiments indicate that IFN-γ is encoded by a single gene (*181, 182*) located in the p1205 → qter region of chromosome 12 (*183*). In contrast to the α and β IFN genes, the γ gene is interrupted by three introns, the first of which contains a repetitive sequence. The transcription unit is preceded by familiar promoter signals, i.e., a TATAA box 28 bases upstream and a sequence similar to the CCAAT box 89 bases upstream from the cap site. A polyadenylation signal (AATAAA) is found 16 bases upstream from the 3' end of the mRNA sequence.

The gene encodes a "preprotein" (a protein carrying a signal sequence) of 166 amino acids (Fig. 8). By analogy with the type I IFNs, a signal sequence cleavage site was assumed to exist between Gly(S)^{20} (footnote 3) and Cys^1, which would lead to the formation of a mature protein of 146 residues starting with Cys-Tyr-Cys-Gln; this is

---

[3] Gly(S)^{20} indicates residue 20, a glycine, in the signal sequence.

the basis for the numbering system used by most authors and in Fig. 8. However, natural IFN-γ from induced human lymphocytes lacks the Cys-Tyr-Cys sequence and starts with pyroglutamate (pGlu) (89); in addition, it has a heterogeneous C-terminal end with 9 to 16 C-terminal amino acids missing. It is not clear whether the amino terminus of the natural product represents the site of signal sequence cleavage or arises from subsequent proteolytic degradation. In any event, biologically active IFN-γ contains no cysteine residues, in contrast to type I IFNs. There are two potential N-glycation sites, at positions 28 and 100; differences in glycation are held responsible for the heterogeneity of natural IFN-γ; two species of 20 and 25 kDa (184) contain identical protein moieties (89).

Several sequence variants of IFN-γ were prepared by recombinant DNA techniques. A molecule with a Cys-Tyr-Cys-Gln N-terminus, that is, with 3 more amino-terminal residues than natural IFN-γ, was reported to have only 10% of the specific antiviral activity of the protein in which Cys-Tyr-Cys was replaced by Met, i.e., the methionyl derivative of (unglycated) natural IFN-γ (109), but more recent work suggests that the specific activities are actually quite similar (N. Stebbing, personal communication). Deletion of 18 C-terminal amino acids yielded a protein with about 1% residual antiviral activity; deletion of residues 58–60 completely abolished antiviral activity (185). Using synthetic genes, Alton et al. (109) introduced a number of amino-acid substitutions, all resulting in more or less reduced antiviral activity.

B. Other Mammalian IFN-γ Genes

The IFN-γ genes of mouse (186, 187), rat (188, 189), and cattle (68, 89) have been cloned and sequenced. All species examined had a single IFN-γ gene containing three introns in identical positions as the human gene; in the case of the rat gene (for which sequence data are available), the intron lengths are very similar to those of the human gene.

The aligned amino-acid sequences of human, bovine, mouse, and rat IFN-γ are shown in Fig. 8 (residue Asp[34] given in 68 for bovine IFN-γ should be Glu[34] as in 89; D. Goeddel, personal communication). Whereas the bovine and human coding sequence are of the same length (Table II), both the mouse and the rat coding sequences are shorter by nine codons at the C-terminus. In addition, both have one less codon in the signal sequence, and codon 26 of the mature coding sequence in the mouse IFN-γ is deleted. Bovine IFN-γ contains no cysteine residues; in contrast, mouse and rat IFN-γ have two cysteines at the same positions as the human protein and a third one at the C-terminus. The two potential N-glycation sites of the animal γ IFNs have locations different from those of human IFN-γ (Table II). Using genetic engineering techniques, modifications involving the putative N-terminal Cys-Tyr-Cys residues of mouse IFN-γ have been made; these had no effect on the antiviral activity (*185*). In addition, two human/mouse hybrid proteins have been constructed (*185*). One, with 15 amino acids of the human sequence at the N-terminus, retained activity on mouse cells, but not on human cells; the other, with 44 amino acids of the human sequence at the C-terminus, had no activity on either mouse or human cells.

VI. The Evolution of the IFN Gene Family

A. The Origin of the IFN Genes

IFN genes probably derived from ancestral genes whose products had functions other than antiviral activity in early evolutionary history. The slight similarity between the lymphocyte function-associated glycoprotein 1 (LFA-1; *190*) and HuIFN-α1 has been taken to mean that the two genes derive from a common, early ancestor; however, the statistical significance of such marginal similarities is questionable, especially if one considers the finding that within segments of 98 residues of T_4-induced DNA β-glucosyltransferase (EC 2.4.1.27) and HuIFN-α there are 24 identical and 21 related residues (*125a*).

If the suggestion of Erickson *et al.* (*126*) that the genes for IFN-α and IFN-β show traces of an internal repetition is confirmed by statistical analysis, the IFN genes as we now know them may have arisen from an about half-sized gene by duplication.

The similarities between the IFN-α and IFN-β genes (*9*) make it very likely that these diverged from a common ancestor. Can we calculate how long ago the ancestral gene was duplicated to give rise to the α and β lineages? It has been proposed that the sequence divergence [corrected for multiple events (*133*)] between two homologous genes in different species is proportional to the time elapsed since the separation of the species (*191*); the proportionality factor is the mutation fixation rate (corrected nucleotide exchanges per residue per year). This rate differs very considerably at replacement sites of different genes, probably because of different evolutionary constraints, but is thought to be the same for homologous genes in different species; our data suggest that this assumption is not correct (see below). It is also thought that silent sites and noncoding sequences of all genes show

TABLE IV
MUTATION FIXATION RATES AT REPLACEMENT SITES FOR IFN-α AND IFN-β GENES IN DIFFERENT SPECIES[a]

	Man	Cattle	Mouse
IFN-α genes	$1.0(\pm 0.2) \times 10^{-9}$	$1.6(\pm 0.3) \times 10^{-9}$	$2.1(\pm 0.3) \times 10^{-9}$
IFN-β genes	$1.7(\pm 0.5) \times 10^{-9}$	$2.5(\pm 0.5) \times 10^{-9}$	$3.1(\pm 0.6) \times 10^{-9}$

[a] Values (mutations/site/year) were calculated according to ref. *193* from the average divergence values for replacement sites given in Table III.

the same mutation fixation rates in the more recent evolutionary past, but, because of saturation with silent site mutations, not over longer periods.

The proposed constancy of the mutation fixation rate is the basis for using the "molecular clock" to determine the time elapsed since two genes went their own evolutionary ways. Even if applicable (under appropriate provisos) to homologous genes in different species, the molecular clock is not valid for homologous genes in one and the same species, because of the unknown frequency of gene rectification. This process time and again counteracts the divergence resulting from nucleotide changes; therefore, for genes within a single species, only the *minimum* time since gene duplication can be estimated. We have estimated this value to be 250 ± 60, 180 ± 34, and 155 ± 24 million years for human, bovine, and murine IFN-α and -β, respectively, using divergence at replacement sites [determined according to Perler *et al.* (*133*)] and mutation fixation rates (Table IV) calculated for each species as proposed by Miyata and his colleagues (*192, 193*). These values are lower than those resulting from calculations carried out at the amino-acid sequence level by Miyata (*193*), namely 310 million years for man; even higher estimates had been obtained earlier using mutation fixation rates of globin genes (*9*). As mentioned above, these are all minimal value because of the possibility of gene rectification; the finding that *HuIFN-α$_{II}$1* and *HuIFN-α2* appear more closely related to HuIFN-β than the other α HuIFNs (Table III; *122*) suggests that such sequence interchanges may have taken place even in more recent times. A more reliable dating should be obtained by determining whether both IFN-α and -β genes are present in species that diverged in earlier evolutionary times.

From Table III, it is evident that *IFN-γ* is considerably less conserved between different species than *IFN-β*, which in turn is less conserved than the *IFN-α* family. This first suggests that there are stronger constraints on the IFN-α than on the IFN-γ proteins; however, the ratios of silent to replacement site mutations (which are considered a measure for such constraint; see below) are about the same (about two) for all these genes.

The relationship between *IFN-γ* on the one hand and *IFN-α* and *-β* on the other is unclear. The genes differ in that the former has three introns while type I IFNs have none. There are precedents for evolutionarily related genes that differ as regards content (*194, 195*) and position of introns (*196*). Although no significant homologies were detected between *IFN-γ* and α or β *IFNs* at the nucleotide sequence level, appropriate alignments suggest conservation of several amino-acid clusters and single residues (*11, 179, 197, 198*); the most striking similarity is between positions 60 and 98 of IFN-γ and positions 67 to 106 of IFN-β (33% match) or positions 65 to 104 of the IFN-α consensus sequence (16% match). After introduction of multiple gaps, 47 amino acids are aligned and identical between IFN-γ and IFN-α1 as well as between IFN-γ and IFN-β, and 24 residues are aligned and identical in all three sequences, but the statistical significance of this finding could not be determined (*198a*).

IFN-$β_2$, although neutralizable by antibodies against IFN-β, shows little sequence homology to IFN-β or IFN-α; the gene encoding it contains several introns, and the evolutionary relationship to the other IFNs is obscure.

In conclusion, whereas IFN-α and -β genes are clearly derived from a common ancestor, at least some 200–300 million years ago, the IFN-γ and IFN-$β_2$ gene could, by the same criteria, have originated from a common ancestor only very much earlier, if at all.

B. The Origin of the IFN Gene Families

All mammalian species examined have large IFN-α gene families, but only one IFN-γ and one IFN-β gene; exceptions to this are the ungulates, which have a large IFN-β gene family (*79, 80*).

Gene families could arise either by a series of sequential duplications spread over a long evolutionary time period (*199*) or by a "big bang" event giving rise to many gene copies in one generation. In sequential duplication, the first duplication is probably a rare event, perhaps from unequal crossing-over at flanking repeated sequences, while subsequent duplications may come about more frequently by unequal crossing-over within the duplicated segments (*200*). "Big bang" amplification could be the consequence of an aberration of replication in germ line cells, as proposed for the amplification of the dihydrofolate reductase (EC 1.5.1.3) and other genes in somatic cells (*201*). This phenomenon may come about when a DNA segment of some 10 kilobases replicates more frequently than the rest of the genome, giving rise to multiple gene copies that are eventually integrated in one or more loci (*202*).

Members of gene families accumulate various changes that may result in genes encoding products with modified and even novel functions (*199*) or in nonfunctional pseudogenes. In addition, genes corresponding to one and the same locus may acquire different mutations, giving rise to different alleles (polymorphic variants). Allelic genes (with notable exceptions, see above; *130, 131*) usually do not diverge very much; this is ascribed to a high rate of allele loss.

Because of its ubiquity, the IFN-α gene family was most probably established prior to the mammalian radiation 85 million years ago (*114, 134*). That the divergence of the individual IFN-α genes seems to point to a recent origin of the IFN-α gene family (*203*) does not argue against this conclusion, because gene rectification counteracts divergence of genes within the same species. Certainly the *IFN-α_I* and *-α_{II}* subfamilies diverged prior to mammalian radiation, because the IFN-α_{II} genes of cattle and of man are more closely related to each other than to the IFN-α_I genes of the cognate species (*122*). As both perissodactyls (horse) and artiodactyls (pig, cow), which diverged about 55 million years ago, have a multigene IFN-β family, and other mammals do not, the amplification of the β-gene probably occurred between 55 and 85 million years ago.

Whatever the mechanism leading to the IFN-α gene family, gene amplification seems to have occurred recently in at least one region, namely, within the DNA segment comprising ΨIFN-α10 (Ψ*LeIF-L*), *IFN-α7* (*LeIF-J*), and *IFN-α4* (Fig. 1, linkage group VI), as argued in *120*.

How does sequence divergence come about? *Point substitutions* are generally the most frequent mutations; in the IFN genes transitions (i.e., purine–purine and pyrimidine–pyrimidine exchanges) and transversions (purine–pyrimidine exchanges) are about equally frequent, both in the coding and the noncoding regions. A similar ratio was also found for the β-globin gene (*204*), whereas in an RNA phage (*205*) almost only transition mutations were found. It has been suggested that natural point mutations arise because the wrong pyrimidine (or purine) nucleotide can be incorporated when it or the template nucleotide is in its unusual tautomeric form (*206*). As this mechanism predicts a predominance of transition mutations, a different mechanism, such as mutagenesis via apurinic sites, as described for prokaryotes (*207*), must account for a substantial fraction of the point mutations in eukaryotic genes.

Within the coding regions, mutations at silent sites are usually more frequent than at replacement sites. On average, the ratio of divergence of silent to that of replacement sites for human IFN-α genes is 2.8 ± 0.7, the corresponding value for the murine IFN-α genes is 2.7 ± 0.9, and between murine and human IFN-α genes 2.2 ± 0.3. For *IFN-β*, similar values are found. These values are lower than those for the insulin C chain (5 ± 1), the α- and β-globin genes of different mammalians (7 ± 4 and 6 ± 2, respectively), and the insulin A and B chains (16 ± 2) [all values calculated from Table II in Perler *et al.* (*133*)]. It is thought that the rate of silent substitutions is approximately the same for all genes (*208*), at least for the more recent evolutionary past [about 85 million years (*133*)], whereas the rate of replacement substitution depends on the degree of constraint imposed on the protein molecule. By these criteria, the IFNs are less constrained than are the globin and the insulin chains.

C. Mutation Fixation Rates

If the mutation fixation rates were the same in different species, it would be possible to determine them by simply dividing the corrected divergence values by the time since the species separated. Man, cattle, and rodent lineages are assumed to have separated 85 million years ago (*209*); the separation of rat and mouse is thought to have occurred 2–10 million years ago (K. Hünerman, personal communication, *210*, *211*), but Wilson *et al.* (*212*) give a range of 5 to 35 million years. Contrary to expectation, the mutation fixation rates for both α and β IFN genes differ significantly when computed from man/cattle, man/rodent, or cattle/rodent nucleotide divergence (Table III). We therefore calculated the mutation fixation rates for each species separately, as described by Miyata *et al.* (*193*, *203*) (Table IV). The values for the IFN-α genes differ very considerably between man [about $(1 \pm 0.2) \times 10^{-9}$ site^{-1} year^{-1}], cattle [$(1.6 \pm 0.3) \times 10^{-9}$ site^{-1} year^{-1}], and rodents [$(2.1 \pm 0.3) \times 10^{-9}$ site^{-1} year^{-1}]. Miyata (personal communication) has calculated mutation fixation rates of 1.6×10^{-9} for primates, 3×10^{-9} for mammals (including rodents), and 5.5×10^{-9} for rodents, using a different set of primary data.

It is worth remarking that the values increase with decreasing generation time of the species, though by no means in proportion, as would be expected from the generation time hypothesis (*211*). A similar pattern was found for mutation fixation rates computed from IFN-β gene divergence; however, except for man, the values were significantly higher. The reason for this is not obvious, but it is clear that the estimation of evolutionary age from mutation fixation rates must be exercised with due caution. Using the IFN-α data, it would seem that mouse and rat diverged about 26 ± 7 $(0.11/(2 \times 2.1 \times 10^{-9})$ million years ago, a value far higher than that accepted by palaeontologists on the basis of the fossil record (K. Hünermann, personal communication).

D. Deletions and Insertions

Deletions and insertions of one or several nucleotides are quite frequent in noncoding, but relatively rare in the coding regions of functional genes, because the strong disruptive effects they usually entail generate pseudogenes. *HuIFN-α2* has a 3-nucleotide deletion that precisely removes triplet 44, without any known deleterious effect. *MuIFN-α4* has a 15-nucleotide deletion removing codons 103 through 107, which decreases the antiviral activity to 1% of what is thought to be its normal level (*213*). The coding sequences of *HuIFN-α8* and *HuLeIF B* differ by 4 contiguous amino-acid changes in positions 98 to 101 of *IFN-α8* which could be explained by a nucleotide insertion (position 1291; Fig. 2 in ref. *120*) and a deletion (following position 1303) giving rise to a short frame-shifted region (*120*). Beyond the putative point insertion and deletion mentioned above, there are only four single nucleotide substitutions, two of which, at the 5' end of LeIF-B cDNA, could be due to a cloning artifact (*214, 215*). It is astonishing that of six, or more probably four, nucleotide changes, two are an insertion and a deletion, because insertions and deletions occur 5–10 times less frequently than substitutions; this can be seen by comparing IFN-α with ΨIFN-α genes, or with noncoding regions such as introns or flanking regions of related genes.

A similar situation is found in the case of the chromosomal genes *I-II* (*111*) and *IFN-α7* (*120*). If such insertion and deletion mutations were independent events, the gene should have concomitantly accumulated 20 or more nucleotide substitutions. The fact that in both cases the two mutations lie close to each other suggests that they were introduced by a single, quite recent event, perhaps by a gene conversion of limited extent; the donor sequence could be one of the large number of IFN genes or pseudogenes, in which the insertion/deletion pair may have arisen during a long period of time (along with additional mutations in other parts of the gene). Gene conversions affecting regions of less than 100 nucleotides have been described (*216, 217*). The IFN-α gene family was probably subjected to extensive rectification events in the recent evolutionary past, as suggested by the sequences of *HuIFN-α14* and *GX-1* (*120*). Whereas these two cases are as well explained by unequal crossing-over as by gene conversion, the relationship of *IFN-α1* and *IFN-α13* is best explained by gene conversion (*132*).

Finally, it is of interest to consider that although *HuIFN-α$_{II}$1* is more closely related to the *HuIFN-α$_I$* subfamily than to *HuIFN-β*, a *HuIFN-β* probe hybridizes more readily to the former than to the latter (*122*). This suggests that after divergence of the HuIFN-α subfamilies, a gene conversion event may have transferred *HuIFN-β* sequences to the *HuIFN-α$_{II}$1* progenitor, or vice versa.

E. Conservation of Sequences

A comparison of all human IFN-α amino-acid sequences shows that 36% of the positions are invariant and in part clustered. When the comparison is extended to the human IFN-β sequence, conserved regions (about 20% of the residues) become even more evident: in region A (positions 29–50) 45% of the residues are conserved, in region B (positions 58 to 74) 38%, and in region C (positions 126 to 150) 48%. When IFN-αs of all species are compared there are 25 conserved positions (15%) (labeled with asterisks in Fig. 6) of which 13 are clustered between positions 29–41 and 131–142. Nine positions are conserved in all type I IFNs of all species examined; four of these lie between positions 130–141 and four between positions 123–141. It is commonly thought that positions conserved to such a degree are essential for the function of the molecule, however, when Gln^{62} of *HuIFN-α2*, which is conserved in all known α and β IFNs of man, mouse, rat, and cattle (Fig. 6) was substituted by lysine using *in vitro* site-directed mutagenesis there was no effect on the specific antiviral activity of the resulting molecule (*106*).

Two other positions that had been considered invariant in the human IFN-α gene family have also been substituted without causing a detectable difference in the biological activity of the IFN (*106*). Perhaps the changes that were introduced affect the biological activity to such a small degree that they cannot be detected biochemically, but yet may influence the fitness of the carrier over long evolutionary periods. Alternatively, mutations at these sites could lead to inadequate expression of the gene by impairing transcription, translation, or posttranslational events, or impair some as yet unrecognized function of IFN.

The other region of the IFN genes that shows striking sequence conservation is within the 110-nucleotide 5' flanking region preceding the "cap" site. As even genes with intact reading frames are not necessarily transcribed efficiently or at all (for example, IFN-α6 mRNA has never been detected in induced human cells, and *IFN-α5*,

```
               -111       -101       -91        -81        -71        -61        -51
                !          !          !          !          !..        !...       !
Hu-αI cons.   gAGtgcataAAgGAAAGCaAAAAcAGAaaTaGAAAgtaa.aCa.ggaggcaTTtaGAAAaT
Hu-αII        GGGAGAACACACAAATGAAAACAGTAAAAGAAACTGAAAGTACAGAGAAATGTTCAGAAAA

Bo-αI cons.   tTGgAGAGTgCAAAcTGAAAAaCAAAAACAAAAGTAGAAAgcAAGAGGGAACTTTCAgAAA
Bo-αII        CGGAATACATAAATGAAAATCAAAAAAGGAAGTACAAGTACACAGAAATGACTAGAAAATG
                                           ....
Ro-α cons.    AGAaTgAgtTAAAGaaagTGAAAaGAcAatTgGaAAGt.AggGgAGGGcATtcaGAAAgta

Hu-β          TAAAATGTAAATGACATAGGAAAACTGAAAGGGAGAAGTGAAAGTGGGAAATTCCTCTGAA
                                    .
Bo-β cons.    TtAAatgACA.aGGAaAAcTgAAAggGAgAactGAAAGTGGGAAATtcCTCtc.aatagaa
                                                                      ......
Hu-γ          AAACTATCATCCCTGCCTATCTGTCACCATCTCATCTTAAAAAAACTTGTGAAAATACGTAA

Ra-γ          ACTAGCTCCCGCCACCTATCTTTCACCATCTTAACTTAAAAAAAAACCTGTGAAAATACGT


                         -41        -31        -21        -11        -1      Base comp.*
                          !          !          !          !          !      (%)
                                                                    ↦        A   G   C   T
Hu-αI cons.   GgAAAttaGTATGTtcccTaTTTAAGaccTatGCAcaaAgCAAgGtCTtCAG        51  23   7  16
Hu-αII        TGAAAACCATGTGTTTCCTATTAAAAGCCATGCATACAAGCAATGTCTTCAG        56  19  13  13

Bo-αI cons.   aTGGAAACCATGGgCTCCTATTTAAgACACAGgCCTGAAGGAAGGTCTTCAG        53  23  13  11
Bo-αII        AAAATTACTGTGTTCCCTATTTAATGGCCTTGCTTAGAAAGCATGGCATCAG        56  17  10  17
                     ...       .
Ro-α cons.    aAAacTagtgTtTgtcCCTATTTAAGAcAcAtccacacaGgAtGgtcttCAG        46  30   6  17

Hu-β          TAGAGAGAGGACCATCTCATATAAATAGGCCATACCCACGGAGAAAGGACAG        46  30   7  17
              ......
Bo-β cons.    agaatgGAGGGCCATgCTgTATAAGTAGCCCACACTcAAGgAgGAAGGcCAG        44  27  10  16

Hu-γ          TCCTCAGGAGACTTCAATTAGGTATAAATACCAGCAGCCAGAGGAGGTCAG        33  11  27  29

Ra-γ          AATCCCAAGAAGCCTTCGGTCATGTATAAAACTGGAAGCAAGAGAGGTCAG        37   9  30  24
```

FIG. 9. Promoter sequences of interferon genes. The numbering refers only to *HuIFN-α1* (position −1 is the position preceding the cap site); all other sequences were aligned by their presumed cap sites, more or less as indicated by the original authors (for references, see table footnotes). Hu-αI is a plurality consensus of the (expressed) human IFN-α genes *HuIFN-α1, -α2, -α4b, -α5, -α7, -α13,* and *-α14;* Hu-αII is *HuIFN-α_{II}1.* Bo-α_I is a consensus of bovine *BoIFN-α_I1, -α_I2,* and *-α_I3* genes; Bo-αII is *BoIFN-α_{II}1.* Ro-α is a consensus of murine *MuIFN-α2, -α4, -α5, -α6(P), -αA,* and rat *RaIFN-α1.* Hu-β is human *HuIFN-β;* Bo-β is a consensus of bovine *BoIFN-β1, -β2,* and *-β3.* Hu-γ and Ra-γ are the human and the rat IFN-γ genes, respectively. Upper case letters indicate completely conserved bases; lower case letters are used where only pluralities of sequences have the indicated base. Where no plurality could be established, the base is represented by a dot. Dots above base symbols indicate nucleotides that are absent in some sequences of a consensus. The repetitive GAAA sequences and the TATA boxes are underlined. *, The base composition refers to the region from nucleotide −41 to −110.

IFN-α7, and IFN-α8 are transcribed only at a low level), only the 5′ flanking regions of genes known to be expressed efficiently in at least some cell types [*IFN-α1, -α2, -α13, -α14, IFN-α_{II}1;* (7, 8, 48, 122)] were used to compile the plurality sequences in Fig. 9. The 5′ flanking region comprises the so-called Hogness box, which usually has the sequence TATA(A/T)A(A/T) (*218*), but in all IFN-α_I genes of all species examined has the uncommon structure TATTTAAG (except for the *IFN-α5*, which has the even more unusual TTTTTAAG). The type I IFN promoters show a high purine content (on average 75%) in the region between positions -140 and -110, with a 4- to 6-fold recurrence of the sequence GAAA. The expected frequency of this tetrameric sequence is two at the given base composition. The 5' flanking regions of the IFN-γ genes have a quite different overall pattern, with a purine content of 44%, but two oligo(A) clusters are conserved in the two species examined. It has been shown experimentally that 5' deletions extending downstream to position -109 relative to the cap site do not affect induced expression of the human IFN-α1 gene, whereas expression is abolished stepwise by a successive deletion extending to position -79 (*98, 99*). Similar results have been found for the human IFN-β gene; however, somewhat different 5' boundaries were determined by Fujita *et al.* (*102*) (-117 to -105), Maniatis *et al.* (*103, 104*) (-77 to -73), and Dinter *et al.* (*219*) (-107 to -67). The requirement for the region downstream at about position -110 for induced transcription of type I IFN genes can account for the sequence conservation in this region.

VII. Conclusions

Interferons were early classified into two types, I and II, on the basis of distinct biological, biochemical, and serological properties. Type II IFN, which is acid-labile, is secreted by T lymphocytes after immune or mitogenic stimulation. Analyses at the genomic level have revealed only one gene for type II IFN, namely *IFN-γ*, which shows no clear relationship to genes of type I IFN (but see refs. *12, 54*) and is located on a different chromosome than the latter. Type I IFNs are induced by viral infection or treatment of a variety of cells with poly(I)·poly(C) and are usually acid-stable.

The genes encoding type I IFNs are members of a superfamily. All are located on one chromosome (chromosome 9 in man; chromosome 4 in mouse), and comprise the IFN-β and IFN-α families, which diverged 200–300 million years ago or more. In ungulates the IFN-β family comprises several genes, in man and mouse only one. The recently described IFN-β_2, while cross-reacting with IFN-β antibodies, is apparently not related to the IFN-β at the genetic level.

The IFN-α family comprises at least two subfamilies, IFN-α_I and IFN-α_{II}, which diverged 120 million years ago or more, clearly before the mammalian radiation. In mammalians the IFN-α_I subfamily comprises many loci; in man, there are at least 15 loci encoding potentially functional IFNs, 9 loci encoding pseudogenes, and not less than 17 additional alleles. The IFN-α_{II} subfamily consists of at least one functional gene and several pseudogenes. Some additional pseudogenes may belong to other, as yet not characterized subfamilies. The time at which the IFN-α_I subfamily diverged is difficult to determine, as there has evidently been considerable recombination and/or gene conversion among related genes, but it is probable that it occurred before the mammalian radiation. The divergence of the IFN genes in different species suggests that the mutation fixation rates differ for IFN-α, -β, and -γ, and are higher in the mouse and cattle than in man.

What is the biological significance of the multiplicity of IFN-α genes? Is it simply an evolutionary accident that brings no selective advantage to the organism, or have the distinct species evolved to exercise specific functions? The fact that cattle but not other mammalians have evolved a multigene IFN-β family in the recent evolutionary past argues that accidental gene amplification and fixation of gene families can occur as an essentially neutral event. The situation may well be different in more ancient gene families. There are several reports suggesting that the different IFN-α species show quantitatively distinct patterns of antiviral, growth inhibitory, and killer cell stimulatory activities. It is clear that in man at least 10 IFN-α loci are expressed in normal and leukemic leukocytes; the relative degree of expression in response to induction varies considerably in the different cell types. The fact that a battery of IFN genes is not coordinately turned on in response to induction, but that different genes are turned on to varying extents in different cells invites the speculation that the individual IFN species may have further, as yet unrecognized activities of importance, not only in the antiviral and the immune response, but perhaps also in physiological growth control and differentiation.

Acknowledgments

This work was supported by the Schweizerische Nationalfonds (3.147.81 and 3.515.83) and the Kanton of Zürich. We thank John David Weissmann for writing the program for divergence calculation and Rose-Marie von Rotz for her invaluable secretarial help. We are very grateful to G. Allen, E. DeMayer, L. Epstein, D. Goeddel, A. von Gabain, K. Henko, T. Miyata, P. M. Pitha, P. B. Sehgal, N. Stebbing, T. Taniguchi, J. Trapman, and J. Vilček for many suggestions, corrections, and unpublished information.

References

1. A. Isaacs and J. Lindenmann, *Proc. R. Soc. Lond. B* **147**, 258 (1957).
2. J. Taylor, *BBRC* **14**, 447 (1964).
3. H. Rubin, *Virology* **13**, 200 (1961).
4. E. A. Havell, B. Berman, C. A. Ogburn, K. Berg, K. Paucker and J. Vilček, *PNAS* **72**, 2185 (1975).
5. K. Berg, C. A. Ogburn, K. Paucker, K. E. Mogensen and K. Cantell, *J. Immunol.* **114**, 640 (1975).
6. E. A. Havell, T. G. Hayes and J. Vilček, *Virology* **89**, 330 (1978).
6a. E. Saksela, I. Virtanen, T. Hovi, D. S. Secher and K. Cantell, *Prog. Med. Virol.* **30**, 78 (1984).
7. J. Hiscott, K. Cantell and C. Weissmann, *NARes* **12**, 3727 (1984).
8. J. Hiscott, J. Ryals, P. Dierks, V. Hofmann and C. Weissmann, *Phil. Trans. R. Soc. Lond. B* **307**, 217 (1984).
9. T. Taniguchi, N. Mantei, M. Schwarzstein, S. Nagata, M. Muramatsu and C. Weissmann, *Nature* **285**, 547 (1980).
10. E. F. Wheelock, *Science* **149**, 310 (1965).
11. L. B. Epstein, *Nature* **295**, 453 (1982).
12. L. B. Epstein, in "Interferon: Interferons and the Immune System" (J. Vilček and E. De Mayer, eds.) Vol. **2**, p. 185. Elsevier, Amsterdam, 1984.
12a. A. Zilberstein, R. Ruggieri and M. Revel, in "The Interferon System" (F. Dianzani and G. B. Rossi, eds.), p. 73. Serono Symposia, Raven, New York, 1985.
13. W. E. Stewart II, I. Gresser, M. G. Tovey, M.-T. Bandu and S. Le Goff, *Nature* **262**, 300 (1976).
14. M. G. Masucci, R. Szigeti, E. Klein, G Klein, J. Gruest, L. Montagnier, H. Taira, A. Hall, S. Nagata and C. Weissmann, *Science* **209**, 1431 (1980).
15. A. Adams, H. Strander and K. Cantell, *J. Gen. Virol.* **28**, 207 (1975).
16. D. Santoli, G. Trinchieri and H. Koprowski, *J. Immunol.* **121**, 532 (1978).
17. S. Einhorn, H. Blomgren and H. Strander, *Int. J. Cancer* **22**, 405 (1978).
18. R. B.Herberman, J. R. Ortaldo, A. Mantovani, D. S. Hobbs, H.-F. Kung and S. Pestka, *Cell. Immunol.* **67**, 160 (1982).
19. M. Hokland, I. Heron, K. Berg and P. Hokland, *Cell. Immunol.* **72**, 40 (1982).
20. S. H. S. Lee and L. B. Epstein, *Cell. Immunol.* **50**, 177 (1980).
21. S. Keay and S. E. Grossberg, *PNAS* **77**, 4099 (1980).
22. G. J. Jonak and E. Knight, Jr., *PNAS* **81**, 1747 (1984).
22a. E. Knight, Jr., E. D. Anton, D. Fahey, B. K. Friedland and G. J. Jonak, *PNAS* **82**, 1151 (1985).
22b. M. Einat, D. Resnitzky and A. Kimchi, *Nature* **313**, 597 (1985).
23. I. Gresser, in "The Biology of the Interferon System" (E. De Maeyer, G. Galasso and H. Schellekens, eds.) p. 141. Elsevier, Amsterdam (1981).
23a. J. L. Pace, S. W. Russell, R. D. Schreiber, A. Altman and D. H. Katz, *PNAS* **80**, 3782 (1983).
24. T. Y. Basham and T. C. Merigan, *J. Immunol.* **130**, 1492 (1983).
25. V. E. Kelley, W. Fiers and T. B. Strom, *J. Immunol.* **132**, 240 (1984).
26. H. P. Koeffler, J. Ranyard, L. Yelton, R. Billing and R. Bohman, *PNAS* **81**, 4080 (1984).
27. J. Weil, Ch. J. Epstein, L. B. Epstein, J. van Blerkom and N. H. Xuong, *Antiviral Res.* **3**, 303 (1983).
28. R. L. Friedman and G. R. Stark, *Nature* **314**, 637 (1985).

29. A. C. Larner, G. Jonak, Y.-S. E. Cheng, E. Knight and J. E. Darnell, Jr., *PNAS* **81**, 6733 (1984).
30. J. Weil, C. J. Epstein, L. B. Epstein, J. J. Sedmak, J. L. Sabran and S. E. Grossberg, *Nature* **301**, 437 (1983).
31. J. Weil, Ch. J. Epstein and L. B. Epstein, *Nat. Immun. Cell Growth Regul.* **3**, 51 (1983/84).
32. W. E. Stewart II, in "The Interferon System," 2nd ed. Springer-Verlag, Berlin and New York, 1981.
33. M. Streuli, A. Hall, W. Boll, W. E. Stewart II, S. Nagata and C. Weissmann, *PNAS* **78**, 2848 (1981).
34. E. Rehberg, B. Kelder, E. G. Hoal and S. Pestka, *JBC* **257**, 11497 (1982).
35. P. K. Weck, S. Apperson, L. May and N. Stebbing, *J. Gen. Virol.* **57**, 233 (1981).
36. E. Yelverton, D. Leung, P. Weck, P. W. Gray and D. V. Goeddel, *NARes* **9**, 731 (1981).
37. S. Pestka, J. McInnes, E. A. Havell and J. Vilček, *PNAS* **72**, 3898 (1975).
38. B. Berman and J. Vilček, *Virology*, **57**, 378 (1974).
39. M. Wiranowska-Stewart and W. E. Stewart II, *J. Gen. Virol.* **37**, 629 (1977).
40. G. Shaw, W. Boll, H. Taira, N. Mantei, P. Lengyel and C. Weissmann, *NARes* **11**, 555 (1983).
40a. W. E. Stewart II and E. A. Havell, *Virology* **101**, 315 (1980).
41. W. E. Stewart II, F. H. Sarkar, H. Taira, A. Hall, S. Nagata and C. Weissmann, *Gene* **11**, 181 (1980).
42. J. Werenne, C. Vanden Broecke, A. Schwers, A. Goossens, L. Bugyaki, M. Maenhoudt and P.-P. Pastoret, *J. Interferon Res.* **5**, 129 (1985).
43. K. Cantell, in "Interferon 1" (I. Gresser, ed.), p. 1. Academic Press, New York, 1979.
43a. M. Rubinstein, S. Rubinstein, P. Familletti, M. S. Gross, A. Waldman and S. Pestka, *Science* **202**, 1289 (1978).
44. E. Knight, Jr. in "Interferon 2" (I. Gresser, ed.), p. 1. Academic Press, New York, 1980.
44a. H. Taira, R. J. Broeze, B. M. Jayaram, P. Lengyel, M. W. Hunkapiller and L. E. Hood, *Science* **207**, 528 (1980).
45. J. Vilček, in "Interferon 4" (I. Gresser, ed.), p. 129. Academic Press, New York, 1982.
46. T. Taniguchi, S. Ohno, Y. Fujii-Kuriyama and M. Muramatsu, *Gene* **10**, 11 (1980).
47. N. Mantei, M. Schwarzstein, M. Streuli, S. Panem, S. Nagata and C. Weissmann, *Gene* **10**, 1 (1980).
48. D. V. Goeddel, D. W. Leung, T. J. Dull, M. Gross, R. M. Lawn, R. McCandliss, P. H. Seeburg, A. Ullrich, E. Yelverton and P. W. Gray, *Nature* **290**, 20 (1981).
49. G. Allen and K. H. Fantes, *Nature* **287**, 408 (1980).
50. D. S. Secher and D. C. Burke, *Nature* **285**, 446 (1980).
51. W. E. Stewart II and J. Desmyter, *Virology* **67**, 68 (1975).
52. M. Rubinstein, S. Rubinstein, P. C. Familletti, R. S. Miller, A. A. Waldman and S. Pestka, *PNAS* **76**, 640 (1979).
53. K. Berg and I. Heron, *Scand. J. Immunol.* **11**, 489 (1980).
54. W. P. Levy, S. Shively, M. Rubinstein, U. Del Valle and S. Pestka, *PNAS* **77**, 5102 (1980).
55. H. C. Kelker, J. Le, B. Y. Rubin, Y. K. Yip, C. Nagler and J. Vilček, *JBC* **259**, 4301 (1984).
56. E. Rinderknecht, B. H. O'Connor and H. Rodriguez, *JBC* **259**, 6790 (1984).
57. E. Knight, Jr., M. W. Hunkapiller, B. D. Korant, R. W. F. Hardy and L. E. Hood, *Science* **207**, 525 (1980).

57a. NCIUB and JCBN, Newsletter 1984, *EJB* **138**, 7 (1984).
58. M. Rubinstein, *in* "Interferon: 25 years on" (D. A. J. Tyrell and D. C. Burke, eds.), p. 39. Royal Society London, 1982.
59. W. P. Levy, M. Rubinstein, S. Shively, U. Del Valle, C.-Y. Lai, J. Moschera, L. Brink, L. Gerber, S. Stein and S. Pestka, *PNAS* **78**, 6186 (1981).
60. S. Pestka, *ABB* **221**, 1 (1983).
60a. J. E. Labdon, K. D. Ginson, S. Sun and S. Pestka, *ABB* **232**, 422 (1984).
61. S. Pestka, J. E. Labdon, A. Rashidbaigi, X.-Y. Liu, J. A. Langer, J. R. Ortaldo, R. B. Herberman and V. Jung, *in* "The Biology of the Interferon System 1984" (H. Kirchner and H. Schellekens, eds.), p. 3. Elsevier, Amsterdam, 1985.
62. M. Revel, *in* "Interferon 5" (I. Gresser, ed.), p. 205. Academic Press, New York, 1983.
63. M. Revel, *in* "Antiviral Drugs and Interferons: The Molecular Basis of Their Activity" (Y. Becker, ed.), p. 357. Martinus Nijhoff, Boston, Massachusetts, 1984.
64. J. Collins, *in* "Interferon 3: Mechanisms of Production and Action" (R. M. Friedman, ed.), p. 33. Elsevier, Amsterdam, 1984.
65. S. Pestka, *Sci. Am.* **249**, 29 (1983).
66. R. Dijkmans and A. Billiau, *in* "Interferons: Their Impact in Science and Medicine" (J. Taylor-Papadimitriou, ed.), p. 1. Oxford Univ. Press, London and New York, 1985.
67. K. C. Zoon and R. Wetzel, *in* "Handbook of Experimental Pharmacology" (P. E. Came and W. A. Carter, eds.), p. 79. Springer-Verlag, Berlin and New York, 1984.
68. R. Derynck, *in* "Interferon 5" (I. Gresser, ed.), p. 181. Academic Press, New York, 1983.
69. E. Knight, Jr., *in* "Interferon 1" (N. B. Finter, ed.), Vol. 1, p. 61. Elsevier, Amsterdam, 1984.
70. F. H. Reynolds, Jr., E. Premkumar and P. M. Pitha, *PNAS* **72**, 4881 (1975).
71. P. B. Sehgal, B. Dobberstein and I. Tamm, *PNAS* **74**, 3409 (1977).
72. R. L. Cavalieri, E. A. Havell, J. Vilček and S. Pestka, *PNAS* **74**, 4415 (1977).
73. F. H. Reynolds, Jr. and P. M. Pitha, *BBRC* **59**, 1023 (1974).
74. T. P. St. John and R. W. Davis, *Cell* **16**, 443 (1979).
75. J. H. J. Hoeijmakers, P. Borst, J. van den Burg, C. Weissmann and G. A. M. Cross, *Gene* **8**, 391 (1980).
76. M. M. Harpold, P. R. Dobner, R. M. Evans and F. C. Bancroft, *NARes* **5**, 2039 (1978).
77. S. Nagata, H. Taira, A. Hall, L. Johnsrud, M. Streuli, J. Ecsödi, W. Boll, K. Cantell and C. Weissmann, *Nature* **284**, 316 (1980).
78. E. M. Southern, *JMB* **98**, 503 (1975).
79. V. Wilson, A. J. Jeffries, P. A. Barrie, P. G. Boseley, P. M. Slocombe, A. Easton and D. C. Burke, *JMB* **166**, 457 (1983).
80. D. W. Leung, D. J. Capon and D. V. Goeddel, *Biotechnology* **2**, 458 (1984).
81. D. Owerbach, W. J. Rutter, T. B. Shows, P. Gray, D. V. Goeddel and R. M. Lawn, *PNAS* **78**, 3123 (1981).
82. T. B. Shows, A. Y. Sakaguchi, S. L. Naylor, D. V. Goeddel and R. M. Lawn, *Science* **218**, 373 (1982).
83. M. Lovett, D. R. Cox, D. Yee, W. Boll and C. Weissmann, *EMBO J.* **3**, 1643 (1984).
84. F. Dandoy, K. A. Kelley, J. DeMaeyer-Guignard, E. DeMaeyer and P. M. Pitha, *J. Exp. Med.* **160**, 294 (1984).
85. D. L. Slate, P. D'Eustachio, D. Pravtcheva, A. C. Cunningham, S. Nagata, C. Weissmann and F. H. Ruddle, *J. Exp. Med.* **155**, 1019 (1982).
86. J. M. Trent, S. Olson and R. M. Lawn, *PNAS* **79**, 7809 (1982).

87. D. V. Goeddel, E. Yelverton, A. Ullrich, H. L. Heyneker, G. Miozzari, W. Holmes, P. H. Seeburg, T. Dull, L. May, N. Stebbing, R. Crea, S. Maeda, R. McCandliss, A. Sloma, J. M. Tabor, M. Gross, P. C. Familletti and S. Pestka, *Nature* **287**, 411 (1980).
88. H. Schellekens, A. de Reus, R. Bolhuis, M. Fountoulakis, C. Schein, J. Ecsödi, S. Nagata and C. Weissmann, *Nature* **292**, 775 (1981).
89. E. Rinderknecht and L. E. Burton, in "The Biology of the Interferon System 1984" (H. Kirchner and H. Schellekens, eds.), p. 397. Elsevier, Amsterdam, 1985.
90. D. V. Goeddel, H. M. Shepard, E. Yelverton, D. Leung, R. Crea, A. Sloma and S. Pestka, *NARes* **8**, 4057 (1980).
91. E. Knight, Jr., *PNAS* **73**, 520 (1976).
92. J.-I. Fujisawa, Y. Iwakura and Y. Kawade, *JBC* **253**, 8677 (1978).
93. J. Haynes and C. Weissmann, *NARes* **11**, 687 (1983).
94. S. J. Scahill, R. Devos, J. Van der Heyden and W. Fiers, *PNAS* **80**, 4654 (1983).
95. F. McCormick, M. Trahey, M. Innis, B. Dieckmann and G. Ringold, *MCBiol* **4**, 166 (1984).
96. R. Fukunaga, Y. Sokawa and S. Nagata, *PNAS* **81**, 5086 (1984).
96a. D. G. Knorre and V. V. Vlassov, *This Series* **32**, 292 (1985).
97. U. Weidle and C. Weissmann, *Nature* **303**, 442 (1983).
98. H. Ragg and C. Weissmann, *Nature* **303**, 439 (1983).
99. J. Ryals, P. Dierks, H. Ragg and C. Weissmann, *Cell* **41**, 497 (1985).
100. L. Maroteaux, C. Kahana, C. Mory, Y. Groner and M. Revel, *EMBO J.* **2**, 325 (1983).
101. U. Nir, B. Cohen, L. Chen and M. Revel, *NARes* **12**, 6979 (1984).
102. T. Fujita, S. Ohno, H. Yasumitsu and T. Taniguchi, *Cell* **42**, 489 (1985).
103. K. Zinn, D. DiMaio and T. Maniatis, *Cell* **34**, 865 (1983).
104. S. Goodbourn, K. Zinn and T. Maniatis, *Cell* **41**, 509 (1985).
105. H. Hauser, H. Dinter and J. Collins, in "The Biology of the Interferon System 1984" (H. Kirchner and H. Schellekens, eds.), p. 21. Elsevier, Amsterdam, 1985.
106. D. Valenzuela, H. Weber and C. Weissmann, *Nature* **313**, 698 (1985).
107. D. F. Mark, S. D. Lu, A. A. Creasey, R. Yamamoto and L. S. Lin, *PNAS* **81**, 5662 (1984).
108. H. M. Shepard, D. Leung, N. Stebbing and D. V. Goeddel, *Nature* **294**, 563 (1981).
109. K. Alton, Y. Stabinsky, R. Richards, B. Ferguson, L. Goldstein, B. Altrock, L. Miller and N. Stebbing, in "The Biology of the Interferon System 1983" (E. De Maeyer and H. Schellekens, eds.), p. 119. Elsevier, Amsterdam, 1983.
110. M. D. Edge, A. R. Green, G. R. Heathcliffe, V. E. Moore, N. J. Faulkner, R. Camble, N. N. Petter, P. Trueman, W. Schuch, J. Hennam, T. C. Atkinson, C. R. Newton and A. F. Markham, *NARes* **11**, 6419 (1983).
110a. M. D. Edge, A. R. Greene, G. R. Heathcliffe, V. E. Moore, N. J. Faulkner, R. Camble, N. N. Petter, P. Trueman, W. Schuch, J. Hennam, T. C. Atkinson, C. R. Newton and A. T. Markham, *NARes* **11**, 6419 (1983).
111. S. Maeda, R. McCandliss, T.-R. Chiang, L. Costello, W. P. Levy, N. T. Chang and S. Pestka, *ICN-UCLA Symp. Mol. Cell. Biol.* **23**, 85 (1981).
112. S. Nagata, N. Mantei and C. Weissmann, *Nature* **287**, 401 (1980).
113. M. Streuli, S. Nagata and C. Weissmann, *Science* **209**, 1343 (1980).
114. C. Weissmann, S. Nagata, W. Boll, M. Fountoulakis, A. Fujisawa, J.-I. Fujisawa, J. Haynes, K. Henco, N. Mantei, H. Ragg, C. Schein, J. Schmid, G. Shaw, M. Streuli, H. Taira, K. Todokoro and U. Weidle, *Philos. Trans. R. Soc. London Ser. B* **299**, 7 (1982).
115. A. Ullrich, A. Gray, D. V. Goeddel and T. J. Dull, *JMB* **156**, 467 (1982).

116. S. Maeda, R. McCandliss, M. Gross, A. Sloma, P. C. Familletti, J. M. Tabor, M. Evinger, W. R. Levy and S. Pestka, *PNAS* **77**, 7010 (1980).
117. C. Brack, S. Nagata, N. Mantei and C. Weissmann, *Gene* **15**, 379 (1981).
118. R. M. Lawn, J. Adelman, T. J. Dull, M. Gross, D. V. Goeddel and A. Ullrich, *Science* **212**, 1159 (1981).
119. R. M. Lawn, M. Gross, C. Houck, A. E. Franke, P. V. Gray and D. V. Goeddel, *PNAS* **78**, 5435 (1981).
120. K. Henco, J. Brosius, A. Fujisawa, J.-I. Fujisawa, J. R. Haynes, H. Hochstadt, T. Kovacic, M. Pasek, A. Schamböck, J. Schmid, K. Todokoro, M. Wälchli, S. Nagata and C. Weissmann, *JMB* **185**, 227 (1985).
121. B. Lund, T. Edlund, W. Lindenmaier, T. Ny, J. Collins, E. Lundgren and A. von Gabain, *PNAS* **81**, 2435 (1984).
121a. K. Hauptmann and P. Swetly, *NARes* **13**, 4739 (1985).
122. D. J. Capon, H. M. Shepard and D. V. Goeddel, *MCBiol* **5**, 768 (1985).
123. H. M. Shepard, D. Eaton, P. Gray, S. Naylor, P. Hollingshead and D. Goeddel, in "The Biology of the Interferon System 1984" (H. Kirchner and H. Schellekens, eds.), p. 147. Elsevier, Amsterdam 1985.
124. R. M. Lawn, J. Adelman, A. E. Franke, C. M. Houck, M. Gross, R. Najarian and D. V. Goeddel, *NARes* **9**, 1045 (1981).
125. S. Ohno and T. Taniguchi, *NARes* **10**, 967 (1982).
125a. J. Tomaschewski, H. Gram, J. W. Crabb, and W. Rüher, *NARes* **13**, 7551 (1985).
126. B. W. Erickson, L. May and P. B. Sehgal, *PNAS* **81**, 7171 (1984).
127. R. Wetzel, *Nature* **289**, 606 (1981).
127a. R. Wetzel, H. L. Levine, D. A. Estell, S. Shire, J. Finer-Moore, R. M. Stroud and T. A. Bewley, *UCLA Symp. Mol. Cell. Biol.* **25**, 365 (1982).
127b. IUPAC-IUB Joint Commission on Biochemical Nomenclature, *EJB* **138**, 9 (1984).
128. N. Mantei and C. Weissmann, *Nature* **297**, 128 (1982).
129. D. L. Slate and F. H. Ruddle, *Cell* **16**, 171 (1979).
130. R. Ollo and F. Rougeon, *Cell* **32**, 515 (1983).
131. G. Widera and R. A. Flavell, *EMBO J.* **3**, 1221 (1984).
132. K. Todokoro, D. Kioussis and C. Weissmann, *EMBO J.* **3**, 1809 (1984).
133. F. Perler, A. Efstratiadis, P. Lomedico, W. Gilbert, R. Kolodner and J. Dodgson, *Cell* **20**, 555 (1980).
134. D. Gillespie and W. Carter, *J. Interferon Res.* **3**, 83 (1983).
135. F. S. Collins and S. M. Weissman, *This Series* **31**, 317 (1984).
136. S. I. Feinstein, Y. Mory, Y. Chernayovsky, L. Maroteaux, U. Nir, V. Lavie and M. Revel, *MCBiol* **5**, 510 (1985).
137. P. K. Weck, S. Apperson, N. Stebbing, P. W. Gray, D. Leung, H. M. Shepard and D. V. Goeddel, *NARes* **9**, 6153 (1981).
138. T. Goren, A. Kapitkovsky, A. Kimchi and M. Rubinstein, *Virology* **130**, 273 (1983).
139. J. R. Ortaldo, R. B. Herberman, C. Harvey, P. Osheroff, Y.-C. E. Pan, B. Kelder and S. Pestka, *PNAS* **81**, 4926 (1984).
140. S. Pestka, B. Kelder, E. Rehberg, J. R. Ortaldo, R. B. Herberman, E. S. Kempner, J. A. Moschera and S. J. Tarnowski, in "The Biology of the Interferon System 1983", (E. De Maeyer and H. Schellekens, eds.), p. 535. Elsevier, Amsterdam, 1983.
141. N. Stebbing and P. K. Weck, in "Recombinant DNA Products: Insulin, Interferon and Growth Hormone" (P. Bollon, ed.), p. 75. CRC Press, Boca Raton, Florida, 1984.
142. R. Dijkema, P. Pouwels, A. de Reus and H. Schellekens, *NARes* **12**, 1227 (1984).

143. K. A. Kelley, C. A. Kozak, F. Dandoy, F. Sor, D. Skup, J. D. Windass, J. DeMaeyer-Guignard, P. M. Pitha and E. De Maeyer, *Gene* **26**, 181 (1983).
143a. F. Dandoy, E. De Mayer, F. Bonhomme, J. L. Guenet and J. De Mayer, *J. Virol.* **56**, 216 (1985).
144. J. A. G. M. Van Der Korput, J. Hilkens, V. Kroezen, E. C. Zwarthoff and J. Trapman, *J. Gen. Virol.* **66**, 493 (1985).
144a. Z. Y. Cheng, M. Lovett, L. B. Epstein and C. J. Epstein, in press (1986).
145. B. Daugherty, D. Martin-Zanca, B. Kelder, K. Collier, T. C. Seamans, K. Hotta and S. Pestka, *J. Interferon Res.* **4**, 635 (1984).
146. K. A. Kelley and P. M. Pitha, *NARes* **13**, 825 (1985)
147. E. C. Zwarthoff, A. T. Mooren, and J. Trapman, *NARes* **13**, 791 (1985).
147a. I. Seif and J. De Mayer-Guignard, in press (1986).
148. T. Taniguchi, M. Sakai, Y. Fujii-Kuriyama, M. Muramatsu, S. Kobayashi and T. Sudo, *Proc. Jpn Acad. Ser. B* **55**, 464 (1979).
149. T. Taniguchi, L. Guarente, T. M. Roberts, D. Kimelman, J. Douhan III and M. Ptashne, *PNAS* **77**, 5230 (1980).
150. R. Derynck, J. Content, E. DeClercq, G. Volckaert, J. Tavernier, R. Devos and W. Fiers, *Nature* **285**, 542 (1980).
151. L. T. May and P. B. Sehgal, *J. Interferon Res.* **5**, 521 (1985).
152. S. Ohno and T. Taniguchi, *PNAS* **78**, 5305 (1981).
153. M. Houghton, I. J. Jackson, A. G. Porter, S. M. Doel, G. H. Catlin, C. Barber and N. H. Carey, *NARes* **9**, 247 (1981).
154. Y. Mory, Y. Chernajovsky, S. I. Feinstein, L. Chen, U. Nir, J. Weissenbach, Y. Malpiece, P. Tiollais, D. Marks, M. Ladner, C. Colby and M. Revel, *EJB* **120**, 197 (1981).
155. G. Gross, U. Mayr, W. Bruns, F. Grosveld, H.-H. M. Dahl and J. Collins, *NARes* **9**, 2495 (1981).
156. Y. K. Yip, R. H. L. Pang, C. Urban and J. Vilček, *PNAS* **78**, 1601 (1981).
157. E. A. Havell, S. Yamazaki and J. Vilček, *JBC* **252**, 4425 (1977).
158. Y. H. Tan, F. Barakat, W. Berthold, H. Smith-Johannsen and C. Tan, *JBC* **254**, 8067 (1979).
159. M. Ohlsson, J. Feder, L. L. Cavalli-Sforza and A. von Gabain, *PNAS* **82**, 4473 (1985).
160. A. Meager, H. Graves, D. C. Burke and D. M. Swallow, *Nature* **280**, 493 (1979).
161. A. Meager, H. E. Graves, J. R. Walker, D. C. Burke, D. M. Swallow and A. Westerveld, *J. Gen. Virol.* **45**, 309 (1979).
162. J. Weissenbach, Y. Chernajovsky, M. Zeevi, L. Shulman, H. Soreq, U. Nir, D. Wallach, M. Perricaudet, P. Tiollais and M. Revel, *PNAS* **77**, 7152 (1980).
162a. M. Revel, R. Ruggieri and A. Zilberstein, *in* "The Biology of the Interferon System 1985" (W. E. Stewart II and H. Schellekens, eds.), Elsevier, Amsterdam, in press.
163. Y. H. Tan, R. P. Creagan and F. H. Ruddle, *PNAS* **71**, 2251 (1974).
164. D. L. Slate and F. H. Ruddle, *Ann. N.Y. Acad. Sci.* **350**, 174 (1980).
165. P. M. Pitha, D. L. Slate, N. B. K. Ray and F. H. Ruddle, *MCBiol.* **2**, 564 (1982).
166. P. B. Sehgal, L. T. May, A. D. Sagar, K. S. LaForge and M. Inouye, *PNAS* **80**, 3631 (1983).
167. A. D. Sagar, P. B. Sehgal, L. T. May, M. Inouye, D. L. Slate, L. Shulman and F. H. Ruddle, *Science* **223**, 1312 (1984).
168. P. B. Sehgal and L. T. May, *in* "The Biology of the Interferon System 1984" (H. Kirchner and H. Schellekens, eds.), p. 27. Elsevier, Amsterdam, 1985.

169. J. Van Damme, A. Billiau, M. De Ley and P. De Somer, *J. Gen. Virol.* **64**, 1819 (1983).
170. J. Van Damme, G. Opdenakker, A. Billiau, P. De Somer, L. De Wit, P. Poupart and J. Content, *J. Gen. Virol*, in press.
171. J. Van Damme, M. De Ley, H. Claeys, A. Billiau, C. Vermylen and P. De Somer, *Eur. J. Immunol.* **11**, 937 (1981).
172. J. Van Damme, M. De Ley, G. Opdenakker, A. Billiau, P. De Somer and J. Van Beeumen, *Nature* **314**, 266 (1985).
173. Y. Higashi, Y. Sokawa, Y. Watanabe Y. Kawade, S. Ohno, C. Takaoka and T. Taniguchi, *JBC* **258**, 9522 (1983).
174. J.-I. Fujisawa and Y. Kawade, *Virology* **112**, 480 (1981).
175. D. Skup, J. D. Windass, F. Sor, H. George, R. G. Williams, H. Fukuhara, J. De Maeyer-Guignard and E. De Maeyer, *NARes* **10**, 3069 (1982).
176. D. R. Gewert, M. Castellino, D. Skup and B. R. G. Williams, in "The Biology of the Interferon System 1984" (H. Kirchner and H. Schellekens, eds.), p. 49. Elsevier, Amsterdam, 1985.
177. P. W. Gray, D. W. Leung, D. Pennica, E. Yelverton, R. Najarian, C. C. Simonsen, R. Derynck, P. J. Sherwood, D. M. Wallace, S. L. Berger, A. D. Levinson and D. V. Goeddel, *Nature* **295**, 503 (1982).
178. R. Devos, H. Cheroute, Y. Taya, W. Degrave, H. Van Heuverswyn and W. Fiers, *NARes* **10**, 2487 (1982).
179. P. W. Gray and D. V. Goeddel, *Nature* **298**, 859 (1982).
180. Y. Taya, R. Devos, J. Tavernier, H. Cheroutre, G. Engler and W. Fiers, *EMBO J.* **1**, 953 (1982).
181. R. Derynck, D. W. Leung, P. W. Gray and D. V. Goeddel, *NARes* **10**, 3605 (1982).
182. T. Taniguchi, H. Sugano, T. Nishi, J. T. Vilček and Y. K. Yip, Eur. Pat. Appl. 83109788.6. (1983).
183. S. L. Naylor, A. Y. Sakaguchi, T. B. Shows, M. L. Law, D. V. Goeddel and P. W. Gray, *J. Exp. Med.* **157**, 1020 (1983).
184. Y. K. Yip, B. S. Barrowclough, C. Urban and J. Vilček, *PNAS* **79**, 1820 (1982).
185. L. E. Burton, P. W. Gray, D. V. Goeddel and E. Rinderknecht, in "The Biology of the Interferon System 1984" (H. Kirchner and H. Schellekens, eds.), p. 403. Elsevier, Amsterdam, 1985.
186. P. W. Gray and D. V. Goeddel, *PNAS* **80**, 5842 (1983).
187. R. Dijkmans, G. Volckaert, A. Billiau and P. De Somer, *J. Interferon Res.* in press.
188. R. Dijkema, P. H. van der Meide, P. H. Pouwels, M. Caspers, M. Dubbeld and H. Schellekens, *EMBO J.* **4**, 761 (1985).
189. R. Dijkema, P. H. van der Meide, M. Dubbeld, M. Caspers, J. Wubben and H. Schellekens, *Methods Enzymol.*, in press.
190. T. A. Springer, D. B. Teplow and W. J. Dreyer, *Nature* **314**, 540 (1985).
191. E. Zuckerkandl and L. Pauling, in "Evolving Genes and Proteins" (V. Bryson and H. J. Vogel, eds.), p. 97. Academic Press, New York, 1965.
192. T. Miyata and H. Hayashida, *PNAS* **78**, 5739 (1981).
193. T. Miyata, H. Hayashida, R. Kikuno, H. Toh and Y. Kawade, in "Interferon 6" (I. Gresser, ed.), p. 1. Academic Press, New York, 1985.
194. R. P. Harvey, J. A. Whiting, L. S. Coles, P. A. Krieg and J. R. E. Wells, *PNAS* **80**, 2819 (1983).
195. D. Wells and L. Kedes, *PNAS* **82**, 2834 (1985).
196. R. Zakut, M. Shani, D. Givol, S. Neumann, D. Yaffe and U. Nudel, *Nature* **298**, 857 (1982).

197. W. F. DeGrado, Z. R. Wassermann and V. Chowdhry, *Nature* **300**, 379 (1982).
198. J. Tavernier and W. Fiers, *Carlsberg Res. Commun.* **49**, 359 (1984).
198a. P. Sehgal, L. T. May and F. R. Landsberger, *in* "The Biology of the Interferon System 1985" (H. Schellekens and W. E. Stewart II, eds.). Elsevier, Amsterdam, in press.
199. S. Ohno, "Evolution by Gene Duplication." Springer Verlag, Berlin and New York, 1970.
200. T. Edlund and S. Normark, *Nature* **292**, 269 (1981).
201. M. Botchan, W. Topp and J. Sambrook, *CSHSQB* **43**, 709 (1978).
202. J. M. Roberts, L. B. Buck and R. Axel, *Cell* **33**, 53 (1983).
203. T. Miyata and H. Hayashida, *Nature* **295**, 165 (1982).
204. J. van den Berg, A. van Ooyen, N. Mantei, A. Schamböck, G. Grosveld, R. A. Flavell and C. Weissmann, *Nature* **275**, 37 (1978).
205. E. Domingo, D. Sabo, T. Taniguchi and C. Weissmann, *Cell* **13**, 735 (1978).
206. J. W. Drake and R. H. Baltz, *ARB* **45**, 11 (1976).
207. T. A. Kunkel, *PNAS* **81**, 1494 (1984).
208. T. Miyata, T. Yasunaga and T. Nishida, *PNAS* **77**, 7328 (1980).
209. A. E. Romero-Herrera, H. Lehmann, K. A. Joysey, and A. E. Friday, *Nature* **246**, 389 (1973).
210. N. R. Rice, *in* "Evolution of Genetic Systems" (H. H. Smith, ed.), p. 44. Gordon & Breach, New York, 1972.
211. C. D. Laird, B. L. McConaughy and B. J. McCarthy, *Nature* **224**, 149 (1969).
212. A. C. Wilson, S. S. Carlson and T. J. White, *ARB* **46**, 573 (1977).
213. K. A. Kelley and P. M. Pitha, *NARes* **13**, 825 (1985).
214. G. Volckaert, J. Tavernier, R. Derynck, R. Devos and W. Fiers, *Gene* **15**, 215 (1981).
215. S. Fields and G. Winter, *Gene* **15**, 207 (1981).
216. J. F. Ernst, J. W. Stewart and F. Sherman, *PNAS* **78**, 6334 (1981).
217. D. L. Bentley and T. H. Rabbitts, *Cell* **32**, 181 (1983).
218. R. Breathnach and P. Chambon, *ARB* **50**, 349 (1981).
219. H. Dinter, H. Hauser, U. Mayr, R. Lammers, W. Bruns, G. Gross and J. Collins, *in* "The Biology of the Interferon System 1983" (E. De Maeyer and H. Schellekens, eds.), p. 33. Elsevier, Amsterdam, 1983.
220. M. O. Dayhoff, "Atlas of Protein Sequence and Structure." National Biomedical Research Foundation, Washington, D.C. **5** Suppl. 3, (1978).
221. P. H. van der Meide, R. Dijkema, M. Caspers, K. Vijverberg and H. Schellekens, *Methods Enzymol.*, in press.
222. R. M. Lawn, *Genet. Eng.* **4**, 199 (1982).
223. E. Gren, V. Berzin, I. Jansone, A. Tsimanis, Y. Vishnevsky and U. Apsalons, *J. Inerferon Res.* **4**, 609 (1984).
224. R. M. Torczynski, M. Fuke and A. P. Bollon, *PNAS* **81**, 6451 (1984).
225. A. Sloma, *Eur. Pat. Appl.* 83102893.1. (1983).
226. D. V. Goeddel and S. Pestka, *Eur. Pat. Appl.* 82107337.6 (1983).
227. A. Ovchinnikov, E. D. Sverdlov, S. A. Tsarev, E. M. Khodkova, G. S. Monastyrskaya, V. A. Efimov, O. G. Chakhmakhcheva, V. D. Solov'ev, V. P. Kuznetsov and V. M. Kavsan, *Dokl. Akad. Nauk SSSR* **262**, 725 (1982).

Addendum to The Ubiquitin Pathway for the Degradation of Intracellular Proteins

Following the completion of this chapter, a brief review on ubiquitin by Finley and Varsharsky appeared (1), a method for the large scale purification of ubiquitin has been described (2), and immunochemical methods for the quantitative determination of free and protein-conjugated ubiquitin have been worked out (3).

We found that E_3 contains the protein substrate binding site of the ubiquitin ligase system, and that proteins with free α-amino groups or with oxidized methionine residues selectively bind to this site (4). Rechsteiner and co-workers provided furhter evidence that proteins conjugated to ubiquitin are degraded by an ATP-dependent process (5) and partially purified an ATP-dependent protease that degrades ubiquitin-conjugated lysozyme (6). Since we observed (unpublished) that three factors are required for this process, it remains to be seen whether the "ATP-dependent protease" is a multienzyme complex. Rechsteiner and colleagues also showed that the degradation of guanidinated lysozyme (which has a free α-NH_2, but no ε-NH_2 groups) is stimulated by ubiquitin (7). The formation of high-molecular-weight ubiquitin conjugates of guanidinated lysozyme was observed, but only at 5% of the level obtained with unmodified lysozyme (7). The high-molecular-weight derivatives might be polyubiquitin chains linked to the α-amino group of guanidinated lysozyme. It remains unexplained why the levels of these conjugates are so low relative to the rates of proteolysis of guanidinated lysozyme.

An interesting function of ubiquitin was discovered by the observation that the lymphocyte "homing" receptor contains a ubiquitin moiety (8, 9). The presence of ubiquitin in other cell surface proteins was detected (8), suggesting that it might be a widespread phenomenon. It is not yet known whether ubiquitin ligation modifies receptor function. Another interesting question is at what stage of the biosynthesis of membrane proteins ubiquitin ligation occurs.

Addendum References

1. D. Finley and A. Varsharsky, *TIBS* **10**, 343 (1985).
2. A. L. Haas and K. D. Wilkinson, *Prep. Biochem.* **15**, 49 (1985).
3. A. L. Haas and P. M. Bright, *JBC* **260**, 2464 (1985).
4. A. Hershko, H. Heller, E. Eytan and Y. Reiss, *JBC*, in press.
5. R. Hough and M. Rechsteiner, *JBC* **261**, 2391 (1986).
6. R. Hough, C. Pratt and M. Rechsteiner, *JBC* **261**, 2400 (1986).
7. D. T. Chin, N. Carlson, L. Kuehl and M. Rechsteiner, *JBC* **261**, 3883 (1986).
8. M. Siegelman, M. W. Bond, W. M. Gallatin, T. St. John, H. T. Smith, V. A. Fried and I. L. Weissman, *Science* **231**, 823 (1986).
9. T. St. John, W. M. Gallatin, M. Siegelman, H. T. Smith, V. A. Fried and I. L. Weissman, *Science* **231**, 845 (1986).

Index

A

N^α-Acetylation, degradation by ubiquitin and, 37–38, 39
Actinomycin D, DNA polymerase and, 80
Activated ubiquitin, transfer for conjugation, 27–30
Adenosine triphosphate
  conjugation of activated ubiquitin to protein and, 27–28
  degradation of ubiquitin–protein conjugates and, 30–33
Adenosine triphosphate
  DNA polymerse-α processivity and, 69, 71
  ubiquitin activation and, 25–26
Amines, transfer of ubiquitin to, 34
Amino acids, analogs, ubiquitin and, 23
Amino acid sequences
  of bovine α-IFNs, 272, 275
  of human α interferons, 260–261, 275
  of interferons, conservation of, 289–291
  of mammalian β-IFNs, 277, 278
  of mammalian γ-IFNs, 281, 282
  of rodent α-IFNs, 273, 274, 275
  of ubiquitin, conservation of, 20, 21
Aminoacyl-tRNA synthetases
  free from core complex, 212–214
  phosphorylation of, 216–219
  characteristics of synthetases, 210–216
  effects at molecular level, 219–222
α-Amino group, specificity of ubiquitin and, 36
Antibody
  to aminoacyl-tRNA synthetases, 217–218
Antibody
  monoclonal
    chromosomal localization of gene for DNA polymerase-α and, 83–84
    localization of DNA polymerase-α and, 79, 80, 81
    purification of DNA polymerase-α and, 61, 62
    to ubiquitin–protein conjugates, 44, 47
Aphidicolin, DNA polymerase-α and, 67, 73, 80, 81–83, 95–97
Arabinocytidine, DNA polymerase-α and, 96–97
Aromatic residues, of gene-V protein, intercalation with ssDNA, 151–152
Auxilliary proteins, for oriC, 118

B

Bacteriophages, see also Phage
  overlapping genes in, 149–150
Barriers, to DNA elongation, 70–71
Base-pairing, energies, fidelity of DNA synthesis and, 84–85
Beta-protein, of phage λ, initiation by cos and, 177
Biological activity, of human α interferons, 270–271
P^1,P^4-Bis(adenosine-5′)tetraphosphate
  aminoacyl-tRNA synthetases and, 215, 222
  control of DNA replication and, 75–77, 215
  as primer, 76–77
Bovine
  IFN-α genes of, 271–272
  IFN-β genes of, 280
Break-copy model, mechanism of cos-increased exchange rate and, 175–176
Bromodeoxyuridine, DNA polymerase-α mutants and, 96–97, 98
Butylanilinouracil, DNA polymerase-α and, 80–81
Butylphenylguanine, DNA polymerase-α and, 80–81

C

Calcium ions, protein synthesis and, 223
Calmodulin, ubiquitin and, 38
Carbachol, phosphorylation of ribosomal protein S6, 199, 200
Carbamoylation, protein degradation by ubiquitin and, 36
Casein kinase I, aminoacyl-tRNA synthetases and, 221
Catalytic core component, of DNA polymerase-α
 fidelity of, 85–86
 mutational spectra, 89–90
 single base substitutions using DNA templates, 88–89
 use of polynucleotide templates, 86–88
Cell-cycle mutant, with thermolabile ubiquitin-activating enzyme, temperature-sensitive protein breakdown in, 46–48
Cell membrane, filamentous phage assembly and, 157
Chi
 discovery of, 186–188
 rules of activity in phage λ, 188–191
Chicken, ubiquitin genes of, 22, 23
C kinase, phosphorylation of ribosomal protein S6 and, 204–205
Cloning, of interferon genes, 253–254
 human IFN-α1, 255
 human IFN-β, 276, 279
 human IFN-γ, 280–281
Coat proteins, filamentous phage assembly and, 156–157
Codon, recognition, phosphorylation of aminoacyl-tRNA synthetases and, 221–222
Complexes, supramolecular, of aminoacyl-tRNA synthetases, 210–213
Concatemers, phage λ lytic cycle and, 172–173
Conformational model, DNA polymerase-α fidelity and, 93
cos
 detailed model for initiation by, 177–179
 DNA synthesis at, 180–181
 interaction with Chi, 189–191
 involvement in recombination by λ's Red system, 174–175
 break-copy model, 175–176
 distinguishing the models, 176–177
 initiation model, 176
 nonreciprocal exchange at, 180
Crossing over, human IFN-α genes and, 265, 269
Cruciform structures, superhelical density and, 128
Cyclic adenosine monophosphate, phosphorylation of ribosomal protein S6 and, 196–198, 208–209
Cycloheximide
 DNA polymerase-α and, 80
 phosphorylation of ribosomal protein S6 and, 199, 200, 209
Cytochrome c, ubiquitin and, 37
Cytoskeleton, organization of protein synthesis and, 224

D

dC-dG sequences, DNA methylation and, 231, 234
 effect of spacing on, 238–239
Deletions, in IFN genes, 288–289
Deoxyribonucleic acid
 elongation of, 129–130
 elongation by DNA polymerase-α
  barriers to, 70–71
  gap filling and long-stretch synthesis, 71–72
  processivity, 69
 initiation of replication, 116
  oriC and λ dv, 117–120
  pBR322 DNA, 120–124
  template discrimination, 124–129
 de novo versus maintenance sites of methylation, 239–244
 primary sequence, effect on methylation, 236–238
 recombinant, analysis of interferon system by, 253–254
 replication
  topological constraints and, 112–116
  single-stranded, complex with gene-V protein, 152, 155
  termination and segregation of daughter molecules, 131–137

Deoxyribonucleic acid-A protein, 122, 123
  prepriming complex and, 119
Deoxyribonucleic acid-B protein, priming complex and, 119
Deoxyribonucleic acid gyrase
  separation of daughter DNA molecules and, 135, 137
  superhelical DNA replication and, 117–118, 120, 126, 129
Deoxyribonucleic acid methyltransferases
  characterization of
    biochemical, 233–234
    purification, 232–233
    substrate concentration dependence, 234
    substrate specificity, 234–235
  inhibitors of
    polynucleotides and polydeoxynucleotides, 245–248
    RNA, 244–245
    triplexes, 248
Deoxyribonucleic acid polymerase-α
  auxilliary activities associated with
    exonuclease and DNA primase, 72–74
    primer recognition and Ap$_4$A binding protein, 74–77
    template binding proteins, 77–78
  biochemical characteristics of
    catalytic mechanism, 62–65
    elongation of DNA, 69–72
    interaction with template and primer stem, 65–66
    nucleotide substrates, metal activators and initial order of reactants, 66–69
    structure, 60–62
  gene for
    chromosomal localization of, 83–84
    prospects for cloning of, 99–101
  identification of, 58–59
  mutants, 95–99
  transfection and, 99
  roles, 78–79
    in fidelity of DNA synthesis, 84–95
    in repair, 81–83
    in replication, 79–81
  selective inhibitors of, 82

Deoxyribonucleic acid polymerase-β
  isolation and properties of, 59
  long stretch DNA synthesis and, 72
Deoxyribonucleic acid polymerase-γ,
  isolation and properties of, 59
Deoxyribonucleic acid polymerase-δ,
  isolation and properties of, 59
Deoxyribonucleic acid primase
  DNA polymerase-α and, 73–74
  function of, 60
Deoxyribonucleic acid replication, in f1 phage, 143–145
  need for regulation, 145–147
Deoxyribonucleic acid synthesis
  role of DNA polymerase-α infidelity of, 84–85
  catalytic core component, 85–86
  at cos, 180–181
  enhancement $in\ vivo$, 94–95
  enhancement mechanisms, 91–94
  holoenzyme complexes, 90–91
Deoxynucleoside phosphates, thioredoxin and, 159, 160
Deoxyuridine triphosphate, as substrate for DNA polymerase-α, 67
Dimethylnitrosamine, phosphorylation of ribosomal protein S6 and, 199, 209
Discrimination, between tRNA species, 210
Divergence
  of human α IFNs, 266, 267
  of human β IFNs, 266, 267
  of human γ IFNs, 267
$Drosophila\ melanogaster$, ubiquitin genes of, 22

E

Effectors, promoting incorporation of five phosphates into ribosomal protein S6, 199
Elongation proteins, for $oriC$, 118
Energy-relay model, enhancement of fidelity of DNA polymerase-α and, 92
Epidermal growth factor, phosphorylation of ribosomal protein S6 and, 199, 200, 201, 202, 203, 206, 207
Error corrections, in DNA of animal cells, 95

Escherichia coli
   fip gene, characterization of, 158–159
   RecBC pathway of
      discovery of Chi, 186–188
      roles of Chi activity in phage λ, 188–191
      in vitro studies, 191
   Red-like pathways in, 181
Exonuclease
   DNA polymerase-α and, 72–73
   initiation by *cos* in phage λ and, 177
Expression levels, of human α-interferon genes, 271

F

Fibroblasts, interferon of, 251
Filamentous phage, *see also* Phage
   gene-II protein
      activities, 147–148
      specificity, 148–149
   gene-V protein
      biochemistry, 151–152
      specific repression of gene-II and gene X translation, 153–154
      ssDNA complex and, 152
   gene X
      function of, 150–151
      overlapping genes, 149–150
   introduction to
      f1 DNA replication, 143–145
      need for regulation, 145–147
   regulatory circuit in, 154–156
   role of thioredoxin, 162–163
      in assembly, 156–157
      host mutants in assembly, 157–159
      mutant thioredoxin, 160–162
      thioredoxin, 159–160
fip gene, phage assembly and, 157–159
f1 phage, DNA replication, 143–145
   need for regulation, 145–147

G

Gap filling, DNA polymerase-α and, 71–72
Gene(s)
   localization of DNA polymerase-α and, 83–84
   for ubiquitin, structure and organization of, 21–23

Gene conversion, human IFN-α and, 265, 269
Gene dosage, of mutant rRNA, 5
   effect of, 7–8
Gene expression, ubiquitinated histone and, 49
Gene I, phage assembly and, 162
Gene II
   gene X and, 150–151
   translation, specific repression by gene-V protein, 153–154
Gene-II protein
   activities of, 147–148
   specificity of, 148–149
Gene-V protein
   biochemistry, 151–152
   free, as gauge of phage replication, 155
   specific repression of gene-II and gene-X translation, 153–154
   ssDNA complex with, 152
Gene X
   function, 150–151
   overlapping genes, 149–150
   translation, specific repression by gene-V protein, 153–154
Glucagon
   aminoacyl-tRNA synthetase activity and, 217, 219
   ribosomal protein S6 and, 196
Glutamine synthetase, regulation of degradation of, 43
Glutaredoxin, deoxyribonucleotides and, 160
Glycation, of interferons, 253, 262, 276, 282
Growth rate, deletion mutations in *rrn*B and, 4–5, 7, 12
Guanidination, protein degradation by ubiquitin and, 37
Gyrase, definition of, 142

H

Heat shock, protein synthesis and, 224–225
Heat-shock proteins, ubiquitin and, 23, 47
Helicase, definition of, 142
Hemoglobin, degradation, ubiquitin and, 44

INDEX   307

Heteroduplexes, phage λ Red system
    and, 174
Histone(s)
    modification, possible roles of ubi-
       quitin in, 48–51
    turnover rates of, 48
Histone 2A, conjugate with ubiquitin,
    21, 35
Holoenzyme complexes, of DNA poly-
    merase-α fidelity of, 90–91
Host mutants, filamentous phage assem-
    bly and, 157–159
Human
    interferon-α genes
       biological activity and expression
          level, 270–271
       general description, 255–262
       linkage and loci, 262–264
       other related loci, 270
       pseudogenes, 269–270
       relatedness, 264–269
    interferon-β genes, 276–277
    interferon-γ genes of, 280–282
    ubiquitin genes of, 22, 23
Hydrophobic interactions, of aminoacyl-
    tRNA synthetases, 214
Hydrophobicity, of substrate binding
    site, fidelity of DNA polymerase-α
    and, 92

I

Infections, phosphorylation of ribosomal
    protein S6 and, 199, 200
Inhibitors, of DNA polymerase-α, 82
Initiation model
    mechanism of *cos*-increased exchange
       rate and, 176
    tests of, 179–181
Initiation proteins, for *ori*C, 118
Insertions, in IFN genes, 288–289
Insulin
    aminoacyl-tRNA synthetase activity
       and, 217, 219
    phosphorylation of ribosomal protein
       S6 and, 199, 200, 201, 202, 205,
       207, 208
Interferon(s)
    different types of, 251
    physical properties of, 252–253
    pleiotropic effects of, 252

Interferon-α genes
    human
       biological activity and expression
          level, 270–271
       general description, 255–262
       linkage and loci, 262–264
       other related loci, 270
       pseudogenes, 269–270
       relatedness, 264–269
    in nonhuman species
       bovine, 271–272
       rodent, 273–276
Interferon-β genes
    human, 276–277
    nonhuman, 279–280
    other human, 278–279
Interferon-γ genes
    human, 280–282
    other mammalian, 282–283
Interferon gene family, evolution
    conservation of sequences, 289–291
    deletions and insertions, 288–289
    mutation fixation rates, 287–288
    origin, 283–285
    origin of families, 285–287
Interferon system, analysis by recombi-
    nant DNA technology, 253–254
Iodoacetamide
    conjugation of ubiquitin to protein
       and, 28
    ubiquitin activation and, 26
3-Isobutyl-1-methylxanthine, aminoacyl-
    tRNA synthetase activity and, 217,
    219
Isolation, of DNA polymerase-α, difficul-
    ties with, 60
Isopeptidase, ubiquitin–histone conju-
    gates and, 35

K

Kinetics, of decay of ubiquitin–protein
    conjugates, 45–46

L

Lambda
    O protein, DNA replication and, 117
Lambda-P_L promoter, in pNO2680, con-
    ditional rRNA gene expression
    system and, 8–10

Lethal mutations, in rRNA, 17
  testing for, 9–10
Leukocytes, interferon of, 251
Linkage, of human IFN-α genes, 262–264
Location, of human IFN-α genes, 262–264
Loci, human IFN-α-related, 270
Lysosomes, proteolysis by, 19

M

Magnesium ions, DNA polymerase-α and, 67–68
Manganese ions, fidelity of DNA polymerase-α and, 93
Mapping, of interferon genes, 254
Mating-type, switching of, 183–184
Maxicell-like system
  induced chemically, specific labeling of cloned rDNA genes and, 16–17
  induced by UV, specific labeling of cloned rDNA genes and, 14–16
Meiosis, initiation events, double-chain breaks and, 183
Meiotic recombination, Red and RecBC as models for, 191–192
Metal activator, of DNA polymerase-α, 67–68
Methylation, of DNA, mutations and, 192
Mitosis, ubiquitin–histone H2A and, 49
Mitotic activity, DNA polymerase-α and, 79–80
Mutational spectra, catalytic core component of DNA polymerase-α and, 89–90
Mutation fixation rates, of IFN genes, 284, 287–288

N

Nucleosomes
  DNA replication and, 117
  ubiquitin–histone 2A conjugate and, 21, 50
Nucleotide(s), conformation, binding to DNA polymerase-α and, 94
Nucleotide substrates, of DNA polymerase-α, 66–67

O

Oligodeoxynucleotides, methylation of, 236
  effect of nucleotide spacing between dC-dG sequences, 238–239
  effect of primary DNA sequence, 236–238
oriC, initiation of DNA replication at, 117
  proteins required for, 118
Origin, of IFN genes, 283–285
  gene families, 285–287
oriλ, initiation of DNA replication at, 117, 119
Orthovanadate, phosphorylation of ribosomal protein S6 and, 199, 206

P

Packaging, interaction of Chi with cos and, 189–190
Phage f1, see also Filamentous phage
  export of single-stranded DNA and, 12
Phage T4, recombination in, 182
Phage T7, thioredoxin and, 159, 162–163
Phage T7 late-promoter, in pAR3056, conditional rRNA-gene-expression system and, 10–13
Phage λ
  exchange stimulated by restriction enzyme cuts, 179–180
  glossary of terms peculiar to study of genetic recombinations in, 187–188
  recombination by Red system
    detailed model for initiation by cos, 177–179
    double-chain breaks as initiators elsewhere, 181–186
    heteroduplexes, 174
    involvement of cos, 174–177
    tests of model, 179–181
  lytic cycle of, 172–173
  recombination of nonreplicated chromosomes, 173–174
  role in recombination studies, 169–171
  rules of Chi activity in, 188–191
Phorbol esters, phosphorylation of ribosomal protein S6 and, 199, 200, 201, 202, 204, 205

Phospholipids, protease-activated kinase and, 203
Phosphorylation phosphatases, aminoacyl-tRNA synthetases and, 216
Phosphorylation
 of aminoacyl-tRNA synthetases, 216–219
 coordination of protein synthesis and, 223, 224–225
 of ribosomal protein S6
  molecular analysis of protein synthesis and, 207–209
  in response to cAMP, 196–198
  in response to growth-promoting compounds and insulin, 198–207
 of ribosomal protein S6 and aminoacyl-tRNA synthetases
  historical background, 195–196
Physarum polycephalum, ubiquitinated histone in, 49
Plasmid(s), recombinations in yeast and, 182
Plasmid pBR322, DNA, initiation of replication of, 120–124
Plasmid pEJM007, mutant rRNA expressed from, 5–8
Plasmid pGQ15, construction of, 12
Plasmid pKK3535
 construction of, 3
 mutant rRNA expressed from, 3–5
Plasmid pNO2680, lambda-P_L promoter in, conditional rRNA gene expression system and, 8–10
Platelet-derived growth factor, phosphorylation of ribosomal protein S6 and, 199, 200
Point substitutions, in IFN genes, 286
Poly(A), β-IFNs and, 278, 279
Poly[d(A-T)], as template, error rate and, 87
Polydeoxynucleotides, as inhibitors of DNA methyltransferase, 246–248
Polynucleotides
 as inhibitors of DNA methyltransferase, 245–246
 as templates, fidelity of catalytic core component of DNA polymerase-α and, 86–88
Poly(rG), inhibition of DNA methytransferase by, 245–246
Polyribonucleotides, DNA polymerase-α and, 65, 66
Polysomes, formation, phosphorylation of ribosomal protein S6 and, 207
Primase, definition of, 142
Primer, of DNA polymerase-α, 58
 magnesium and, 68
 recognition and Ap_4A-binding proteins, 74–77
Priming complex, components of, 119–120
Primer stems, DNA polymerase-α and, 66
Processivity
 of DNA polymerase-α, 69
 fidelity and, 92–93
 thioredoxin and, 159, 162–163
Progesterone, phosphorylation of ribosomal protein S6 and, 199, 200
Promoters, inducible, expression of mutant rRNA from plasmids with, 8–13
Promoters P1 and P2, expression of mutant ribosomal RNA from, 3–8
Proofreading exonucleotide activity, of DNA polymerases, 85–86, 91–92, 94–95
Prostaglandin E_1, ribosomal protein S6 and, 196
Prostaglandin $F_{2\alpha}$, phosphorylation of ribosomal protein S6 and, 199, 200, 207
Protease-activated kinase, phosphorylation of ribosomal protein S6 and, 202, 203–204, 208
Protein(s)
 abnormal, conjugation to ubiquitin and degradation, 43–46
 binding Ap_4A, 76
 breakdown, discovery of role of ubiquitin in, 23–24
 conjugation of ubiquitin to, 24
 enzymatic reactions in, 25–30
 conjugated with ubiquitin, breakdown of, 30–33
 microinjected, degradation of, 43–44
 primer recognition factors, DNA polymerase-α and, 74–75
 recognition of structure by ubiquitin system, role of α-amino group, 35–40

template binding, 77–78
ubiquitin-mediated breakdown, involvement of tRNA in, 41–43
Protein kinases
  aminoacyl-tRNA synthetases and, 216, 219–221
  cAMP-dependent, ribosomal protein S6 and, 197–198
Protein synthesis
  coordinate regulation of, 223–225
  phosphorylation of ribosomal protein S6 and, 207–209
Protein-tyrosine kinases, phosphorylation of ribosomal protein S6 and, 206–207
Proteolysis
  of DNA polymerase-α, 87–88
  ubiquitin-dependent, evidence in various cells, 43–48
Pseudogenes, human IFN-α and, 255, 257, 269–270, 288
Pyrimidine nucleotides, mutant DNA polymerase-α and, 97–98
Pyrophosphate, DNA polymerase-α fidelity and, 93

Q

Q fever, phosphorylation of ribosomal protein S6 and, 199, 209

R

RecBC pathway, in *E. coli*
  discovery of Chi, 186–188
  rules of Chi activity in phage λ, 188–191
  in vitro studies, 191
Receptors, for IFN-α, 270–271
Recombination studies, role of phage λ in, 169–171
Red and RecBC, as models for meiotic recombination, 191–192
Regeneration, phosphorylation of ribosomal protein S6 and, 199, 200
Regulation, of f1 DNA replication, 145–147
Regulatory circuit, of filamentous phage, 154–156

Relatedness, of human IFN-α genes, 264–269
Repair synthesis, DNA polymerase-α and, 81–83, 98
Replication, DNA polymerase-α and, 79–81
Restriction enzyme, cuts by, exchange stimulation and 179–180
Ribonucleic acid
  as inhibitor of DNA methyltransferase, 244–245
  messenger, binding to phosphorylation ribosomal subunits, 208
  messenger, of genes II and X gene-V protein and, 155
  as primer for DNA polymerase-α, 65, 66, 73
  replication of pBR322 DNA and, 122–123
  transfer, recognition by aminoacyl-tRNA synthetases, 215
Ribonucleic acid polymerase
  DNA replication and, 117–119, 120, 124, 126, 128–129
  modulation of action of, 124–126
  T7, integration into *lac* UV5 promoter, 11–12
Ribosomal DNA, cloning of, 1–3
Ribosomal DNA genes, cloned, specific labeling of, 13–17
Ribosomal protein, S6, phosphorylation of
  molecular analysis of protein synthesis and, 207–209
  in response to cAMP, 196–198
  in response to growth-promoting compounds and insulin, 198–207
Ribosomal RNA
  conditional gene expression system utilizing lambda-P_L promoter in pN02680, 8–10
  conditional gene expression system utilizing T7 late promoter in pAR3056, 10–13
  mutant
    expressed from high-copy-number plasmid pKK3535, 3–5
    expressed from low-copy-number plasmid pEJM007, 5–8
Rifampicin, specific labeling of cloned rDNA genes and, 16–17

Rodents
  IFN-α genes of, 273–276
  IFN-β genes of, 280
rrnB, structural gene, construction of site-specific mutations on, 12

S

α-Sarcin, construction of mutations and, 12
Serum, phosphorylation of ribosomal protein S6 and, 199, 200, 201, 202, 203, 204, 205, 207
Single-stranded binding protein, DNA elongation and, 70
Specificity
  of gene-II protein, 148–149
  of ubiquitin conjugation to protein, 29, 35–40, 43
Spermine, DNA polymerase-α processivity and, 69
Subfamily, of human IFN-$α_1$, 268
Substrate, for DNA methyltransferase
  concentration dependence, 234
  specificity, 234–235
Subunits, of DNA polymerase-α, 60, 62, 63–64
Superhelical density, topoisomerase and, 126–128

T

Template
  interaction of DNA polymerase-α with, 65, 68
  secondary structure and, 70–71
  natural DNA, single-base substitution using, 88–89
  polynucleotide, fidelity of catalytic core component of DNA polymerase-α and, 86–88
  proteins binding, 77–78
Template discrimination, initiation of DNA replication and, 124–129
Thioacetamide, phosphorylation of ribosomal protein S6 and, 199, 209
Thioredoxin
  characterization of, 159–160
  functions of, 163
  mutant, 160–162
  filamentous phage assembly and, 161–162
  role in phage assembly, 162–163
  filamentous phage, 156–157
  host mutants and, 157–159
  mutant thioredoxin, 160–162
  thioredoxin, 159–160
Thiodoxin reductase, thiodoxin mutants and, 161
T lymphocytes, interferon of, 251
Topoisomerases
  DNA replication and, 114–116
  separation of daughter DNA molecules and, 131, 132, 135–137
  template discrimination and, 124, 126, 129
Transfer RNA, involvement in ubiquitin-mediated protein breakdown, 41–43
Transformation, phosphorylation of ribosomal protein S6 and, 199, 200, 201, 206, 207
Transcription, ubiquitinated histone and, 50–51
Transposase, definition of, 142
Triplexes, inhibition of DNA methyltransferase and, 248
Tryptophanyl-tRNA synthetase, DNA polymerase-α and, 76, 77

U

Ultraviolet, maxicell system induced by, specific labeling of cloned rDNA genes and, 14–16
Ubiquitin
  activation of, 25–27
  breakdown of proteins conjugated with, 30–33
  conjugate with histone, 21
  discovery of role in protein breakdown, 23–24
Ubiquitin
  possible roles in histone modification, 48–51
  proteolytic pathway, proposed sequence of events in, 40–41
  structure of, 19–21

Ubiquitin-activating enzyme
  properties of, 25–26
  thermolabile, temperature-sensitive protein breakdown and, 46–48
Ubiquitin genes, structure and organization of, 21–23
Ubiquitin–protein conjugates, high molecular weight, 32–33, 36–37
Ubiquitin–protein lyases, 33
  function of, 34–35

X

X-chromosome, localization of gene for DNA polymerase-α on, 84
Xenopus laevis, ubiquitin genes of, 22

Y

Yeast
  ubiquitin genes of, 21–22
  recombination in, 182–185